Binary Bullets

Binary Bullets

THE ETHICS OF CYBERWARFARE

EDITED BY FRITZ ALLHOFF,
ADAM HENSCHKE,
and
BRADLEY JAY STRAWSER

OXFORD
UNIVERSITY PRESS

Oxford University Press is a department of the University of
Oxford. It furthers the University's objective of excellence in research,
scholarship, and education by publishing worldwide.
Oxford is a registered trademark of Oxford University Press
in the UK and certain other countries.

Published in the United States of America by
Oxford University Press
198 Madison Avenue, New York, NY 10016

© Oxford University Press 2016

All rights reserved. No part of this publication may be reproduced, stored in
a retrieval system, or transmitted, in any form or by any means, without the prior
permission in writing of Oxford University Press, or as expressly permitted by law,
by license, or under terms agreed with the appropriate reproduction rights organization.
Inquiries concerning reproduction outside the scope of the above should be sent to the
Rights Department, Oxford University Press, at the address above.

You must not circulate this work in any other form
and you must impose this same condition on any acquirer.

Library of Congress Cataloging-in-Publication Data
Binary bullets: the ethics of cyberwarfare / edited by Fritz Allhoff,
Adam Henschke, Bradley Jay Strawser.
pages cm
Includes bibliographical references and index.
ISBN 978–0–19–022108–9 (pbk.: alk. paper)—ISBN 978–0–19–022107–2 (cloth: alk. paper)
1. Cyberspace operations (Military science)—Moral and ethical aspects. I. Allhoff, Fritz, editor.
II. Henschke, Adam, 1976– editor. III. Strawser, Bradley Jay, editor. IV. Title: Ethics of cyberwarfare.
U163.B5525 2016
172'.42—dc23
2015009185

9 8 7 6 5 4 3 2 1
Printed in the United States of America
on acid-free paper

Contents

Foreword: Ethics for the Coming Epoch of Conflict   vii
   JOHN ARQUILLA

Notes on Contributors   xiii

Introduction: Editors' Introduction   1

Part I   FOUNDATIONAL NORMS FOR CYBERWARFARE

1. Emerging Norms for Cyberwarfare   13
   GEORGE R. LUCAS JR.

2. The Emergence of International Legal Norms for Cyberconflict   34
   MICHAEL N. SCHMITT AND LIIS VIHUL

3. Distinctive Ethical Issues of Cyberwarfare   56
   RANDALL R. DIPERT

Part II   CYBERWARFARE AND THE JUST WAR TRADITION

4. Cyber *Chevauchées*: Cyberwar Can Happen   75
   DAVID WHETHAM

5. Cyberwarfare as Ideal War   89
   RYAN JENKINS

6. Postcyber: Dealing with the Aftermath of Cyberattacks   115
   BRIAN OREND

PART III ETHOS OF CYBERWARFARE

7. Beyond *Tallinn*: The Code of the Cyberwarrior? 139
   MATTHEW BEARD

8. Immune from Cyberfire? The Psychological and Physiological Effects of Cyberwarfare 157
   DAPHNA CANETTI, MICHAEL L. GROSS, AND ISRAEL WAISMEL-MANOR

9. Beyond Machines: Humans in Cyberoperations, Espionage, and Conflict 177
   DAVID DANKS AND JOSEPH H. DANKS

PART IV CYBERWARFARE, DECEPTION, AND PRIVACY

10. Cyber Perfidy, Ruse, and Deception 201
    HEATHER M. ROFF

11. Cyberattacks and "Dirty Hands": Cyberwar, Cybercrime, or Covert Political Action? 228
    SEUMAS MILLER

12. Moral Concerns with Cyberespionage: Automated Keyword Searches and Data Mining 251
    MICHAEL SKERKER

Name Index 277
Subject Index 283

Foreword

Ethics for the Coming Epoch of Conflict

Early on in the American involvement in World War II, when the Axis Powers were still on the march and the Allies were suffering sharp reverses, Secretary of the Navy Frank Knox gave a speech in which he noted that "modern warfare is an intricate business about which no one knows everything—and few know very much." He spoke in the wake of German armored blitzkriegs and aerial terror bombings of cities, and of crippling Japanese aircraft carrier strikes from Pearl Harbor to the Philippines—and well beyond. Eventually, American and Allied forces learned to match Axis capabilities; sadly, they mirrored many of their unethical practices as well. Nazi U-boats waged brutal, unrestricted submarine warfare in the Atlantic; Americans did the same in the Pacific. Axis air forces killed civilians with abandon from Shanghai to London; the Allies firebombed from Hamburg and Dresden to Tokyo—and as soon as atomic bombs became available, they used them against the innocent at Hiroshima and Nagasaki. Yes, modern war was indeed "intricate." And in this intricacy mankind's moral compass malfunctioned. In the opinion of J. F. C. Fuller, one of the fathers of modern strategic thought, it was the weaponry itself that proved too enthralling. As he noted when surveying all the carnage at war's end: "It is in this appalling dissolution of morality that armaments have played so great a part."[1]

The shadow cast by nuclear weapons has kept wars smaller, but they have still been waged bitterly and unethically. In Korea, as one historian has put it, "Pyongyang and other major cities had been flattened, hundreds of thousands of North Koreans killed."[2] A decade later, "entire areas of South Vietnam were designated Free Fire Zones which could be pulverized without regard for the inhabitants."[3] And the Iran-Iraq War in the 1980s degenerated into a missile-lobbing "war

---

[1] J. F. C. Fuller, *Armament and History* (New York: Charles Scribner's Sons, 1946), xiii.
[2] Max Hastings, *The Korean War* (New York: Simon & Schuster, 1987), 268.
[3] George Herring, *America's Longest War* (New York: Alfred A. Knopf, 1986), 152.

of the cities." American military doctrine came to be associated with such terms as "overwhelming force" (forget proportionality!) and "shock and awe." With the rise of terrorist networks and the moral quagmires of nation-states that resorted to torture, mass surveillance, and preventive war in response, Fuller's assessment of the "appalling dissolution" of war ethics is clearly still pertinent. Yet there is also a ray of hope: the emerging technologies with which twenty-first-century conflicts will be waged have the potential to set the moral compass aright.

How so? By emphasizing operations in the information domain—especially, but not solely, in cyberspace—conflicts can now be conducted in more disruptive but less destructive ways. A modern field army whose communications links are severed or spoofed can hardly function, meaning that the side with the "information edge" can outmaneuver its opponent in short order, bringing war to a close quickly and with fewer losses. The cyberattacks on Georgian command and control systems during the 2008 war with Russia hamstrung the former. While there was little doubt about who would win that war, the Georgians would almost surely have fought longer and harder than the five days the conflict lasted had their command capabilities not been so seriously disrupted. Russian use of psychological operations in cyberspace—messages designed to spark panic and spur mass refugee flows—put many noncombatants in harm's way, making the conflict much less than the "ideal war" that Ryan Jenkins describes in his chapter. But overall, the Russo-Georgian War did provide a glimpse of how cyberattack can help to avoid lengthy bloodletting. This notion of short, sharp, and less destructive conflict is what my colleague David Ronfeldt and I had in mind when we developed the concept of cyberwar over twenty years ago.[4]

Back then there was little interest in our focus on cyberwar as a battle-oriented concept—a state of affairs that persists today. Instead, there has been and continues to be an unbridled enthusiasm for and a simultaneous fear of "strategic cyberwar" —the mounting of mass disruptive attacks on critical, information-dependent infrastructures. In many ways this mirrors the early discourse about air power back in the 1920s and 1930s, in which proponents of strategic bombing prevailed, for the most part, over those who preferred—for both practical and ethical reasons—to emphasize the role of attack aircraft in land and sea battles against enemy armed forces. Strategic air power advocates still predominate today, suggesting that the proponents of strategic cyberwar are likely to continue to exert a strong influence upon policy and doctrine as well. Given that few strategic bombing campaigns have ever succeeded, and that noncombatants in huge numbers have been killed in them along the way, it is very troubling to think that, even if cyberattacks don't live

---

[4] John Arquilla and David Ronfeldt, "Cyberwar Is Coming!" *Comparative Strategy* 12, no. 2 (April–June 1993): 141–65.

up to their hype, this may be the form of cyberwar—the less ethical, in *jus in bello* terms—that is resorted to more often.[5]

It is especially good, then, that many of the contributors to this volume have focused on ethical questions raised by the specter of strategic cyberwar. Indeed, taken together their insights begin to sketch out what Randall Dipert describes in his chapter as a "full ethics" of cyberwar. Such comprehensive analysis is much needed, and *Binary Bullets* is replete with fresh insights and interesting analyses of existing aspects of ethical debates about cyberwar. In terms of going to war justly, it seems clear that the *jus ad bellum* admonition to resort to force only as a last resort comes under considerable pressure, given the relative ease of use, precision targeting capability, and generally nonlethal effects of strategic cyberattack. Indeed, it is a bit troubling to note, as Randall Dipert does, that there appears to be "tacit international acceptance" of cyberattacks like the preventive strike on an Iranian nuclear facility by whoever launched the Stuxnet worm. If anything, strategic cyberattack is likely to be viewed as an attractive "first resort," something that might be resorted to in lieu of aerial bombing, commando raids, or larger-scale field operations. Thus "early use" of cyberattack may violate one classic tenet of just war theory while at the same time shoring up the norm of lessening the overall amount of harm.

A further factor encouraging not only early use of strategic cyberattack, but also resort to digital attacks more generally throughout a conflict, is the point that "mass disruption" of information systems may impose very significant economic costs but is hardly likely to cause much, if any, loss of life. And in the situations in which cyberattack *is* likely to lead to large numbers of deaths—such as when military communications are compromised and soldiers, sailors, and airmen are killed as a result—the losses are to the combatants, not to civilians. Noncombatant immunity, a key *jus in bello* principle considered by several of the contributors to this volume, thus seems to be left relatively untouched by cyberwar—unless the economic cost inflicted on a whole society by strategic cyberwar is to be viewed as a violation of the ethical norm.

Other themes in military and security affairs that receive thoughtful treatment in *Binary Bullets* have to do with classical notions of deterrence and retaliation, and the relatively new problem of "attribution"—that is, the problem of ambiguity as to the identity of the cyberattacker. The veil of anonymity that often accompanies cyberattacks suggests that a new lease on life may be granted to the secret coup d'état and unattributable assassination techniques that comprised so much of "covert action"—but that went into eclipse—in the latter decades of the Cold War.[6] Against whom is one to retaliate when the identity of the perpetrator cannot be firmly

---

[5] The air power debate about strategic bombing is thoroughly discussed in Robert A. Pape, *Bombing to Win* (Ithaca: Cornell University Press, 1996).

[6] On the waning of this mode of "secret warfare," see Gregory Treverton's powerful critique, *Covert Action: The Limits of Intervention in the Postwar World* (New York: Basic Books, 1987).

established? And how is deterrence to work if the punitive threat of retaliation cannot be accurately aimed? What may happen is that standards of proof required for retaliatory action may be lowered, as seems the case with the Shamoon cyberattack in 2012 on the information systems of the Arabian-American Oil Company.[7] It is believed—but not proved beyond doubt—that Iran perpetrated this attack in retaliation for what it believed was an American-led Stuxnet cyberstrike on an Iranian nuclear facility. Neither attack has been attributed via high standards of verification; nevertheless, retaliatory action was taken.

Ethical reasoning in the form of judgments about good versus harm done, the level of proof required to act, and even the matter of ensuring that a retaliatory act is proportionate will all be of critical importance if an age emerges in which covert action is reenergized by worms, viruses, Trojans, and other tools of cyberwarfare. Thankfully, ethical judgments can to some extent be informed, perhaps even guided, by analogous historical situations. For example, there came a time in the Spanish Civil War (1936–1939) when ships bearing supplies for the Republican government's forces were being torpedoed by unknown submarines. British naval Intelligence felt it knew, but could not prove, that Italian submarines were the perpetrators, given that Mussolini supported Franco and the fascist forces trying to overthrow the government in Madrid. The Italians denied any involvement, but the British sent the warning that Rome would be held responsible for future attacks anyway. The "phantom submarines" ceased their strikes.[8]

While there may be some ways of shoring up deterrence by setting a norm for retaliation based on compelling but imperfect information, it seems very clear that the attribution problem is a grave one—not likely to be solved any time soon by technical measures—that poses the risk of striking at an innocent party while the true perpetrator gets away scot-free. Indeed, it is growing ever more apparent that deterrence in an age of cyberwar is likely to be far less reliable than it was in the long decades of the Cold War nuclear standoff. With this in mind, legal and ethical efforts to cope with the far-reaching consequences of the attribution problem should perhaps extend to consideration of another concept from that earlier era: arms control. In their thoughtful chapter, Michael Schmitt and Liis Vihul do this, exploring the possibilities for a kind of behavior-based form of arms control. Given that virtually all information technology should be seen as "dual use"—that is, can be employed to conduct cyberwar almost as easily as for commercial, social, or other purposes—arms control cannot be pursued by quantitative measures such as those that aim at reducing or limiting missiles, warheads, and fissile material.

---

[7] See Nicole Perlroth, "In Cyberattack on Saudi Firm, U.S. Sees Iran Firing Back," *New York Times*, October 24, 2012.

[8] Hugh Thomas, *The Spanish Civil War* (New York: Harper & Brothers, 1961), 475–76, makes clear that the Italians were indeed the perpetrators and that the retaliatory threats led to Mussolini's decision to suspend the submarine campaign.

Nevertheless, as we know from the general success of the conventions regarding biological and chemical weapons, a behavioral basis for arms control is feasible. Ethicists can and should play a powerful role in encouraging such efforts.

Issues of policy and strategy aside, perhaps the most important conceptual contribution ethicists can make to the discourse on cyberwar is at the definitional level. Definitions, and the current state of the debates about definitions, of cyberwar are carefully considered throughout *Binary Bullets*. Yes, there are distinctions to be made between cyberspace-based crime, terrorism, acts of espionage, and sabotage (or as I like to call it, *cybotage*). But war lurks here as well. Cyberwar is disruptive in costly ways rather than physically destructive. It is far less lethal than aerial bombing as a form of strategic attack. But cyberwar techniques used in close support of field operations caused many deaths in the Russo-Georgian War, and will no doubt do so in future conflicts as well. It may turn out that the side in a shooting war that has the cyberedge will win lop-sided victories, much as, when the Germans were still pioneering the blitzkrieg concept, their early campaigns were won at very low cost. In the 1940 Battle of France, German casualties amounted to less than one-fifth those of their opponents—*not* counting prisoners taken. In 1941, the Germans conquered Yugoslavia and its million-man army in less than two weeks, the Wehrmacht suffering only 151 battle deaths.[9] It is this sort of edge that David Ronfeldt and I thought of when we developed the cyberwar concept in the first place.

And if we were right about the potential of cyberwar techniques being able to lower the cost and increase the decisiveness of military operations on land, at sea, and in the air and space, we should also have been concerned, even back in the early 1990s, that we might be contributing to developments that would make war more thinkable once again. In the 1970s, George Quester articulated the argument that, when attacking others seems easier than defending against attacks, war is more likely.[10] At present, it seems that taking the cyberoffensive is far easier—for networked nonstate as well as for nation-state actors—whether in support of "actual wars" or in virtual strikes on critical infrastructures. However one chooses to look at cyberwar there seems to be a world of opportunity for going on the offensive. In this regard, the paradox of cyberwar is that, while it may allow some virtually bloodless conflicts to be waged in cyberspace, and it may reduce the carnage wrought by and enable the swift resolution of more traditional wars, it will make the whole notion of resorting to war more attractive. Ronald Reagan put it well when he spoke of war in the atomic age: "A nuclear war cannot be won and must never be fought." The sad cyber corollary is that cyberwar can be waged all

---

[9] See John Keegan, *The Second World War* (New York: Viking Penguin, 1989), 154–57. Despite the ease of the initial conquest, the Germans would face their most nettlesome partisan resistance in Yugoslavia.

[10] George Quester, *Offense and Defense in the International System* (New York: John Wiley & Sons, 1977).

too easily, too anonymously, and with every prospect of achieving one's aims at low cost.

Thus an age of cyberwar looms ahead. No doubt new forms of arms racing will emerge, and fresh conflicts will emerge along with them. But given the still-embryonic state of cyberweaponry—relative rates of progress are swift, but in absolute terms cyberwar remains for now in a very early stage of development—there is the blessing of time to think through the ways that traditional notions of going to war and waging war justly might apply to this new era of conflict. *Binary Bullets* provides exactly the sort of foundational building blocks needed for parsing, and ultimately making informed judgments about, the ethics of cyberwar. And its contributors have done so at just the right time: when concerns have risen, around the world, among mass publics and their governments—but before cyberwar has had time to "go viral."

John Arquilla
Monterey, California
Fall 2014

Notes on Contributors

Fritz Allhoff is associate professor in the Department of Philosophy at Western Michigan University and senior research fellow at the Centre for Applied Philosophy and Public Ethics, Charles Sturt University, Australia. He has held visiting appointments at the University of Oxford, the University of Michigan, and the University of Pittsburgh. His research principally involves the ethics of war and the ethics of technology, including military technologies. Allhoff is author or editor of nearly thirty books; two of his most recent are *Terrorism, Ticking Time-Bombs, and Torture* (University of Chicago Press, 2012) and *The Routledge Handbook of Ethics and War* (2013). He was a founding member of the International Intelligence Ethics Association, serves on the advisory board for the International Committee of Military Medicine (Switzerland), and is active in the Consortium for Emerging Technologies, Military Operations, and National Security (CETMONS). He recently graduated from the University of Michigan Law School, magna cum laude, and is spending his sabbatical as a clerk for the Alaska Supreme Court.

Matthew Beard is a military ethicist and philosopher with the University of Notre Dame, Australia. He has served as managing editor for *Solidarity: The Journal for Catholic Social Thought and Secular Ethics*, and is currently research associate with the Centre for Faith, Ethics and Society at Notre Dame. Matthew submitted his doctoral thesis in 2014, titled "War Rights and Military Virtues: A Philosophical Reappraisal of Just War Theory," and was the inaugural recipient of the Morris Research Scholarship from Notre Dame. He has discussed subjects including military ethics, moral injury and PTSD, cyberwar, torture, and medical ethics among others in book chapters, scholarly articles, radio interviews, public opinion pieces, and at academic conferences both domestically and internationally.

Daphna Canetti is an associate professor in the School of Political Science at the University of Haifa. Her research focuses on the psychological mechanisms underlying the politics of terrorism and political violence, with a particular interest in individual-level exposure to kinetic and cybernetic security threats.

Methodologically, she uses controlled randomized experiments, spatial analysis, survey experiments, and biopolitical approaches. She has served on editorial boards of *Democracy and Security* and *Political Psychology*, was a Fulbright Fellow at Notre Dame University, and was a Rice Family Foundation Visiting Professor at Yale University. She has received over $3 million in grants and published in journals such as *American Journal of Political Science, The Lancet, Journal of Conflict Resolution, Political Behavior, Political Psychology, Journal of Consulting and Clinical Psychology,* and the *British Journal of Political Science*. Her work has been featured in various media outlets including the *Washington Post, Times of Israel,* and *Psychology Today*.

David Danks is professor of Philosophy and Psychology, and Head of the Department of Philosophy, at Carnegie Mellon University. His primary research is in computational cognitive science, with a particular focus on the representations and processes that underlie complex human cognition. He has also worked on a range of problems in machine learning, including the development of algorithms for automated causal inference from multiple sources and from time-series data. Dr. Danks received his PhD from the University of California, San Diego in 2001. He is the author of *Unifying the Mind: Cognitive Representations as Graphical Models* (MIT Press, 2014) and has published widely in philosophy, psychology, and computer science.

Joseph H. Danks is research professor and Area Director for Performance and Analysis at the University of Maryland Center for Advanced Study of Language (CASL). His research has focused on how people comprehend sentences and text, especially across languages; the cognitive processes involved in translation; and how elderly patients communicate their life-sustaining treatment preferences. At CASL, he has investigated using a cultural lens and social media to forecast the plans and intentions of a country's leadership and populace, and also how to conduct remote psychological assessment of cyberadversaries. Dr. Danks received his PhD from Princeton University and taught for many years at Kent State University, serving as Chair of Psychology and as Dean of Arts and Sciences. He also has taught at Princeton, Stanford, the University of Warsaw, and Aristotle University of Thessaloniki. Dr. Danks has authored and edited several books and published extensively in psycholinguistics and cognitive psychology.

Randall R. Dipert has taught at the (SUNY) University at Buffalo since 2000, where he is the C.S. Peirce Professor of American Philosophy. He was a professor of philosophy from 1995 to 2000 at the US Military Academy (West Point, NY) and did research for a year on cyberwar for the Stockdale Center at the US Naval Academy (Annapolis, MD). His most active current interests are military ethics (the philosophy of war and peace), logic, and applied ontology. He is a founding member of the National Center for Ontological Research (NCOR). His past research includes logic, the history and philosophy of logic, Peirce and Early American Pragmatism, the philosophies of logic and mathematics, the philosophy of artifacts, aesthetics, action theory, and metaphysics, especially the metaphysics and logic of relations. He also has interests in ethics and political philosophy and, most recently, in the philosophy of war and peace and value-theoretic implications of game theory. Recent publications and papers include the ethics of cyberwarfare and the ethics of preemptive/preventive war, including the seminal paper "The Ethics of Cyberwarfare" published in the *Journal of Military Ethics* in 2010.

Michael L. Gross is Professor and Head of the School of Political Science at the University of Haifa, Israel. He completed his BA at the Hebrew University of Jerusalem, MA at Northwestern University (philosophy), and PhD in political science at the University of Chicago. Gross has published widely in medical ethics, military ethics, military medical ethics, and related questions of medicine and national security. His articles have appeared in the *New England Journal of Medicine, American Journal of Bioethics, Journal of Military Ethics, Cambridge Quarterly of Healthcare Ethics, Hastings Center Report, Journal of Medical Ethics, Journal of Applied Philosophy, Social Forces*, and elsewhere. His books include *Ethics and Activism* (Cambridge University Press, 1997), *Bioethics and Armed Conflict* (MIT Press, 2006), *Moral Dilemmas of Modern War* (Cambridge University Press, 2010), *Military Medical Ethics for the 21st Century* (Ashgate, 2013), and a forthcoming book: *The Ethics of Insurgency: A Critical Guide to Just Guerrilla Warfare* (Cambridge University Press). He has been a visiting fellow at the University of Chicago, MacLean Center for Clinical Medical Ethics, and the European University Institute, Department of Political and Social Sciences in Florence, Italy. He serves on regional and national bioethics committees in Israel and has led workshops and lectured on battlefield ethics, medicine, and national security for the Dutch Ministry of Defense, the US Army Medical Department at Walter Reed Medical Center, the US Naval Academy, the International Committee of Military Medicine, and the Medical Corps and National Security College of the Israel Defense Forces.

Adam Henschke is a research fellow at the Australian National University's National Security College, an adjunct research fellow with the Centre of Applied Philosophy and Public Ethics, Charles Sturt University, and was recently a visiting assistant professor in the Department of Politics and Public Administration, University of Hong Kong. In 2012 Adam received a Brocher Foundation Research Fellowship (Geneva, Switzerland) to look at the ethical, legal, and social implications of "Open Health" technologies and programs, and in 2009 was a visiting researcher at the Delft University of Technology (Delft, The Netherlands). He has published in areas that include information theory, ethics of technology, and military ethics. He coedited the *International Journal of Applied Philosophy*'s symposium on *War in the 21st Century* (Fall 2012) and *The Routledge Handbook of Ethics and War* (July 2013). He is currently researching, among other things, ethics and norms in security.

Ryan Jenkins earned a PhD in Philosophy from the University of Colorado Boulder in 2014 and began an assistant professorship at California Polytechnic

State in January 2015. His recent publications include "Is Stuxnet Physical? Does It Matter?" in the *Journal of Military Ethics*, and "Moral Imperialism," "Consumer Society," and "Oxfam" in the *Springer Encyclopedia of Global Justice*. His interests include normative ethics (especially rule-consequentialism), applied ethics (especially military ethics and emerging technologies), and social and political philosophy.

George R. Lucas Jr. recently retired from the Distinguished Chair in Ethics in the Vice Admiral James B. Stockdale Center for Ethical Leadership at the US Naval Academy, and he is currently Professor of Ethics and Public Policy at the Graduate School of Public Policy at the Naval Postgraduate School in Monterey, California. He has taught at Georgetown University, Emory University, Randolph-Macon College, the French Military Academy (Saint-Cyr), and the Catholic University of Leuven in Belgium. His main areas of interest are applied moral philosophy and military ethics, and he has written on such topics as: irregular and hybrid warfare, cyberconflict, military and professional ethics, and ethical challenges of emerging military technologies. His most recent book is *Anthropologists in Arms: The Ethics of Military Anthropology* (AltaMira Press, 2009); he has a commissioned work on military ethics in preparation for Oxford University Press and is currently editing the *Routledge Handbook on Military Ethics* for publication in 2015.

Seumas Miller is a professorial research fellow at the Centre for Applied Philosophy and Public Ethics (CAPPE) (an Australian Research Council Special Research Centre) at Charles Sturt University (CSU) (Canberra) and the 3TU Centre for Ethics and Technology at Delft University of Technology (The Hague). He is the Foundation Director of CAPPE, Foundation Professor of Philosophy at CSU, and served two terms as a head of the School of Humanities and Social Sciences at CSU. He is the author or coauthor of over 200 academic articles and 15 books, including *Social Action* (Cambridge University Press, 2001), *Corruption and Anti-Corruption* (Pearson, 2005), *Terrorism and Counter-terrorism* (Blackwell, 2009), *The Moral Foundations of Social Institutions* (Cambridge University Press, 2010), and *Investigative Ethics* (Blackwell, 2014). He has also been awarded numerous competitive grants and consultancies.

Brian Orend is the Director of International Studies, and a Professor of Philosophy, at the University of Waterloo in Canada. His PhD is from Columbia University in New York City. He has taught at Columbia, Waterloo, and the University of Lund in Sweden (where he was recently Distinguished Visiting Professor of Human Rights). His research concentrates on three areas: the ethics of war and peace (especially postwar reconstruction); human rights; and happiness. He is the author of six books, including three used widely as required texts at colleges and universities around the world: *Human Rights: Concept and Context* (Broadview, 2002); *Introduction to International Studies* (Oxford University Press, 2012); and *The Morality of War*

(Broadview, 2nd ed., 2013). He is currently completing a book on happiness, as well as a new edition of Kant's *Perpetual Peace*.

Heather M. Roff is a visiting professor of Security Studies at Josef Korbel School of International Studies at the University of Denver, and a research associate at the Eisenhower Center for Space and Defense Studies at the US Air Force Academy. Dr. Roff has held faculty positions at the University of Waterloo and the US Air Force Academy. She is author of *Global Justice, Kant and the Responsibility to Protect: A Provisional Duty* (Routledge, 2013), and has published various articles in academic journals, such as the *Journal for Military Ethics* and *Global Responsibility to Protect*. She also blogs for the *Huffington Post* and *Duck of Minerva* on issues of international ethics and politics.

Michael N. Schmitt is the Charles H. Stockton Professor and Director of the Stockton Center for the Study of International Law at the US Naval War College in Newport, Rhode Island, and Professor of Public International Law at the University of Exeter Law School in the United Kingdom. He is also a Senior Fellow at the NATO Cyber Defence Centre of Excellence, Fellow at Harvard Law School's Program on International Law and Armed Conflict, and editor-in-chief of *International Law Studies*. He directed the project leading to completion of the *Tallinn Manual on the International Law Applicable to Cyber Warfare*, and is directing its follow-on project, *Tallinn 2.0*.

Michael Skerker is an associate professor in the Leadership, Ethics, and Law department at the US Naval Academy (USNA). He received his BA in Comparative Literature from Brown University (1997) and his MA and PhD in Ethics from the University of Chicago Divinity School (1999, 2004). Before joining the faculty of USNA, he taught at the University of Chicago and DePaul University. His academic interests include professional ethics, just war theory, moral pluralism, theological ethics, and religion and politics. Publications include works on ethics and asymmetrical war, moral pluralism, intelligence ethics, and the book *An Ethics of Interrogation* (University of Chicago Press, 2010). He is currently working on a book *The Moral Status of Combatants*, which defends the post-Westphalian idea of the moral equality of combatants. The manuscript won the 2013 Charles Sharp Memorial Prize for best unpublished work on military ethics.

Bradley Jay Strawser is an assistant professor of Philosophy in the Defense Analysis Department at the US Naval Postgraduate School in Monterey, California and a research associate with Oxford University's Institute for Ethics, Law, and Armed Conflict. Prior to his current appointments, Dr. Strawser was a Resident Research Fellow at the Vice Admiral James B. Stockdale Center for Ethical Leadership in Annapolis, Maryland. Previously he taught philosophy and ethics at the US Air Force Academy in Colorado and the University of Connecticut. Before his academic career, Strawser served as an active duty officer in the US Air Force for nearly eight years. His research focus is primarily ethics and political philosophy, though he has also written on metaphysics, ancient philosophy, and human rights. His work has appeared in such journals as *Analysis, Philosophia, Public Affairs Quarterly, Journal of Military Ethics, Journal of Human Rights,* and *Epoché.* He recently published *Killing by Remote Control: The Ethics of an Unmanned Military* (New York: Oxford University Press, 2013), an edited volume on the many moral issues raised by drone warfare, and *Killing bin Laden: A Moral Analysis* (New York: Palgrave Macmillan, 2014).

Liis Vihul serves as a scientist in the Law and Policy Branch at the NATO Cooperative Cyber Defence Centre of Excellence. She holds an MA in law from the University of Tartu and an MSc in information security from the University of London. Her research is focused on the interplay between public international law and cyberoperations, in particular the legal requirements for attributing cyberoperations to states and the obligation of states to stop third-party malicious cyberactivities from their territories. Ms. Vihul is a coauthor of the book *International Cyber Incidents: Legal Considerations.* She is also actively involved in planning and executing training on cyberissues for legal professionals. Ms. Vihul was the Project Manager for the *Tallinn Manual on the International Law Applicable to Cyber Warfare,* published by Cambridge University Press in 2013, and has taken up the same role for its follow-on project *Tallinn 2.0.*

Israel Waismel-Manor (PhD in Government, Cornell University) is senior lecturer at the School of Political Science, University of Haifa. His research focuses on biopolitics, new media, political behavior, public opinion, and political communication. His work has been published in such journals as the *Journal of Communication, Political Communication, Political Behavior, International Journal of Press and Politics, New Political Science, International Journal of Public Opinion Research, Political Science and Politics,* and *European Neuropsychopharmacology.* Waismel-Manor has been a visiting professor at Stanford University, Cornell University, and an exchange scholar at the Annenberg School for Communication at the University of Pennsylvania. He received the NSF Time-Sharing Experiments in the Social Sciences (TESS) award, a commission by the Voting Rights Research Initiative at the University of California, Berkeley, and the Outstanding Teaching Award—Faculty of Social Sciences, University of Haifa (2007). His work has been featured in various media outlets such as the *New York Times, Washington Times, Huffington Post,* and *Jerusalem Post.*

David Whetham is a senior lecturer in the Defence Studies Department of King's College London, based at the Joint Services Command and Staff College at the U.K. Defence Academy. David's main research interests are focused on the ethical dimensions of warfare. In Spring 2011, David was a Resident Fellow at the Stockdale Center for Ethical Leadership at the US Naval Academy, Annapolis, and in 2009 he was a visiting fellow with the Centre for Defence Leadership and Ethics at the Australian Defence College, Canberra. He is also a visiting lecturer in military ethics at the Baltic Defence College and for the Royal Brunei Armed Forces. David was a cofounder of the European Chapter of the International Society for Military Ethics and sits on the editorial advisory board of the *Journal of Military Ethics.*

Introduction

Editors' Introduction

The histories of technology and of military practice are intimately entwined. Think of the well-worn examples of the English longbow's dominance in the fifteenth century at the Battle of Agincourt, the close relationship between the advent of aviation technology and military airpower, or the rapid development of nuclear physics in the twentieth century leading directly to the deployment of the atomic bomb in Hiroshima and Nagasaki. The rise of computers is likewise similarly married to the military. Alan Turing is known equally for his work as part of the British code-breaking efforts during World War II and his contributions to computers and the philosophy of computing. In the United States, it was the Advanced Research Projects Agency (ARPA, since renamed DARPA) that developed ARPANET, the genesis of what is now the Internet. During the Cold War, the complexities of overseeing the Strategic Air Command pushed the development of ever more advanced computational systems to monitor, control, and integrate multiple strategic bombers and nuclear missiles, leading to the explosion of computing power we see today. To say that "cyber" and "military" go together is an understatement.

Moving from the command of strategic bombers to our own lives, consider that you most likely used an Internet-connected device today—perhaps a computer at your workplace, a portable laptop, or, increasingly, a smart phone or tablet. Such devices have infiltrated our personal and professional lives. Indeed, most people reading these words are online every day. We are—nearly all of us—deeply enmeshed in the cyberrealm. We too quickly forget just how precipitous this change has been. As former President Bill Clinton once noted, "When I took office, only high energy physicists had ever heard of what is called the World Wide Web. . . . Now even my cat has its own page."[1] Although Clinton exaggerates in attributing early Internet access only to physicists, his broader point is on target. Only twenty years ago, less than half of one percent of the world's population had Internet access.

[1] White House Office of the Press Secretary, "Excerpts from Transcribed Remarks by the President."

Today over 40 percent of the globe gets online regularly. Taking into account the starkly differing levels of development across the globe, as well as considering the many children and elderly who do not get online, that number as a portion of total global population is shocking. In the so-called developed world, the speed with which the Internet has penetrated every aspect of our lives is astonishing. In Iceland, over 96 percent of the population is online; in Norway it is over 97 percent. There is no hyperbole in claiming that this constitutes a worldwide revolution in the ways and means in which people communicate, taking place over merely two decades. That kind of rapid change, this cyberrevolution, may well be unprecedented in human history.

And let us not get lost thinking of the cyberrealm as merely "being online." Beyond widespread (and growing) Internet dependency, many basic social functions are now reliant upon computer-mediated communications. Your electricity, water, and telephone are all dependent upon computers operating reliably. And, with this dependence comes certain vulnerabilities: even though an industrial control system might not be online, it is still be a target for a well-designed cyberweapon. This increasingly wired world has created an array of new channels for communication, services, and functions interwoven into our everyday lives. Perhaps not surprisingly, this revolution has also brought with it an array of new channels for waging war. And just as the changes to information, communication, and commerce have been new and unprecedented, so too does this new form of warfare—cyberwarfare—bring with it original and unique challenges to our understanding of war. The questions raised by firing binary bullets instead of lead bullets are many, and they weigh heavily upon some of the central tenets of the normative frameworks that have been traditionally used for thinking about warfare.

But just what do we mean by "cyberwarfare"? To understand "cyberwarfare," let us start with our conceptions of "war" and "warfare," and then add the "cyber" dimension. Brian Orend writes that "[w]ar should be understood as an *actual, intentional,* and *widespread* armed conflict between political communities. Thus, fisticuffs between individual persons do not count as a war, nor does a gang fight, nor does a feud on the order of the Hatfields versus the McCoys."[2] Prussian military general and strategist Carl von Clausewitz famously said that war is *"an act of violence intended to compel our opponent to fulfill our will."*[3] So, war is something violent, organized, and done for some second-order purpose—we do not go to war simply to kill people, but to bring about the fulfillment of our will. Warfare is simply the means by which we carry out this thing called "war."

In our efforts to understand how adding "cyber" to warfare changes matters, we might first look to ways of conceptualizing cyberspace. The 2014 report on cybersecurity by the National Academies in the United States offers this "rough definition":

---

[2] Orend, "War."
[3] Clausewitz, *On War*, 101, emphasis in original.

> [C]yberspace consists of artifacts based on or dependent on computing and communications technology; the information that these artifacts use, store, handle, or process; and the interconnections among these various elements. But the reader should keep in mind that this is a rough and approximate definition and not a precise one.[4]

Thus, bringing the concepts of war and cyber together, we can think of cyberwarfare as "an *actual, intentional,* and *widespread* armed conflict between political communities," conducted by means of, or targeted at, "computing and communications technology; the information that these artifacts use, store, handle, or process; and the interconnections among these various elements." Or, to adapt Clausewitz's concept, "cyberwar is an act of violence conducted by, or targeted at, information communication technologies intended to compel our opponent to fulfill our will." Such a simple, straightforward definition can serve as a rough starting-point for the investigations in this book.[5]

This construction may seem sensible to some. But given that the cyberrealm necessarily involves the so-called virtual realm, some question whether any cyberact, no matter how malicious in intent, is even properly thought of as violent. Thomas Rid makes this argument most forcefully and contends that we should reject the concept of cyberwar as a meaningful notion.[6] In contrast to Rid, however, we may think that the virtual nature of a cyberattack is irrelevant. What matters, on this view, is the actual results that are brought about in the world: the measured outcome, the damage (physical or otherwise), or, of course, lives lost.[7]

The cyberweapon known as Stuxnet, for example, caused significant physical damage. Stuxnet was

> a hostile cyber operation that targeted the computer-controlled centrifuges of the uranium enrichment facility in Natanz, Iran. . . . After taking control of these centrifuges, Stuxnet issued instructions to them to

---

[4] Clark, Berson, and Lin, *At the Nexus of Cybersecurity and Public Policy*, 8–9.

[5] We should note at the outset of this volume that even given our rough attempt here at getting a handle on the terms "cyberwar" and "cyberwarfare," the concepts themselves are still very much ill-defined and highly contentious in the present debate. This is even more the case for the various derivative terms that come from it, some of which we have already used, such as "cyberweapon," "cyberconflict," "cyberattack," and all the rest. The vocabulary in this debate is still in flux at this early stage and in need of good analysis and clarification. Indeed, getting clear on some of these concepts is a principal question tackled by some of the chapters in this book, as discussed below.

[6] Rid, *Cyber War Will Not Take Place*.

[7] The *Tallinn Manual*, for example, takes this analogical reasoning—whether an attack is virtual or not is irrelevant. What matters is the damage caused. Schmitt, *Tallinn Manual*. Also see David Whetham's chapter in this volume, which directly challenges some of Rid's claims.

operate in ways that would cause them to self-destruct, unbeknownst to the Iranian centrifuge operators.[8]

In more traditional military attacks, cyberweapons have been used as force multipliers in individual skirmishes between Israel and Syria in 2007[9] and in larger conflicts such as the Russian invasion of Georgia in 2008.[10] Perhaps the most controversial event occurred in 2007 in the Estonian capital, Tallinn. In this case, military, civil, and private actors located in Tallinn were the target of highly disruptive cyberattacks.[11] This event is particularly instructive as it raises some of the key ethical issues around cybewarfare. Firstly, in terms of attributing the attack, while Russian patriots are widely believed to be the source of the cyberattack, it is still unconfirmed if elements in the Russian government or military should rightly be held accountable. Further, these attacks were mostly sustained distributed denial of service (DDoS) attacks, which did not damage the targets so much as merely disrupted their normal capability function. Despite some calls for a kinetic military response by North Atlantic Treaty Organization (NATO) at the time, most hold that a cyberattack that inflicts no lasting physical damage does not rise to the level of sufficient harm to justify a traditional military response, yet the harm inflicted by the disruption was real.

Adding further complexity to these discussions is the moral responsibility for escalation and the risk of conflict between nations. For instance, one might agree with Rid that "cyberwarfare" is something of a misnomer. However, given that international relations in cyberspace are still emerging, the potential for nonviolent cyberactions to possibly bring on or escalate physical conflict brings with it its own moral concerns. In one such example the American cybersecurity company Mandiant released a report accusing the People's Liberation Army (PLA) of China of being deeply involved in cyberespionage.[12] This report exposed and increased tensions between the United States and China, and in May 2014, the US Justice Department charged five members of the PLA with cyberattacks.[13] While no military force has been used, the cyberespionage has certainly had a significant impact on relations between the United States and China.[14]

In a bid to reduce uncertainty about the legal status of cyberattacks, a group of experts looked at cyberattacks in relation to international law, producing what is known as the

---

[8] Clark, Berson, and Lin, *At the Nexus of Cybersecurity and Public Policy*, 35.
[9] Rid, *Cyber War Will Not Take Place*, 42.
[10] Rohan, "Georgia Says Russian Hackers Block Govt Websites."
[11] Rid, *Cyber War Will Not Take Place*, 6–7, 31–32.
[12] China has never officially confirmed these accusations. For the full report, see "APT 1: Exposing One of China's Espionage Units."
[13] Schmidt and Sanger, "5 in China Army Face U.S. Charges of Cyberattacks."
[14] Seumas Miller's and Michael Skerker's chapters in this volume both discuss different aspects of responsibility and states in and around cyberespionage.

Tallinn Manual.[15] The *Tallinn Manual* sought to develop a series of "black letter rules" in which the laws of international humanitarian law are applied to the cybercontext, with an accompanying commentary by the authors. Though no international consensus on cyberwarfare, cyberattacks, or cyberespionage is likely to emerge in the near future, efforts like the *Tallinn Manual* seek to give clarity to a developing set of practices.[16]

That said, it bears repeating that we are still at the earliest stages of answering the conceptual and the many other ethical, legal, policy, and normative questions raised by cyberwar. Indeed, it is fair to say that we are still only on the cusp of even fully *understanding* what the most important questions for this new form of warfare will be. This scholarship immaturity[17] is surprising given the widespread impact the cyberrealm already commands in our lives and the speed with which modern militaries have adopted cybertactics. Yet we are still unsure what to make of the fact that our ever-expanding cybercapabilities bring with them new means of waging war. How should we think about the moral implications of a cyberweapon that can "attack" a computer system on the other side of the world in an instant? Should such a thing be properly considered an act of war like a physical attack would be? Or, alternatively, is this simply another advanced form of espionage, or perhaps something altogether different still? The ways in which this form of warfare diverges from conventional violence are multitudinous. Consider that a target in cyberspace is in many ways more appealing than physical targets since the aggressor would not need to incur the expense and risk of transporting equipment and deploying troops into enemy territory, not to mention the political risk of casualties. Cyberweapons could be used to attack anonymously at a distance while still causing much mayhem, on targets ranging from banks to media to military organizations. These kinds of advantages suggest that cyberweapons would be an excellent choice for an unprovoked surprise strike. Yet many wonder whether cyberconflict and the harms it can inflict are weighty enough to be considered a legitimate *casus belli*. Others wrestle with the ways in which this form of conflict changes some of the very terms of the debate that are traditionally taken for granted. For example, simply knowing *who* is attacking *who*—the so-called attribution problem—or even what the weapon is that is being used, or even *when* or *if* one is being attacked at all, are all questions that may not be so easily ascertained in the world of cyberwarfare.[18]

---

[15] Schmitt, *Tallinn Manual*.

[16] The chapters by George R. Lucas Jr., Michael N. Schmitt and Liis Vihul, and Randall Dipert all cover related and overlapping issues of moral, legal, and social norms around cyberwarfare.

[17] As recognised by both George R. Lucas Jr. and Randall Dipert in this volume, some areas of scholarship are a little more advanced—considerations and concerns had been raised from the early 1990s by practitioners in computer science, international relations, and special operations. Sustained ethical analyses, they suggest, have entered the dialogue more recently.

[18] Some of these points have been raised in Allhoff and Jenkins, "When Is a Real-World Response to a Cyberattack Justifiable?"; and Henschke and Lin, "Cyberwarfare Ethics, or How Facebook Could Accidentally Make Its Engineers into Targets."

Some hold that these and other challenges should compel us to reexamine the lenses through which we view cyberconflict. Today, many nations have acquired the capability to strike in cyberspace, but whether this new means of state force is a legitimate form of coercion remains an open question. After all, the laws of armed conflict (LOAC), or international humanitarian law (IHL), were not written with cyberspace in mind. The world community thus faces a significant policy gap with regards to cyberwar. Several organizations and experts have tried to address these gaps in recent years, yet international consensus and legal authority both remain elusive.[19] Further, though some have focused on the philosophical framework for information warfare, there remains a significant gap in our understanding of how the ethics of cyberwar should be understood.[20] Are the traditional confines of just war theory adequate to critically engage cyberwar or are new approaches needed? Can just war theory accommodate the new means and methods of cyberwarfare, or do these new ways of fighting render its application obsolete?

These questions strike at the very center of contemporary intellectual discussion over the ethics of war. This book engages these questions head-on with contributions from the top scholars working in this field today. Intended for anyone trying to find their way through the maze of work on the topic, this volume's chapters aim to strike a balance between deep normative questions cyberwarfare entails for just war theory and the more pragmatic and policy questions raised by conflict in the cyberrealm. The volume is designed to present the leading edge of philosophical research on the ethics of cyberwarfare, and our hope is that it will help chart the course for future debate as this technology continues to evolve.

The book begins with a foreword by John Arquilla, a professor and chair of the Defense Analysis Department at the Naval Postgraduate School. Arquilla is himself a pioneer in thinking about cyberwarfare. His contribution helps show us the ways in which a familiarity with recent historical developments and uses of military technologies can inform our understanding of this emerging set of practices. Following the foreword, the book is composed of four principal parts.

Part one, "Foundational Norms for Cyberwarfare," explores the normative framework—both moral and legal—that attaches to this realm. George R. Lucas Jr.'s chapter contends that new moral norms have begun to emerge regarding the use of cyberwarfare. Lucas also provides a useful history of cyberconflict. Michael N. Schmitt and Liis Vihul countenance that legal norms are emerging,

---

[19] See, e.g., Owens, Dam, and Lin, "Technology, Policy, Law, and Ethics Regarding U.S. Acquisition and Use of Cyberattack Capabilities"; Liebenthal and Singer, *Cybersecurity and U.S.-China Relations*; Libicki, *Cyberdeterrence and Cyberwar*; Lin, "Cyber Conflict and International Humanitarian Law"; and Arquilla, "Cyberwar Is Already Upon Us: But Can It Be Controlled?"

[20] See, e.g., Floridi, "Get Ready For Cyberwar," *Philosopher's Magazine*, 46; Taddeo, "Analysis for a Just Cyber Warfare."

and they track their recent historical development and their impact on regulating cyberwarfare. In their essay, they survey the roles of treaty and soft law, arguing that the tools needed for good cybergovernance are already substantially in place, but need to be augmented by other instruments of international law. The next chapter in this section comes from Randall R, Dipert, who considers what makes cyberwarfare distinctive from more traditional forms of warfare. One of the key differences, he suggests, is that given that a cyberattack will typically not kill or cause widespread physical destruction, cyberwarfare causes us to rethink our normative frameworks and thresholds and consider other, lesser forms of intentional harm, such as economic damage and espionage.

Part two of the book, "Cyberwarfare and the Just War Tradition," considers the ways in which cyberwarfare trades on historical questions and issues in the just war tradition. The most fundamental question therein is whether cyberwar should properly be understood as war. David Whetham begins this unit by proposing that—contra thinkers like Rid—cyberwar is indeed a plausible concept. To make his argument he looks back into medieval history, to the strategic role played by *chevauchées*. These *chevauchées* involved mounted soldiers who would deliberately spread across an enemy's territory, plundering and destroying everything in their path. They did this to weaken rivals, undermining their political legitimacy and forcing negotiation and compromise, perhaps forgoing the need to fight at all. By comparing some forms of cyberattacks to these *chevauchées*, Whetham suggests that they can be conceptualized within a broader notion of war as social phenomenon. In the next chapter, Ryan Jenkins postulates a significant upside to cyber-, as opposed to kinetic, warfare, namely the possibility of mitigating collateral damage. Whereas kinetic warfare invariably puts noncombatants at risk, Jenkins argues that cyberwarfare could, at least in principle, be more discriminatory and, therefore, morally preferable. Brian Orend has made an important contribution to the just war tradition in arguing that, as wars have beginnings, middle, and ends, so should *jus post bellum* complement the traditional categories of *jus ad bellum* and *jus in bello*. In his contribution in this book, he applies his *jus post bellum* framework to cyberwarfare, arguing for associative rights and obligations following cyberattacks.

The "Ethos of Cyberwarfare" frames part three of the book. Here, our authors examine the roles of various agents who participate in cyberwarfare. Matthew Beard takes up the operational side, looking at what he dubs "cyberwarriors." He argues that serious conversations about the morality of cyberwarriors are lacking from the broader dialogue of cyberwarfare. Building from notions of honor and virtue, Beard carves out a distinct conceptual and moral space for how a cyberwarrior should act. Daphna Canetti, Michael L. Gross, and Israel Waismel-Manor bring an interdisciplinary perspective to bear on the effects of cyberwarfare on noncombatants in their chapter. In particular, they offer fascinating empirical results exploring the notion of the harm that a cyberattack causes psychologically. They evaluate the psychological effects of cyberterrorism, drawing from laboratory-simulated cyberattacks, ultimately concluding that cyberterrorism causes significant anxiety and substantially influences rational political thinking, so much so that cyberterrorism violates the principle of noncombatant immunity even when it does not cause physical harm. Drawing on the human targets of cyberattack, David Danks and Joseph H. Danks investigate the cognitive constraints, biases, and heuristics of human agents across a range of roles: developer of a cyberaction; target of that cyberaction; defender against some cyberaction; and third-party observer, whether neutral nations or even the public within a nation engaged in cyberactions. Closing off the section, and in line with Beard, Canetti, Gross, and Waismel-Manor, Danks and Danks remind us that a full understanding of the conceptual distinctions and ethical dimensions of cyberwarfare must incorporate the human actors with all of their cognitive, conceptual, and cultural biases, tendencies, and foibles.

The final section, "Cyberwarfare, Deception, and Privacy," tackles what many view to be lacunae of the cyberrealm. From recent high-profile cases peeled from the headlines, to more abstract academic discourse, nobody doubts the capacity of cyber to bear on the issues of deception and privacy. The just war tradition offers a distinction between licit deception (e.g., ambush) and illicit perfidy (e.g., feigning protected status), and Heather M. Roff queries how to extend this distinction into cyberwarfare. In particular, there is a way in which all cyberwarfare could appear perfidious, and therefore proscribed by the laws of war. Roff, though, considers and argues for a more moderate approach. Seumas Miller's chapter asks us to think of differences between cyberwar, cyberterrorism, cybercrime, cyberespionage, and what he refers to as covert political cyberaction—a species of covert political action. He argues that many, if not most, cyberattacks perpetrated by nation-states on other nation-states are best understood neither as acts of war, nor as crimes, but rather as a new form of covert political action. He gives a preliminary ethical analysis of covert political cyberaction, arguing that much covert political cyberaction is best understood as a species of dirty hands action. The book closes with an essay from Michael Skerker on the morality of various types of national security electronic surveillance programs, which seek to analyze communication "metadata," and the targeted intercepts of communication by particular foreigners. Skerker argues that though moral consistency permits foreign governments to engage in intelligence operations that are necessary for their national security, it requires such intelligence collection abroad to use the best technology and tactics at a state's disposal.

The book's aims are not to offer a comprehensive set of answers about the ethics of cyberwarfare. Rather, they are to advance the discussion and to ensure that a range of the important conceptual and ethical concerns are brought together to provide a more expansive view of the debate than that found in heretofore isolated places. The revelations surrounding Stuxnet, the role of the People's Liberation Army as exposed by the Mandiant Report, and the recent whistleblowing regarding the US surveillance program PRISM each give considerable weight to concerns that cyberincidents are getting ever closer to standard conceptions of warfare. This is shocking news to many, and it reveals the precipitous nature that this transformation has had and will continue to have for the foreseeable future. *Binary Bullets* presents an opportunity for academically rigorous ethical discussion of what is almost certain to be a continuing problem for warfare into the future.

In the production of this book, we have incurred many debts. Those are hardly discharged with thanks, but we would like to acknowledge those who have helped. The National Science Foundation (NSF) has generously supported our research through a multiyear project, "Developing a Normative Framework for Cyberwarfare," funded under grants #1318126, #1317798, and #1318270. We have also received generous research support from our institutions, the Australian National University, the Naval Postgraduate School, and Western Michigan University.

Earlier drafts of some of these papers were presented at a workshop at the Brocher Foundation in Hermance, Switzerland, and we thank it for its hospitality. That conversation was continued in a multiday event at the International Committee for the Red Cross (ICRC), under the title "Cyberwarfare, Ethics, and International Humanitarian Law." We thank the ICRC for hosting us as well as Patrick Lin—also a principal on the NSF grant—for his organizational efforts. Rapporteurs for the ICRC event included Shannon Ford and Dusty VanPelt, who we also thank.

The project itself had its germination in a grant from the National Security College (NSC) at the Australian National University, and some of the papers found their inception at a 2013 workshop on the ethics of cybersecurity, jointly hosted by the NSC and the Centre for Applied Philosophy and Public Ethics (CAPPE) at Charles Sturt University. We also wish to thank CAPPE for helping to financially support the workshops in Geneva.

At Oxford University Press, we owe much gratitude to our commissioning editor, Peter Ohlin. Peter's support for this project has been unwavering from the outset, and it has been such a pleasure to work with him. We also thank our production manager Sunoj Sankaran and our copyeditor Kaykodner, without whose excellent work we would have a much lesser book. Our penultimate thanks goes to our contributors, without whose excellent work we would have no book at all. They endured much feedback and produced myriad revisions, and we are grateful for both their final products and good spirits. Finally, we thank you, the reader, for picking up the book. We think it houses critical discussions and appreciate your becoming part of the conversation.

Fritz Allhoff
Anchorage, Alaska
Adam Henschke
Canberra, Australia
Bradley Jay Strawser
Monterey, California
July 2015

Bibliography

Allhoff, Fritz, and Ryan Jenkins. "When Is a Real-World Response to a Cyberattack Justifiable?" *Slate*, June 11, 2014, available at: http://www.slate.com/articles/technology/future_tense/2014/06/cyberwar_ethics_when_is_a_real_world_response_to_a_cyberattack_justifiable.html, accessed October 2, 2014.

Arquilla, John. "Cyberwar Is Already Upon Us: But Can It Be Controlled?" *Foreign Policy*, February 27, 2012, available at: http://www.foreignpolicy.com/articles/2012/02/27/cyberwar_is_already_upon_us, accessed October 16, 2014.

Clark, David, Thomas Berson, and Herbert S. Lin, eds. "At the Nexus of Cybersecurity and Public Policy: Some Basic Concepts and Issues." In *Committee on Developing a Cybersecurity Primer: Leveraging Two Decades of National Academies Work*. Washington, DC: National Academies Press; Computer Science and Telecommunications Board; Division on Engineering and Physical Sciences; National Research Council, 2014.

von Clausewitz, Carl. *On War*. London: Penguin Classics, 1968.

Floridi, Luciano. "Get Ready for Cyberwar." *Philosopher's Magazine* 46 (2009): 12–13.

Henschke, Adam, and Patrick Lin. "Cyberwarfare Ethics, or How Facebook Could Accidentally Make Its Engineers into Targets." *Bulletin of the Atomic Scientists*, August 25, 2014, available at: http://thebulletin.org/cyberwarfare-ethics-or-how-facebook-could-accidentally-make-its-engineers-targets7404, accessed October 2, 2014.

Libicki, Martin. *Cyberdeterrence and Cyberwar*. N.p.: RAND Corporation, Prepared for the United States Air Force, 2009.

Lieberthal, Kenneth, and Peter W. Singer. *Cybersecurity and U.S.-China Relations*. 21st Century Defense Initiative. N.p.: Brookings Institution, 2012.

Lin, Herb. "Cyber Conflict and International Humanitarian Law." *International Review of the Red Cross* 94 (2012): 886.

Mandiant. "APT 1: Exposing One of China's Espionage Units." *Mandiant Inc.*, http://intelreport.mandiant.com/Mandiant_APT1_Report.pdf, accessed October 17, 2014.

Orend, Brian. "War." In *The Stanford Encyclopedia of Philosophy*, ed. Edward N. Zalta. Fall 2008: http://plato.stanford.edu/archives/fall2008/entries/war/, accessed September 26, 2014.

Owens, William, Kenneth Dam, and Patrick Lin, eds. "Technology, Policy, Law, and Ethics Regarding U.S. Acquisition and Use of Cyberattack Capabilities." Committee on Offensive Information Warfare. Washington, DC: National Academies Press; National Research Council, 2009.

Rid, Thomas. *Cyber War Will Not Take Place*. London: Hurst, 2013.

Rohan, Brian. "Georgia Says Russian Hackers Block Govt Websites." *Reuters*, August 11, 2008. Accessible at: http://uk.reuters.com/article/2008/08/11/us-georgia-ossetia-hackers-idUKLB2050320080811, accessed October 16, 2014.

Schmidt, Michael S., and David E. Sanger. "5 in China Army Face U.S. Charges of Cyberattacks." *New York Times*, May 19, 2014, available at: http://www.nytimes.com/2014/05/20/us/us-to-charge-chinese-workers-with-cyberspying.html?_r=0, accessed October 17, 2014.

Schmitt, Michael N., ed. *Tallinn Manual on the International Law Applicable to Cyber Warfare*. Cambridge: Cambridge University Press, 2013.

Taddeo, Mariarosaria. "An Analysis for a Just Cyber Warfare." In *4th International Conference on Cyber Conflict*, ed. C. Czosseck, R. Ottis, K. Ziolkowski. 2012.

White House Office of the Press Secretary. "Excerpts from Transcribed Remarks by the President and the Vice President to the People of Knoxville on Internet for Schools," October 10, 1996. Accessible at: http://govinfo.library.unt.edu/npr/library/speeches/101096.html, accessed October 16, 2014.

Part I

FOUNDATIONAL NORMS FOR CYBERWARFARE

1

Emerging Norms for Cyberwarfare

GEORGE R. LUCAS JR.

Cyberconflict appears to confront nations and industries with a new form of unrestricted, relentless, and indiscriminate conflict, with attacks on military, industrial, and civilian infrastructure and objects that violate conventional norms of war.[1] Much of this is due to the fact that the conflict represents both criminal theft and vandalism, coupled with sophisticated espionage and intelligence operations. Neither category of operational conflict has heretofore been considered as, nor risen to the level of, the kind of use of force and armed conflict that is governed by existing legal and moral regimes. As a result, cyberconflict has been portrayed as a sort of war without rules, a form of warfare that obviates existing norms and calls for an entirely new legal regime. While recognizing these challenges, I will nonetheless argue that:

(1) there are, or could be considered to be, a set of norms governing cyberconflict practices (apart from criminal activities) that are presently emerging from the engagement in these practices by adversaries;

(2) the set of emergent norms I will sketch resembles the underlying moral (if not necessarily legal) underpinnings of conventional armed conflict, sufficiently so as to provide reasonable assurances of the possibility of responsible and accountable future behavior in this domain; and

(3) perhaps most importantly, this so-called soft law approach, grounded in consensus, agreement, and the advocacy of "best practices," affords a greater probability of widespread acceptance and compliance within the anarchistic and hegemony-averse community of *cyberspace citoyens* than do attempts at imposing formal legislation.

---

[1] This paper is derived from revisions of talks given at Yale University, at Monash University (Melbourne), and at the Australian National University and the National Security College (Canberra). I am grateful to Wendell Wallach, Rob Sparrow, Shannon Ford, and Adam Henschke, as well as to the editors of this volume, for thoughtful discussions of the main ideas, as well as to the Centre for Applied Philosophy and Public Ethics (CAPPE) at Charles Sturt University (Canberra) for the invitation to participate in these symposia.

1.1 Historical Background

Cyberconflict did not attract much attention as a problem in ethics until the beginning of the current decade. Apart from straightforward criminal activities (such as robbery, vandalism, and identity theft), malevolent actions in cyberspace appeared simply to constitute a new kind of information warfare that—alongside propaganda, deception, misinformation, and other more familiar measures preceding or accompanying the onset of war—merely constituted another form of low-intensity conflict between states, more on the order of espionage than of conventional war itself. Apart from safeguarding individual privacy, accordingly, these cyberactivities did not seem to present any genuinely new or serious moral conundrums for consideration.

Explicit *ethical* concerns about cyberweapons and proposed tactics for genuine, full-scale cyberwarfare were first raised, not by ethicists or moral philosophers, but (to their immense credit) practitioners working directly on the problem of information warfare and cyberconflict, including researchers and practitioners in computer science, international relations, and special operations.[2] Despite the decided preference of practitioners to address the emerging concerns initially as a matter of ethics rather than of law, the principal subsequent efforts to develop suitable regimes of oversight and accountability for forms of cybermalevolence (distinct from criminal activity) came from scholars and practitioners of international law, especially following the intense distributed denial of service (DDoS) attacks on Estonian civil infrastructure during the summer of 2007.[3] In point of fact, going all the way back to the path-breaking Budapest Convention on Cyber

---

[2] From the standpoint of special operations in international relations, see Arquilla and Ronfeldt, "Cyberwar Is Coming!" 141–65; Arquilla, "Ethics and Information Warfare," 379–401; and Arquilla, "Conflict, Security, and Computer Ethics," 133–49. See also Libicki, *Conquest in Cyberspace*.

From the perspective of computer science, see Denning, *Information Warfare and Security*; Denning, "Cyberwarriors," 70–75; and Denning, "Ethics of Cyber Conflict." See also Rowe, "War Crimes from Cyberweapons," 15–25; Rowe, "Ethics of Cyber War Attacks," 105–11; and Rowe, "Ethics of Cyberweapons in Warfare," 20–31.

[3] A comprehensive IT report was issued by the Arbor Networks IT Security Blog in 2013. See Dan Holden, "Estonia Six Years Later," at: http://www.arbornetworks.com/asert/2013/05/estonia-six-years-later/.

An excellent summary of the circumstances leading up to the attack on Estonia and its consequences can be found in season 1, episode 2, of the PBS program *Wired Science* from shortly after the incident in 2007, titled "Technology: World War 2.0," at http://xfinitytv.comcast.net/tv/Wired-Science/95583/770190466/Technology%3A-World-War-2.0/videos?skipTo=189&cmpid=FCST_hero_tv. See also Charles Clover, "Kremlin-Backed Group behind Estonia Cyber Blitz," *Financial Times* (London), 11 March 2009; and Tim Espiner, "Estonia's Cyberattacks: Lessons Learned a Year On," *ZD NET UK* (1 May 2008). There is also a summary factual account in Wikipedia: http://en.wikipedia.org/wiki/2007_cyberattacks_on_Estonia.

Crime (CCC) in 2001,[4] the principal proposal for a new normative framework for all forms of cyberconflict came from proponents of international law, rather than ethicists.[5]

That situation changed rather dramatically with the publication of the first ethical analysis of cyberwarfare by a moral philosopher, Randall Dipert, in 2010.[6] Dipert's impact stemmed not just from his status as a well-regarded ethicist but also from his keen grasp of information technology and what he termed the "unique ontology" of cyberspace. The extremely odd and vexing ontological features of the cyberdomain were brilliantly cataloged in that initial essay. Specifically, Dipert demonstrated how these ontological characteristics tended to problematize the very foundations of any legal regime that presupposed well-defined physical locations, clearly delineated geographical boundaries, and concrete personal property or state territory. They likewise challenged conventional legal reasoning addressed toward agents (whether individual or collective) who possessed a clear identity for purposes of legal accountability or the unequivocal attribution of rights (such as privacy in the case of individuals, or sovereignty in the case of states). The peculiarities rife in this virtual, nonphysical, geographically nonlocalized domain presented serious challenges not only for legal analysis, according to Dipert, but also to conventional modes of moral analysis (such as "just war" reasoning). It seemed clear from this initial philosophical analysis, in sum, that simply adhering to the presuppositions entailed in the law of nations, and demanding that the "agents" and events in this new domain conform to its conventional regulatory authority and jurisdiction, was not likely to prove successful.

The field has blossomed substantially in the intervening years, with a bewildering array of new insights almost too numerous to chart. Mariarosaria Taddeo and Luciano Floridi, both eminent philosophers of technology in the United Kingdom, obtained UNESCO sponsorship for a preliminary caucus of leading ethicists focused on the ethical dimensions of cyberconflict immediately following

---

[4] The International Convention on Cybercrime, Council of Europe, "Convention on Cybercrime," Budapest, November 23, 2001: http://conventions.coe.int/Treaty/EN/Treaties/html/185.htm).

[5] For a small representative sample of this considerable literature, see Dunlap, "Perspectives for Cyber Strategists on Law for Cyberwar," 81–99; Lin et al., *Technology, Policy, Law, and Ethics*; Schmitt, "Computer Network Attack and the Use of Force in International Law," 885–937; "Wired Warfare: Computer Network Attack and *Jus in Bello*," 365–99; "Cyber Operations and the Jus in Bello: Key Issues," 89–110; and Wingfield, *Law of Information Conflict*.

[6] See "Ethics of Cyber Warfare," 384–410. A more detailed account of the ontological features of cyberspace is offered by Dipert in "Essential Features for an Ontology for Cyberwarfare," 35–48. Finally, an account of cyberobjects and events not limited to the Internet ("other-than-internet" elements) is offered in "Other Than Internet Warfare," 34–53.

the discovery of Stuxnet (section 1.4).[7] Researchers at the US Air Force Research Institute in Montgomery, Alabama shortly thereafter convened leading military strategists, industrial experts, ethicists, and lawyers to analyze the most urgent ethical issues attendant upon the United States' rapidly developing cyberstrategy,[8] shortly after the formal publication of the new US cyberstrategy in two documents during the summer of 2011.[9] A number of important conferences, symposia, and publications rapidly followed, including a special issue of the *Journal of Military Ethics*, edited by Bradley J. Strawser in 2013, and culminating most recently in the contributions to the present volume.

1.2 From International Law to Ethics

Those prior publications and, indeed, the contributions to this volume amply demonstrate that moral philosophers are not shy about expressing their views on this topic. One might ask, however, what added value ethics and moral philosophy might contribute to these discussions, beyond what has already been achieved over years of normative deliberations from a legal perspective, finally disseminated in summative form in the *Tallinn Manual* of 2013?[10] In fact, the belated arrival of moral philosophy and ethics to the cyberconflict roundtable in recent years apparently elicited a pointedly skeptical reception by those already well along in their efforts, which occurred at an international summit in Rome in November 2013 intended to introduce and discuss the findings published in the *Tallinn Manual*.[11] If these concerted efforts to bring existing international law to bear on cyberconflict failed to persuade nations or individuals to restrain their activities or submit to some form of governance in the cyberdomain, then what hope might moral philosophy hold out for faring any better?

---

[7] Floridi and Taddeo, *Ethics of Information Warfare*.

[8] Yannakogeorgos and Lowther, *Conflict and Cooperation in Cyberspace*.

[9] Somewhat contrasting policy guidance on the cyberdomain was nearly simultaneously by the US Departments of State and Defense. See: "US Department of Defense Strategy for Operating in Cyberspace," http://www.defense.gov/news/d20110714cyber.pdf; and "International Strategy for Cyberspace," http://www.whitehouse.gov/sites/default/files/rss_viewer/international_strategy_for_cyberspace.pdf.

[10] Schmitt, *Tallinn Manual on the International Law Applicable to Cyber Warfare*. Available at: https://www.ccdcoe.org/249.html.

[11] Mariarosaria Taddeo, Assistant Professor of Philosophy and Technology at the University of Warwick (U.K.) and President of the International Society for Philosophy and Computers, reported on a heated exchange concerning the relevance (or irrelevance) of ethics and moral philosophy to adequate legal analysis of the cyberdomain, during a November 2013 international symposium on the Tallinn Manual in Rome, organized by the NATO Center for Cybersecurity Excellence (Tallinn).

In order to address this question, we must attend first to peculiarities inherent in the Tallinn process or effort itself. Contributors to the *Tallinn Manual* chose to focus their efforts on the interpretation of *extant legislation*, rather than advocating new law or international treaties. They collectively attempted to demonstrate how existing international legislation pertaining to the conventional domains of air, sea, land, and space might provide guidance, governance, and accountability in the "fifth domain" of cyberactivities as well. One might therefore have expected the success of that sustained, years-long community effort to mirror the success enjoyed by a similar initiative regarding problems encountered through private military and security contracting: namely, the Montreaux Document of 2008, which quickly achieved widespread acceptance and endorsement, thereby largely silencing decades of prior controversy over the advent of Private Military Contractors (PMCs) and Private Security Companies (PSCs).[12]

Thus far, however, the *Tallinn Manual* has not met with anything like that kind of widespread acknowledgment, adoption, or endorsement. Indeed, United Nations (UN) staff members and nongovernmental organizations (NGOs) at a recent convention on cyberconflict hosted at the International Committee of the Red Cross (ICRC) in Geneva characterized the *Tallinn Manual* (by way of explicit comparison to the Montreaux effort) as instead constituting "a spectacular failure."[13]

As it turns out, there are interesting differences that might be examined between the process and the eventual output of Montreaux, as opposed to the Tallinn deliberations, which might have been attributed to comparatively limited success on the part of the latter. Both were "unofficial" publicist efforts aimed at building consensus for future governance, for instance, but the former was far more widely inclusive of diverse stakeholders in the PMC controversy, including representatives from organizations representing military contracting, while the NATO-centered Tallinn group was more narrowly restricted to representatives of aggrieved or concerned NATO-alliance countries. Hence, the Tallinn deliberations did not merely exclude moral philosophers (to the intense aggravation of the latter) but, far more significantly, also omitted representatives from China, the Russian Federation, and other nations and cultures that might have been thought to have a strong vested interest in

---

[12] The document, and information on its contents and participating nations, may be found on the website on the International Committee of the Red Cross: http://www.icrc.org/eng/resources/documents/misc/montreux-document-170908.htm.

[13] This specific description was offered by a senior staff member [name withheld under Chatham House rules] of the United Nations Institute for Disarmament Research (UNIDIR), during an international workshop at the Humanitarium of the International Committee of the Red Cross headquarters in Geneva: "Cyberwar, Ethics and International Humanitarian Law," 21–22 May 2014, (http://www.icrc.org/eng/resources/documents/event/2014/04-25-panel-cyber-warfare.htm). However, the assessment was, in general, widely shared by most of the remaining UN, ICRC, and NGO participants, to my astonishment.

debating the underlying hermeneutical principles, as well as the results of attempts to extrapolate existing international law to the cyberdomain.

The Montreaux group process was likewise far more open to participation and expert testimony from relevant practitioners, in comparison to the Tallinn group. And, perhaps most fatally, the Tallinn group's very decision to extrapolate from existing formal ("black-letter") legislation differed substantially from the decidedly less ambitious and aggressive "soft law" approach taken by the Montreaux group. Montreaux delegates chose instead to underscore areas of commonality; to endorse emerging "best practices," as discerned and agreed to by otherwise-competing stakeholders; and then to carefully "socialize" these proposals among representatives of the wider international community (ultimately resulting in a favorable hearing on the document by the UN General Assembly in 2009, with an endorsement to date by over fifty signatory nations and several international organizations).[14]

Following the lead of Professor Dipert, philosophers (had they, too, been allowed a voice in the Tallinn deliberations) would almost surely have worried (much more than did lawyers) about the vexing ontological features of the cyberdomain, particularly about their likely damaging impact on some of the most basic assumptions underlying international law. In particular, they would have noted that international law is inherently state-centric, while the cyberdomain is without discernible or meaningful state boundaries or jurisdictions (despite more recent, and what I cannot help but think misguided, attempts among Western European nations at present to reassert national boundaries in cyberspace).[15]

In point of fact, philosophers might well have asserted that, notwithstanding the self-professed expertise of the legal scholars involved, and notwithstanding the sometimes ingenious hermeneutical gymnastics employed to adopt existing law to the new reality, as contained in the *Tallinn Manual, international law itself simply has no obvious jurisdictional authority in this matter.*[16] Indeed, hacktivists and

---

[14] For a list of current signatories to the Montreaux Document, see: http://www.eda.admin.ch/eda/en/home/topics/intla/humlaw/pse/parsta.html.

[15] Reasserting the identity and jurisdiction of national boundaries, especially in the wake of revelations concerning widespread surveillance by the US National Security Agency, was a prominent theme in most of the programs and presentations at the Sixth Annual European Forum on International Cybersecurity (FIC), sponsored by the French National *Gendarmarie* in Lille, France (28–29 January 2014): http://www.forum-fic.com/2014/en/. As one of the invited speakers on this topic, I expressed concerns that the European desires to distance themselves from US providers of Internet services and hardware would result, at best, in marginally greater security at the cost of considerable inconvenience and inefficiency: http://www.forum-fic.com/2014/en/presentation/a7-what-is-a-national-cyberspace/.

[16] Ryan Jenkins observes that this kind of argument is often adduced against "software" weapons: that, since they are not physical objects, existing international conventions regarding the use or threat of force (e.g., UN Charter 2 (4)) do not apply, since these provisions concern prohibitions against the use of *physical* force. He rightly finds that kind of reasoning implausible: see "Is Stuxnet Physical? Does It Matter?" 68–79.

other Internet anarchists *vociferously deny* the jurisdictional authority of *any* legal regime. Their passionate denial does not, of course, settle the matter (especially if they are found to engage in actions harmful to the persons and interests of others), but it does pose a profound obstacle to effective governance that cannot simply be ignored, let alone overridden.

Indeed, the cyberdomain is rife at present with what might be termed "jurisdictional equivocation." International law applies primarily to the behavior of nation-states—in the phrase widely popular among legal scholars, "what States do, and what they will tolerate being done." Yet actors in the cyberdomain often deny or repudiate any state association (even when, as in the case of activities traced to the Shanghai-based Unit 61398 of the People's Liberation Army (PLA), that denial seems difficult to accept). Certainly it remains true that authentic state agency is notoriously difficult to ascertain and attribute with the degree of specificity required in law, especially when (as philosophers Chris Eberle and Ed Barrett rightly worry) the law contemplates sanctioning some sort of punishment or reprisal for alleged wrongdoing.[17] Likewise, although international humanitarian law aims to restrain conventional armed conflict and afford some protection to its most vulnerable victims in the actual world, cyberconflict is not obviously warfare in the conventional sense, even on a so-called effects-based assessment of malevolent cyberactivity. Indeed, some scholars, like Thomas Rid, go so far as to object that there is no such thing, strictly speaking, nor likely ever to be such thing, as a true "cyberwar"[18]—in which case any findings regarding the law of armed conflict are moot. Others[19] (including myself) believe that, while Rid's extreme view is too restrictive in defining the nature of harm inflicted in cyberspace, or in allowing for an effects-based equivalent to the conventional use of armed force, still it is also true that the threats of a "cyber Pearl Harbor" or of cyberterrorism by popular pundits like Richard Clarke or Joel Brenner are often

---

My point here is somewhat different: since the law in question applies to "states" with geographical boundaries, engaged in armed conflict, and since cyberspace is nonphysical in the geographical sense and lacks discernible state boundaries, and also since the conflict in question is not clearly "armed conflict," the otherwise-relevant body of law does not obviously hold sway. The further point could be made (as I do, elsewhere; see "NSA Management Directive # 424: Secrecy and Privacy in the Aftermath of Snowden," 29–38) that espionage specifically is not governed by international law, and most of the cyberconflict in question is espionage.

[17] See Barrett, "Warfare in a New Domain," 4–17; and Eberle, "Just War and Cyber War," 54–67. Although I summarize and criticize elements of the arguments of both authors, I fully agree with their underlying concern that the "attribution problem," and the degree of certainty attached to concerns for accountability, continues to pose a severe problem when governments and their militaries propose, as a matter of policy, to be willing to target persons for retaliation on account of losses of intellectual property: see "Ethics and Cyber Conflict: A Review of Recent Work," 20–31.

[18] See Rid, "Cyber War Will Not Take Place," 5–32. See also "Think Again: Cyberwar": http://www.foreignpolicy.com/articles/2012/02/27/cyberwar

[19] See, e.g., David Whetham's entry in this collection.

inflated.[20] Finally, especially among adversarial nations (such as China and the United States), there remain legitimate differences of opinion on just what sort of entity the cyberdomain is thought to constitute and, hence, what sort of legal framework (if any) would be most appropriate for it.[21]

In light of all these disputes and fundamental philosophical ambiguities, not only does the recent legal scholarship on cyberconflict seem parochial and NATO-centric, but in this instance the lawyers may have been guilty of pressing ahead too aggressively and uncritically, hampered by a range of deeply held, tradition-bound, but highly questionable background assumptions regarding the very nature of governance itself. One might pointedly object that this entire misguided effort constituted a failure to comply with the most basic principle of "good governance."[22] In any case, this insular and discipline-specific effort to apply what many have come to believe is an aging and seriously outmoded legal paradigm to a wholly new, as-yet poorly understood set of circumstances was, accordingly, bound to fail—as apparently, and spectacularly, it has.

1.3 From Fixed Laws and Rigid Principles to Emergent Norms

I underscored the earlier success of the Montreaux Document and its formative process because it unintentionally illustrates a more reflective, less assertive, and therefore more effective approach to the problem of international governance generally. In particular, the Montreaux Document highlighted a range of "best practices" for national militaries to follow in their engagement with private military

---

[20] See, e.g., Clarke and Kanke, *Cyber War*; and Brenner, *America the Vulnerable*. I discuss the nature of the threat they describe, and why I find it inflated, in some greater detail in: "*Jus in Silico*," 367–80.

[21] Thus, following the revelations from Edward Snowden about widespread Internet surveillance and possible invasions of privacy by the United States, the *China Daily* carried an editorial protesting: "For many Chinese, it is bizarre how Washington can continue to pose as the biggest cyber espionage victim and demand that others behave well," in light of Snowden's revelations. The editorial concluded that "by dividing cyber espionage into 'bad' and 'good' activities, Washington is trying to dictate the rules for global cyber domain, which is a public space." To be certain, it appears from this comment that the Chinese (or, more accurately, whoever composed and approved of this editorial) consider the cyberdomain not merely a "public" space, but a "commons," that is, an arena or resource available to all without constraint or restriction. And they are objecting, not consenting, to the imposition of rules, norms, or values on that space. At face value, this seems to portray the cyberdomain in a very different conception than that advanced by the United States and NATO countries. See Sanger, "Differences on Cybertheft Complicate China Talks," www.nytimes.com/2013/07/11/world/asia/differences-on-cybertheft-complicate-china-talks.html.

[22] See O'Meara, *Governing Military Technologies in the 21st Century*, for a complete discussion of the canons of good and effective international governance, chaps. 5 and 6.

contractors. These practices were hammered out, over time, in the crucible of issues presented through practice by the relevant parties. The best practices highlighted, moreover, did not somehow follow, deductively, from the application of existing regulatory schemes and standard operating procedures. Rather, such practices were, more often than not, developed among the relevant parties largely as a response to the *failure* of existing regulatory regimes and procedures—whether in the very early discovery that PMCs, even armed security contractors, did not qualify as "mercenaries" under prior international treaty definitions, to the horrifying realization in the wake of the Nisoor Square incident in Baghdad (September 2007)[23] that many domestic regulatory schemes failed to encompass, or hold accountable, the activity of PMCs and their personnel. As such problems were addressed by all sides and rectified in specific instances, such instances became, in turn, suitably universalizable as best practices for other parties to adopt in similar circumstances.

Moreover, since these best practices had been hammered out cooperatively by the relevant stakeholders themselves in specific instances, it was relatively easy for all stakeholders to recommend and consent to them as an authentic expression of shared intentions and common interests in the general case. Laws and regulations are stipulative, and are imposed externally, often upon unwilling subjects or agents. Best practices, by contrast, *emerge* from the shared practices of the interested parties, reflecting their shared experience and shared objectives. Who, after all (as Kant might remind us), can object to conforming to standards that they themselves have formulated, and subsequently imposed upon themselves, in order to guide and regulate their shared pursuits?

This is an example of what, in international relations (IR), is commonly referred to as "emergent norms." In IR, "norms" are thought to differ from both formal legislation and prior established moral principles. Norms (from "normative," pertaining to agreed standards, like weights and measures, as well as from *nomos*, or law) instead are widely recognized and generally acknowledged standards of conduct by which individuals and nations evaluate their own and others' behavior. Sometimes in international law these norms are termed *ius gentium*, "customary law," or "the standards and customs of civilized nations and peoples." Admittedly, this is not very clear or very precise, nor does this seem very powerful, especially when there are no treaties or bright-line statutes in international or domestic law to clarify these norms, let alone when there is little in the way of effective sanctions to enforce them. But it is remarkable how, over the centuries—and long before these concepts ever

---

[23] For an excellent summary account of this incident, see: http://en.wikipedia.org/wiki/Blackwater_Baghdad_shootings. Although Wikipedia citations are not usual fare for academic research, in this particular case the Wikipedia account is of high quality. Indeed, it gives a more in-depth account than the case study written by my own military students immediately after the Nisoor Square incident: "War Is Big Business: Blackwater Worldwide, Inc.," in W. R. Rubel and G. R. Lucas, eds., *Case Studies in Military Ethics*, 4th ed. (New York: Pearson, 2014), 53–56.

found their way into written, statutory form—nations and peoples have developed and respected a family of norms in international relations pertaining to things like ambassadorial immunity, treatment of prisoners of war, and respect for other basic principles (even when such "respect" is accorded more in the breach than in the observance).[24]

With respect to the cyberdomain, my own understanding and treatment of this topic is deeply indebted to recommendations by Panayotis Yannakegeorgos and Adam Lowther regarding the prospect of US sponsorship of "global norms" for cyberspace.[25] This is what I see the Montreaux Document as advocating with respect to military contracting. Indeed, this is what a number of us likewise advocated with respect to the gaping lacunae in existing governance for military robotics.[26] And this is also the approach I have advocated for cyberconflict since the discovery of Stuxnet in 2010.[27]

But this approach to governance entails its own difficulties, especially from the standpoint of both ethics and law. As a first consideration, how do norms "emerge" from practice? This seems worth explaining, inasmuch as norms are thought to be wholly distinct from—we might say, more accurately, "orthogonal" to—practice, precisely in order to constrain it. Doesn't a concept of emergent norms, then, involve endorsing some version of the naturalistic fallacy, attempting to derive, in this case, "ought" from "is"?

In domestic and international law and public policy, of course, we do not trouble ourselves with this intriguing philosophical puzzle. Instead, as noted earlier, norms are constructed from the precedents set by our practices, or from the legal judgments reached in the evaluation of specific practices. In public policy and in IR, the status of norms is similar to that of causality in the natural sciences: both have remained (in the eloquent and amusing summary of Alfred North Whitehead) "blandly indifferent to their refutation by Hume."[28] And even if this problem can

---

[24] Diplomatic or ambassadorial immunity is by far the oldest of these norms, cited and honored (more or less) since ancient times, but only codified, finally, and written into international law as Article 43 of the Vienna Convention on Consular Relations, in 1963.

[25] Yannakogeorgos and Lowther, "Prospects for Cyber Deterrence," 49–77.

[26] Marchant et al., "International Governance of Autonomous Military Robots," 272–315. See also Lucas, "Automated Warfare," 1–23, for examples of the governing authority of "soft law" precepts for military robotics.

[27] See "Just War and Cyber Conflict," Annual Stutt Lecture on Ethics, U.S. Naval Academy (9 April 2012): http://www.youtube.com/watch?v=hCj2ra6yzl0. Text at: http://www.usna.edu/Ethics/_files/documents/Just%20War%20and%20Cyber%20War%20GR%20Lucas.pdf. Also see "Can There Be an Ethical Cyberwar?" 195–210; "*Jus in Silico*," 367–80; "Emerging Norms for Cyber Warfare": http://philevents.org/event/show/11114; "Navigation, Aviation, 'Cyberation': Developing New Rules of the Road for the Cyber Domain": http://www.youtube.com/watch?v=cs7RXAPzG84; "Permissible Preventive Cyber Warfare," 73–83.

[28] Whitehead's famous phrase is found in *Science and the Modern World* (New York: Macmillan, 1925), 16. It is Hume's version of the naturalistic fallacy, found in his *Treatise* (rather than G. E.

be addressed, both lawyers and ethicists might, in any case, question the degree of normative force that such emergent norms might actually possess. Each of these two fundamental concerns deserves a response.

1.4 The Methodology of Uncertainty: How Norms Emerge

How do we handle uncertainty, including especially the uncertainty attached to novel developments, new technologies, and contrasting or competing ways of living—how do we grow comfortable with the new and the novel? How, in particular, do we discern the appropriate governing principles or "rules of the game," when we don't know precisely what the "game" is (let alone what its rules are)? It was the eminent moral philosopher, Alasdair MacIntyre, who some decades ago pointed out that we have clues about this procedure, illustrated in works of Aristotle on ethics and political theory, and described in detail in one of his logical treatises, *Posterior Analytics*.[29]

Moore's), to which I refer: the (presumably fallacious) derivation of "ought" from "is." (Moore, who coined the famous term for this, had a very different concern, namely, that moral goodness was not a natural kind, and so could not be defined in terms of something else drawn from the natural world, such as pleasure or happiness, as the utilitarians had attempted). Hume did not take this to be an absolute prohibition. Rather, he argued, one was obliged to show how one might derive an "ought" from an "is" (something that R. M. Hare proceeded to do via deontic logic in the 1950s). In Hume's case, I take his own turn to the histories of England to be a different form of derivation, one that demonstrates precisely what we assert in this article: that reflective analysis of customary behaviors and best practices are the origins of norms and principles.

[29] Interestingly, MacIntyre's clearest explicit accounts of what I am terming "the methodology of uncertainty," as well as of his derivation of norms from concrete practices, comes toward the end of his own discovery and elucidation of that method: e.g., in *Whose Justice? Which Rationality?* (South Bend, IN: University of Notre Dame Press, 1988); and even more in his Aquinas Lecture of 1990 at Marquette University. See: *First Principles, Final Ends, and Contemporary Philosophical Issues* for a complete description of the method of "imperfect" sciences, in which the "first principles" remain undisclosed and subject to discovery and subsequent, provisional elucidation (as in ethics or politics, as well as physics and biology), in sharp contrast to geometry, where the first principles are evident from the start.

With these subsequent insights in hand, one reads with greater insight MacIntyre's earlier works, such as the magisterial *After Virtue*, in which the method of inquiry, and the commentary on the nature of emergent moral norms, in particular, is implicit in his critique of modern conceptions of morality. And, while MacIntyre clearly had no patience with Kant, it is interesting that the emergence of MacIntyre's methodology of uncertainty is very much like the emergence of Kant's methodology of "reflective teleological judgments," a method that is implicit, but formally undisclosed in a number of Kant's significant essays on history, political life, and the quest for peace in the 1780s, prior to being more systematically and explicitly elucidated in the second part of *The Critique of Judgment* (1790).

Basically, the method is this: when we are uncertain about what's going on, or what the rules or principles that properly apply to our actions are, we start by gathering all the relevant data we can about the practice in question—for example, moral customs in different cultures, various kinds of constitutions, and political arrangements (or, in our case, various activities and practices taking shape in the cyberdomain). We begin intuitively to discern *better* from *worse* practices, based in large part upon their relative degree of "operational effectiveness." We also draw comparisons between what we know, and what we are newly confronting. We engage in reflection, dialectic, and sustained philosophical argument, in order to analyze both the practices in question and our comparative experiences with their results. We extrapolate from the known to the unknown, and gradually begin to build what Aristotle termed the *Archai*, the "first principles" governing practice, from our varied experiences of, and reactions to, the variant forms of the practices themselves.

Certainly some such "Aristotelian" process was at work with regard to our own and other nations' reactions to the advent of nuclear weapons. The United States developed them, and used them. Upon subsequent reflection and argument—and while the justification for their specific use in a specific war context against a specific enemy deemed guilty of its own war crimes is still hotly debated—few would otherwise doubt that our reflections on this practice of using such indiscriminate and destructive weapons, especially against civilian targets, was a very bad practice in general, and one not to be encouraged. The fear of the United States one day falling victim to the bad precedent it had set helped generate an unfortunate, dangerous, and costly arms race—but that very precedent and the costly arms race it helped generate *also* led responsible states, in turn, to advocate nuclear nonproliferation and discourage threats involving the "first use of nuclear force" by nuclear powers. The norm of nuclear nonproliferation, while not inviolable, has been an especially strong international norm over the past half-century.

In point of fact, for all the criticism presently leveled against legal scholarship in the wake of the lack of uptake of the *Tallinn Manual*, the evolution of law itself, when confronted with new and novel developments, constitutes an especially good case in point. The complex relationship between law and morality is always of great interest, but it need not detain us now. We frequently tease, as well as complain, that both morality and the law are constantly playing catch-up with technology. But if Aristotle is right, this is not a flaw; it is exactly what is to be expected, and precisely what we are now engaged in carrying out. Legal scholars work by analogy, and with past experience in the form of precedent. In order to cite precedent and draw analogies, however, they need a body of new experiences to work with. Confronted with emerging examples of the new and the unknown, they then turn to precedent to help classify and categorize the new behaviors, and reason by analogy and extrapolation from past practices to present governing principles. Solving such recalcitrant problems, in turn, quite often requires what I characterized above as "hermeneutical gymnastics."[30]

---

[30] E.g., see Jonson and Toulmin, *Abuse of Casuistry*, for an account of case-based legal and moral reasoning that offers accounts of unusual and sometimes invalid application of principles to cases.

To be sure: we don't yet have a great deal of experience with cyberconflict—although our familiarity is increasing daily, in accordance with "Moore's Law."[31] But we can pause and notice that we have now, arguably, experienced at least four instances of cyberconflict that could not properly be classified merely as criminal acts, or even simply as acts of espionage and covert action. Earlier I noted that Thomas Rid of King's College had objected that none of the four instances I am about to describe were acts of war, but that all such past cyber conflicts could be classified instead merely as either espionage or sabotage.

But acts of sabotage *are* currently classified as acts of war under international law. Sabotage constitutes an act that, in effect, crosses the boundaries between ongoing low-level conflict between states (like espionage) and a genuine resort to force, or to the equivalent of an "armed attack," by one nation against another. As these are all we have to work with so far, I've already invested a fair amount of time and work in discussing and analyzing what lessons we might learn from them (e.g., see n. 26). Otherwise, however, Professor Rid is basically correct that virtually all cyberconflict has boiled down to more or less straightforward crime, or else acts of espionage (which always classify as "crime" within the domestic legal regime in which they transpire). There is a controversial discussion presently underway about whether infusing a nation's civil infrastructure with "back-door" booby traps likewise crosses that line between espionage and sabotage (which might arguably constitute a use of armed force).[32]

So far, it appears that cybercrime is what has seemed most familiar (for all the novel new ways of carrying it out in the cyberrealm). Cybercrime is the arena in which we have seemed generally most confident and effective in classifying and responding to threats and challenges, even as the extent of criminal ingenuity and enterprise is vastly magnified by the Internet. Notwithstanding the novelty and ingenuity of cybercriminals, the basic nature, intentional structure, and effects of their criminal activity seem all too familiar, common, and generic: suddenly we feel as if we've seen all this before, especially, when all is said and done, if the crime committed amounts to the theft of other people's money or property. The Internet, it seems

---

[31] Named after Gordon Moore, founder and CEO Intel Corp: that, in keeping with the pace of miniaturization of transistors and the increase of the number that can be included upon a single silicon chip of a given size, Moore's Law now holds that the pace of technological change doubles every 18 to 24 months, helping to explain why we persistently feel so overwhelmed by the pace of change in the cyber domain.

[32] This is the chief difference between myself, and Eberle and Barrett (n. 17 above). Eberle, in particular, thinks such cyberactivities are comparable to the building and positioning of strategic nuclear weapons, which is not itself subject to retaliation (unless they are actually used). I challenge that analogy on the basis of the Cuban missile crises, and also find grounds to compare this more to the actions of foreign agents, planting bombs for possible future detonation on US civil infrastructure, in the event of the outbreak of hostilities. The latter strikes me both as closer to what the cyber trapdoors and Trojan horses really represent, and more serious, and more amenable to retaliation, than the development and placement of nuclear weapons. But this is all obviously far from settled.

(in bank robber Willie Sutton's phrase), is simply "where the money is"! Hence, we discover that we can reasonably well classify the criminal acts (even if some old "cons" take new forms on the Internet) and figure out how to prevent them, combat them, and apprehend the criminals, while striving ourselves to remain respectful of individual rights, liberties, and the boundaries of the law—in short, the age-old dilemma of constabulary forces everywhere.

How do we stand presently with respect to cyberwar? The first ethicists to look at the cyberdomain, like Dipert, concluded that there were just too many dissimilarities between the cyberdomain and the other four more familiar domains to permit us to draw any useful analogies, or to make use of our previous knowledge and practices in the other domains to address and resolve the new conundrums confronted in cyberspace. Even if that is so, however, we are then required first to examine carefully what various nations and people have recently been up to, how they behave, how they react, and what the rest of us, upon reflection, think of it all. Richard Clarke offers detailed and dramatic accounts of four recent instances of cyberconflict that might qualify as acts of war, of which there are other accounts, and more numerous examples.[33] I have focused in the past principally on the following four, which I briefly summarize as follows:[34]

(1) Estonia (2007): the Estonian government decided to move an unpopular Russian war memorial from the center of Tallinn to a military graveyard outside the city. Russians citizens and their government were outraged, as were citizens of Estonia of Russian descent. Subsequently, the government of Estonia reported that the country was "under relentless attack" from outside sources unknown. The attack was a cyberattack, a Distributed Denial of Service (DDoS) attack, flooding many Estonian websites with enormous volumes of traffic and effectively shutting them down. Newspapers, banks, government websites, financial and civic transactions, all were brought to a standstill. Hospitals and the medical system were attacked. In a highly "wired"

---

[33] Jeffrey Carr offers additional counts of these four and a number of others that are clearly espionage, and not war, in his *Inside Cyber Warfare*. Ryan Jenkins offers a decisive rebuttal to the patently absurd claim that Stuxnet, constructed of computer software, cannot constitute a use of physical force that would fall under legal regimes governing the use of force in armed conflict. See "Is Stuxnet Physical? Does It Matter?" 68–79. The more interesting feature is, given that it constitutes the equivalent of the use of force in terms of physical effects, it is a preemptive (actually, a preventive) use of force, which is certainly outside existing legal regimes, but can be morally justified. See the continuing discussion of preventive war generally in Chatterjee, *Ethics of Preventive War*.

[34] My earlier analyses of the significance of these cyber events can be found in "Just War and Cyber Conflict," Annual Stutt Lecture on Ethics, U.S. Naval Academy (9 April 2012): http://www.youtube.com/watch?v=hCj2ra6yzl0. Text at: http://www.usna.edu/Ethics/_files/documents/Just%20War%20and%20Cyber%20War%20GR%20Lucas.pdf; "Can There Be an Ethical Cyberwar?" 195–210; "Jus in Silico," 367–80; "Permissible Preventive Cyber Warfare," 73–82.

tech-savvy nation, commercial and government affairs ground to a halt. The government appealed to NATO to come to its aid under the collective security provisions of the NATO treaty, claiming that the attacks originated in Russia. The government of the Russian Federation, however, denied any involvement or responsibility: "we can't be blamed if individual Patriots take matters into their own hands." NATO declined to become involved, stating that the massive cyberattacks "do not rise to the level of armed conflict."

(2) In September 2007, the Israeli Air Force (allegedly) carried out a nighttime bombing raid in Diaya-al-Sahir in Syria, destroying what was alleged to be a nuclear power and weapons facility under construction there, with technical assistance provided by North Korea. The conventional night-bombing raid appeared to succeed, however, because a prior cyberattack disabled and "spoofed" the Soviet-era air missile defense system in Syria, making the military appear to see clear skies and utterly miss the flight of the Israeli bombers into their airspace. The nuclear facility was destroyed, and six North Korean workers were killed in the attack.

(3) Likewise, Russia preceded its conventional armed intervention in the breakaway province of Ossetia in Georgia in 2008 with DDoS attacks, which were designed to frustrate Georgian command and control systems and interfere with government communication and coordination of a response to the Russian invasion. These attacks appeared to be aimed solely at government and military sites. By contrast, and probably in light of the considerable controversy generated by events (1) and (3), there have been to date no such coordinated or massive cyberattacks in the Ukraine, to my knowledge.

(4) Finally, a computer worm (nicknamed "Stuxnet" by Microsoft security experts who later studied it) apparently took control of an array of nuclear centrifuges operated as part of a nuclear weapons development program in Iran. Deceiving the Iranian operators, the worm gained control of the centrifuge array, causing the individual machines to malfunction and self-destruct, while simultaneously transmitting false data to the operators, assuring them incorrectly that all was functioning well. It was thus several months before the underlying problem was discovered, let alone understood as a deliberate attack. Indeed, it is more than two years before analysis, coupled with security leaks, reveal that this particular computer worm was a cyberweapon allegedly created by either Israel or the United States, or from a collaboration between the two, all part of a larger operation of surveillance and attempted sabotage known as "Olympic Games."

What can we glean from these examples? In earlier discussions of these cases, I have argued (see n. 34) that we can derive a great deal that bears upon the present discussion of cybergovernance. The last three of these incidents appear for all the world to be part of serious, grave conflicts between sovereign states: the sort of things that constitute acts of war, or can lead to the declaration of war. Even though

Thomas Rid does not think any of these instances constitute "wars," I have strongly disagreed: the last three are certainly acts of war. As I first argued at the UNESCO symposium in July 2011, however, the very last (4) of these examples of cyberconflict is unique in that it constituted an act of war, resulting in physical damage and destruction of a military target, solely through use of a cyberweapon, very possibly (as Peter Singer later quotes me exclaiming) the first "purely ethical weapon" ever deployed.[35]

By contrast, I have noted then and since, the first of these examples (1) seems somehow out of proportion, or out of place, with respect to the latter three. The justification for armed conflict or the threat of force is nearly nonexistent. This incident was, at most, a diplomatic matter. The attacks, moreover, were directed indiscriminately at civilians and civilian infrastructure. It does not seem morally justifiable, quite apart from legality, to try to harm hospitals and patients and deny ordinary citizens access to their financial resources in a dispute over the placement of a single bronze war memorial. As we know, the government of the territory from which the massive attacks originated denied any knowledge, involvement, or responsibility—invoking the so-called attribution problem. But these attacks, even taken individually, certainly constituted criminal activities, and the International Convention on Cybercrime (CCC) commits nations to policing criminal activities carried out in cyberspace from within their borders (even though the Russian Federation is technically not a signatory to the Bucharest agreement).

Meanwhile, the rulings of the UN Security Council authorizing the US-led intervention in the matter of the Taliban and Al Qaeda in Afghanistan in late 2001 established that the government of an otherwise-sovereign nation could, in fact, be held liable for failing to attempt, in good faith, to put a stop to, or to expel, nonstate actors involved in international criminal conspiracies arising within its borders. At least some of the leading international lawyers engaged in the Tallinn process (such as Mike Schmitt and David Graham) conclude that, in the case of cyberconflict, at least, NATO would have been justified in holding Russia to account for these attacks.[36]

As we know, NATO judged instead that the nature of the harm inflicted and damage done did not, in its opinion, rise to the level of an *armed* attack (unlike the

---

[35] Singer and Friedman, *Cyber Security and Cyber War*. I am cited there on several occasions, both by name and as "a professor at the Naval Academy." This specific characterization of Stuxnet was offered by me in an exchange with Singer at a meeting of the Society for Philosophy and Technology at North Texas State University in 2011, where he was the keynote speaker. In my 2012 "Stutt Lecture on Ethics" at the U.S. Naval Academy, in addition, I noted that the features exhibited by Stuxnet mysteriously seemed to track very closely the principles for the ethical conflict of cyberwarfare that John Arquilla had outlined a decade earlier ("Ethics and Information Warfare"). See "Just War and Cyber Conflict," Annual Stutt Lecture on Ethics, U.S. Naval Academy (9 April 2012): http://www.youtube.com/watch?v=hCj2ra6yzl0. Text at: http://www.usna.edu/Ethics/_files/documents/Just%20War%20and%20Cyber%20War%20GR%20Lucas.pdf.

[36] See Graham, "Cyber Threats and the Law of War," 87–102; and Schmitt, "Cyber Operations and the Jus in Bello: Key Issues," 89–110, both of whom advance forms of this argument.

other three cases). What, we wonder, might have happened, and what would we have *wanted* to see happen, if the attacks had persisted, so that the cumulative harm done became more than merely a massive inconvenience and resulted instead in widespread deaths, immiseration, and loss of property? Would NATO *then* have been justified in some kind of retaliation? If so, what *sort* of attack would be called for? For example, should any response have been limited to an "in-kind" cyberattack (as Eberle and Barrett maintain; see n. 17), or would a conventional, kinetic response be justified?

At the opposite extreme, the alleged attacks by Israel and the United States on nuclear facilities under construction in Iran, in violation of the UN Nuclear Nonproliferation Treaty and against express orders to cease and desist from the International Atomic Energy Commission, seem to constitute a justified military response to a legitimate military threat. The attacks themselves appear to have been undertaken in response to Iran's refusal to follow a direct order from the international community to cease and desist in its program. The cyberattacks themselves were aimed at purely military facilities or installations, not civilians. The damage done seems proportional to the risk of harm threatened.

Lest this apparent blessing of Stuxnet seem merely a partisan conclusion, I note that similar conclusions can be drawn regarding the Russian cyber- and conventional attacks upon Georgia. This is not a matter of playing political favorites, since it is Georgia, after all, that is a member of NATO. Rather, the conflict seemed legitimate, the aspirations of the Ossetians and their complaints against the government of Georgia at least worthy of expression; and the Russian intervention, directed solely against military targets, seemed far more justified as well as proportional to the harm threatened or inflicted, rather than (as in Estonia) wholesale and indiscriminate (although, at the time of this writing, substantial harm and damage to civilians in Ossetia from the Russian-Georgian conflict remain unaddressed and uncompensated).

Is there any legitimate basis for inferring that both the Russian Federation, the United States, and other participatory nations are stumbling, even if blindly and inadvertently, toward some kind of consensus about what constitutes permissible behavior during a cyberconflict? I think these instances, and others, demonstrate that we *are* converging on a kind of normative policy, both constituting best practices and noticing limitations on acceptable practice during a cyberconflict. I enumerate them as follows:

(1) Cyberattacks, like the conventional use of armed force, ought never to be deliberately directed against civilians, civilian objects, or civilian infrastructure.
(2) Cyberattacks may be, and should only be, directed at legitimate military targets, with the twin aims of minimizing collateral damage or loss of life and inflicting only as much damage as is commensurate with the degree of threat represented by the target itself.

(3) A cyberattack can be deemed equivalent to a use of armed force whenever it inflicts harm or damage equivalent to what would be generated by a conventional attack. For instance, it does not matter whether the Iranian centrifuges were destroyed by bombs from aircraft or by a cyberweapon: we might call this the "principle of equivalence" or the "principle of equivalent harm." In turn, however, this principle of equivalent harm entails:

(4) When faced with the choice of means and methods of force to be directed against a justified military target, the weapon that is capable of neutralizing the target with the least threat of additional, collateral damage, harm, or loss of life is always to be selected whenever feasible, consistent with the principles of "economy of force" and military necessity.

That leaves a lot unsettled: for example, the threshold of cyberharm that would justify use of force in retaliation; and whether and when, and how, to retaliate against a serious cyberattack with a conventional attack (rather than a proportionate cyberattack). Note also, in the last case, Stuxnet: the use of force was preemptive, or really, *preventive*—directed against a possible future threat, rather than a clear and present (imminent) danger. Does the advent of discriminate and relatively nondestructive cyberweapons serve to *lower the threshold* against preventive war? Or do the same prohibitions apply as in the case of the use of conventional force?

However we subsequently choose to answer those additional questions, I submit that much has already been accomplished. I believe that this is a process or method that, while similar, is much less presumptuous, and far more inclusive, than is the practice of international law, at least as that practice was exhibited in formulation of the *Tallinn Manual*. There is much that nations could agree upon, and mutually uphold and enforce, without any of the stakeholders feeling they had been ignored, bullied, or unwillingly painted into an unacceptable corner. Moreover, some of the thorny questions still remaining to be addressed have been considerably clarified, and a more acceptable methodology for future work demonstrated that will succeed as well in other areas, such as military robotics, and the use of large-data surveillance in cybersecurity, in which efforts at formal legal governance have likewise thus far failed.[37]

---

[37] My proposed precepts for governance of lethal autonomous systems, attempting to summarize areas of consensus following nearly a decade of contentious public debate on this topic, can be found at the end of "Automated Warfare," 1–23. The precepts are somewhat lengthy, involving knowledge of, and consent to, certain principles governing research, development, manufacture, and use, and specifying the nature of criminal liability for willful violations. Emerging norms for cybersurveillance on the basis of the recent Snowden/NSA controversy can be found at the end of "NSA Management Directive # 424: Secrecy and Privacy in the Aftermath of Snowden," 29–38. These are similar to those constraining cyberconflicts generally, but contain procedures for obtaining informed consent on the part of the general public surveilled.

As to the "normative force" of these emergent norms: they possess as much authority and sanction as does present-day international law, which is to say, not a great deal, in the end. Most of international law, in contrast to domestic legislation, in fact, consists of little more than a somewhat puffed-up and presumptuous attempt to write down, in treaties or statutes, principles that have been long been implicitly agreed to and practiced as norms. Were it not so, there would be little reason to write them down, and nations would tend to ignore or belittle such statutes as constituting "bad governance." Beyond the codification and often helpful clarification of norms emerging from practice, international law accomplishes little on its own (as the abortive attempt to outlaw "killer robots" also currently attests). Killer robots, for example, will only come to be outlawed, or properly regulated, when and if the nations building them, deploying them, and feeling themselves victimized by them together agree to a set of common principles to govern their use, with which all will willingly comply, and to which all will agree to hold one another to account. The same is true with respect to the cyberdomain. That is precisely how the "methodology of uncertainty" works, and how norms themselves first emerge from practice (and reflection upon practice) and come to set effective standards to guide and govern our collective behavior. We need now, with somewhat more humility, to recognize and build upon the progress we have thus far made.

Bibliography

Arquilla, John, and David Ronfeldt. 1993. "Cyberwar Is Coming!" *Comparative Strategy* 12, no. 2: 141–65.

Arquilla, John. 1999. "Ethics and Information Warfare." *The Changing Role of Information in Warfare*, ed. Z. Khalilzad, J. White, and A. Marshall. Santa Monica, CA: RAND Corporation, 379–401.

Arquilla, John. 2010. "Conflict, Security, and Computer Ethics." In *Cambridge Handbook of Information and Computer Ethics*, ed. Luciano Floridi. New York: Cambridge University Press, 133–49.

Arquilla, John. 2012. "Cyber War Is Already Upon Us." *Foreign Policy* (March–April): http://www.foreignpolicy.com/articles/2012/02/27/cyberwar_is_already_upon_us.

Barrett, Edward. "Warfare in a New Domain: The Ethics of Military Cyber Operations." *Journal of Military Ethics* 12, no. 1 (2013): 4–17.

Brenner, Joel. 2011. *America the Vulnerable: Inside the New Threat Matrix of Digital Espionage, Crime, and Warfare*. New York: Penguin.

Brenner, Joel. 2013. "N.S.A.: 'Not (So) Secret Anymore.'" (10 December). http://joelbrenner.com/blog/.

Carr, Jeffrey. 2011. *Inside Cyber Warfare: Mapping the Cyber Underworld*. 2nd ed. Sebastapol, CA: O'Reilly Media.

CCC. 2001. "International Convention on Cybercrime." Council of Europe, "Convention on Cybercrime," Budapest, November 23, 2001: http://conventions.coe.int/Treaty/EN/Treaties/html/185.htm.

Chatterjee, Deen K., ed. 2013. *The Ethics of Preventive War*. New York: Cambridge University Press.

Clarke, Richard A., and Robert K. Kanke. 2010. *Cyber War: The Next Threat to National Security and What to Do about It*. New York: HarperCollins, 2010.

Denning, Dorothy E. 1998. *Information Warfare and Security*. Camden, NJ: Addison-Wesley.

Denning, Dorothy E. 2001. "Cyberwarriors." *Harvard International Review* (summer): 70–75.
Denning, Dorothy E. 2007. "The Ethics of Cyber Conflict." In *Information and Computer Ethics*, ed. K. E. Himma and H. T. Tavani. New York: Wiley. Available at: http://faculty.nps.edu/dedennin/publications/Ethics%20of%20Cyber%20Conflict.pdf.
Dipert, Randall R. 2010. "The Ethics of Cyber Warfare." *Journal of Military Ethics* 9, no. 4 (December): 384–410.
Dipert, Randall R. 2013a. "The Future Impact of a Long Period of Limited Cyberwarfare on the Ethics of Warfare." In *Ethics and Information Warfare: Proceedings of a UNESCO Conference on Ethics and Cyber Security*, ed. Luciano Floridi and Mariarosaria Taddeo. Amsterdam: Springer, 25–38.
Dipert, Randall R. 2013b. "Other Than Internet Warfare: Challenges for Ethics, Law and Policy." *Journal of Military Ethics* 12, no. 1: 34–53.
Dipert, Randall R. 2013c. "The Essential Features for an Ontology for Cyberwarfare." In *Conflict and Cooperation in Cyberspace*, ed. Panayotis A. Yannakogeorgos and Adam B. Lowther. Boca Raton, FL: CRC Press/Taylor and Francis, 35–48.
Department of Defense. 2011. "US. Department of Defense Strategy for Operating in Cyberspace." Washington, DC: Department of Defense, 1 July, http://www.defense.gov/news/d20110714cyber.pdf.
Department of State. 2011. "International Strategy for Cyberspace: Prosperity, Security and Openness in a Networked World." (Washington, DC: Office of the President, 1 May, http://www.whitehouse.gov/sites/default/files/rss_viewer/international_strategy_for_cyberspace.pdf.
Dunlap, Charles J. 2011. "Perspectives for Cyber Strategists on Law for Cyberwar." *Strategic Studies Quarterly* (spring): 81–99.
Eberle, Christopher. 2013. "Just War and Cyber War." *Journal of Military Ethics* 12, no. 1: 54–67.
Floridi, Luciano, and Mariarosario Taddeo, eds. 2014. *Ethics and Information Warfare: Proceedings of a 2011 UNESCO Conference on Ethics and Cyber Security*. Amsterdam: Springer.
Graham, David E. 2010. "Cyber Threats and the Law of War." *Journal of National Security Law* 4, no. 1: 87–102.
Jenkins, Ryan. "Is Stuxnet Physical? Does It Matter?" *Journal of Military Ethics* 12, no. 1 (2013): 68–79.
Johnson, Albert R., and Stephen E. Toulmin. 2009. *The Abuse of Casuistry: A History of Moral Reasoning*. Berkeley and Los Angeles: University of California Press.
Libicki, Martin C. 2007. *Conquest in Cyberspace: National Security and Information Warfare*. New York: Cambridge University Press.
Libicki, Martin C. 2009. *Cyberdeterrence and Cyberwar*. Santa Monica, CA: Rand Corporation.
Lin, Herbert S., et al. 2009. *Technology, Policy, Law, and Ethics Regarding U.S. Acquisition and Use of Cyberattack Capabilities*. Washington, DC: National Research Council/American Academy of Sciences.
Lucas, George R. Jr. 2012. "Just War and Cyber Conflict." Annual Stutt Lecture on Ethics, U.S. Naval Academy (9 April): http://www.youtube.com/watch?v=hCj2ra6yzl0. Text at: http://www.usna.edu/Ethics/_files/documents/Just%20War%20and%20Cyber%20War%20GR%20Lucas.pdf.
Lucas, George R. Jr. 2013a. "Can There Be an Ethical Cyberwar?" In *Conflict and Cooperation in Cyberspace*, ed. Panayotis A. Yannakogeorgos and Adam B. Lowther. Boca Raton, FL: CRC Press/Taylor and Francis, 195–210.
Lucas, George R. Jr. 2013b. "Jus in Silico: Moral Restrictions on the Use of Cyber Warfare." In *The Routledge Handbook of War and Ethics*, ed. Evans Allhoffand and Adam Henschke. Oxford: Routledge, 367–80.
Lucas, George R. Jr. 2013. "Emerging Norms for Cyber Warfare." Keynote Address, "Conference on Ethics & Cyber Security," Center for Applied Philosophy and Practical Ethics, Australian National and Charles Sturt Universities, Canberra, Australia, 5–6 August: http://philevents.org/event/show/11114.
Lucas, George R. Jr. 2013c. "Navigation, Aviation, 'Cyberation:' Developing New Rules of the Road for the Cyber Domain." Inaugural Public Security Lecture, National Security College,

Australian National University, Canberra, Australia, 7 August: http://www.youtube.com/watch?v=cs7RXAPzG84.

Lucas, George R. Jr. 2014a. "Permissible Preventive Cyber Warfare." In *Ethics and Information Warfare*, ed. Luciano Floridi and Mariarosaria Taddeo. Amsterdam: Springer, 73–82.

Lucas, George R. Jr. 2014b. "NSA Management Directive # 424: Secrecy and Privacy in the Aftermath of Snowden." *Journal of Ethics and International Affairs* 28, no. 1 (spring): 29–38. http://journals.cambridge.org/action/displayAbstract?fromPage=online&aid=9207954&fileId=S0892679413000488

Lucas, George R. Jr. 2014c. "Ethics and Cyber Conflict: A Review of Recent Work." *Journal of Military Ethics* 13, no. 1: 20–31.

Lucas, George R. Jr. 2014d. "Automated Warfare." *Stanford Law and Policy Review* 25, no. 2: 1–23.

MacIntyre, Alasdair. 1981. *After Virtue*. South Bend, IN: University of Notre Dame Press.

MacIntyre, Alasdair, 1988. *Whose Justice? Which Rationality?* South Bend, IN: University of Notre Dame Press.

MacIntyre, Alasdair, 1990. *First Principles, Final Ends, and Contemporary Philosophical Issues*. Milwaukee: Marquette University Press.

Marchant, Gary E., et al. 2011. "International Governance of Autonomous Military Robots." *Columbia Science and Technology Law Review* 12: 272–315.

O'Meara, Richard M. 2015. *Governing Military Technologies in the 21st Century*. London: Palgrave Macmillan (forthcoming).

Rid, Thomas C. 2011. "Cyber War Will Not Take Place." *Journal of Strategic Studies* 35, no. 1 (October): 5–32.

Rid, Thomas C. 2012. "Think Again: Cyberwar." *Foreign Policy* (March–April): http://www.foreignpolicy.com/articles/2012/02/27/cyberwar

Rowe, Neil C. 2007. "War Crimes from Cyberweapons." *Journal of Information Warfare* 6, no. 3: 15–25.

Rowe, Neil C. 2008. "Ethics of Cyber War Attacks." In *Cyber Warfare and Cyber Terrorism*, ed. Lech J. Janczewski and Andrew M. Colarik. Hershey, PA: Information Science Reference, 105–11.

Rowe, Neil C. 2010. "The Ethics of Cyberweapons in Warfare." *Journal of Technoethics* 1, no. 1: 20–31.

Rowe, Neil C. 2011. "Toward Reversible Cyber Attacks." In *Leading Issues in Information Warfare and Security Research*, ed. J. Ryan. Reading, UK: Academic, 145–58.

Sanger, David. 2013. "Differences on Cybertheft Complicate China Talks." *New York Times*, July 10, www.nytimes.com/2013/07/11/world/asia/differences-on-cybertheft-complicate-china-talks.html.

Schmitt, Michael N. 1999. "Computer Network Attack and the Use of Force in International Law: Thoughts on a Normative Framework." *Columbia Journal of Transnational Law* 37: 885–937.

Schmitt, Michael N. 2002. "Wired Warfare: Computer Network Attack and *Jus in Bello*." *International Review of the Red Cross* 84, no. 846: 365–99.

Schmitt, Michael N. 2011. "Cyber Operations and the Jus in Bello: Key Issues." *U.S. Naval War College International Law Studies* 87: 89–110.

Singer, Peter W., and Allan Friedman. 2014. *Cyber Security and Cyber War: What Everyone Needs to Know*. New York: Oxford University Press.

Schmitt, Michael N. 2012. *The Tallinn Manual on the International Law Applicable to Cyber Warfare*. Tallinn: NATO Cooperative Cyber Defence Center of Excellence. Available at: https://www.ccdcoe.org/249.html.

Wikipedia contributors, "Nisour Square massacre," *Wikipedia, The Free Encyclopedia*, 21 August 2015, 17:12 UTC, <https://en.wikipedia.org/w/index.php?title=Nisour_Square_massacre&oldid=677180494> Accessed August 25, 2015.

Wingfield, Thomas, 2000. *The Law of Information Conflict*. Falls Church, VA: Aegis Research Corporation.

Yannakogeorgos, Panayotis A., and Adam B. Lowther, eds. 2013. *Conflict and Cooperation in Cyberspace*. Boca Raton, FL: CRC Press/Taylor and Francis.

2

The Emergence of International Legal Norms for Cyberconflict

MICHAEL N. SCHMITT AND LIIS VIHUL

This essay explores the nature, formation, and evolution of legal norms pertaining to cyberactivities in both *jus in bello* and *jus ad bellum*. When referring to the situations in which such norms apply, it adopts, for reasons to be discussed, the term "cyberconflict" in lieu of "cyberwarfare." Only legal norms, that is, those that are binding as a matter of law, are examined. It must be cautioned that political, social, technical, philosophical, and ethical norms also shape cyberactivities; indeed, in many circumstances, they may do so with greater significance than their legal counterparts.

The inquiry begins with a brief review of relevant terminology and the scope of applicability of *jus in bello* and *jus ad bellum*. This survey is essential because the divergent language employed by legal and nonlegal communities is a source of much confusion in discussions of the relevant norms. Such discourse is also often obfuscated by improper reference to various norms that reside in these distinct bodies of international law, norms that have particular purposes and effects only with respect to one of the two bodies of law. Hence, an understanding of the boundary lying between *jus in bello* and *jus ad bellum* is a precondition to comprehending their scope of applicability in a cybercontext.

This chapter then turns to how such norms emerge, are interpreted, and develop through time. In international law, and as acknowledged in Article 38 of the Statute of the International Court of Justice (ICJ), binding legal norms reside in treaties and are found in customary international law.[1] Therefore, the analysis will proceed by addressing each source of law separately—first in the abstract, and then in its cyberconflict context. The chapter concludes that although cyberconflict is no less subject to extant international legal norms than other forms of conflict, the former

---

[1] Statute of the International Court of Justice, June 26, 1945, 59 Stat. 1055, 33 U.N.T.S. 993 [hereinafter ICJ Statute].

presents unique challenges to the application of existing norms and the emergence of new ones.

2.1 Terminological Context and Issues of Scope

In order to understand how legal norms are created for cyberactivities, it is necessary to first grasp the relevant vocabulary. Indeed, perhaps the greatest hindrance to effective interdisciplinary discourse on cyberwarfare is terminological in nature. For instance, nonlawyers tend to speak of "cyberwarfare" in a generic sense, as encompassing all forms of hostile cyberactivities conducted by or against states, and they use the term "cyberattacks" to refer to any harmful cyberoperations. However, these terms do not formally reside in international law. Instead, international law uses a unique language that has discreet normative significance. Fluency in this language requires an understanding of the bifurcated nature of the law governing conflict, whether cyber or kinetic in nature. Such law is found in *jus in bello* and *jus ad bellum*.

With respect to the former, the international community discarded the term "war" in lieu of "armed conflict" during the four 1949 Geneva Conventions.[2] "Armed conflict" refers to the condition within which *jus in bello*, otherwise known as international humanitarian law (IHL), applies. Absent an armed conflict, human rights law, the law of sovereignty, the law of state responsibility, and other genres of international law lay beyond the reach of IHL to govern cyberoperations. Thus, for international lawyers, the term "cyberwarfare" is best translated as "cyber armed conflict."

International humanitarian law, in particular customary international law and the Geneva Conventions with their 1977 Additional Protocols,[3] deals with *how* force may be employed by the parties to an armed conflict (and any others). It applies to all sides of an armed conflict irrespective of the legality of their initial resort to force under *jus ad bellum*, discussed below. International humanitarian law contains, inter alia, the specific rules governing "attacks," delineates protections to which certain

[2] Convention (I) for the Amelioration of the Condition of the Wounded and Sick in the Armed Forces in the Field, August 12, 1949, 75 U.N.T.S. 31; Convention (II) for the Amelioration of the Condition of the Wounded, Sick, and Shipwrecked Members of Armed Forces at Sea, August 12, 1949, 75 U.N.T.S. 85; Convention (III) Relative to the Treatment of Prisoners of War, August 12, 1949, 75 U.N.T.S. 135; Convention (IV) Relative to the Protection of Civilian Persons in Time of War, August 12, 1949, 75 U.N.T.S. 287 [hereinafter GC I–IV, respectively].

[3] Protocol Additional to the Geneva Conventions of 12 August 1949, and Relating to the Protection of Victims of International Armed Conflicts, June 8, 1977, 1125 U.N.T.S. 3; Protocol Additional to the Geneva Conventions of August 12, 1949, and Relating to the Protection of Victims of Non-International Armed Conflicts, June 8, 1977, 1125 U.N.T.S. 609 [hereinafter Additional Protocol I and II, respectively].

persons and objects are entitled during an armed conflict, and restricts the kinds of weapons that may be employed in order to conduct hostilities.

Armed conflict takes one of two forms: international and noninternational. *International armed conflict* exists when states engage in hostilies with other states.[4] No particular level of intensity is required so long as the activities involved qualify as "attacks," as that term is understood in IHL, a topic explored in sections 2.2 and 2.3. *Noninternational armed conflict*, by contrast, occurs when a state is involved in hostilities with an organized armed group or when organized armed groups are engaged in hostilities with each other.[5] In order to distinguish this latter form of conflict from civil disturbances and other lower-level acts of violence that may take place in a state,[6] noninternational armed conflict requires that the hostilities have attained a high level of intensity and that the groups involved are sufficiently organized.[7]

It is clear that when cyberoperations accompany kinetic hostilities qualifying as either international or noninternational armed conflict—as with the conflict between Russia and Georgia in 2008 or that taking place in Syria, respectively—IHL applies fully to those operations. For instance, in the same way that IHL prohibits injurious or destructive kinetic attacks against civilians and civilian objects, it likewise prohibits cyberattacks against them having the same effects. However, the precise application of IHL norms in particular circumstances, such as whether a non–physically destructive cyberoperation against a civilian object qualifies as a prohibited attack, is the source of some contention.[8] The question of when cyberoperations alone qualify as hostilities for the purpose of initiating an armed conflict also remains unsettled.[9]

Jus ad bellum is an entirely separate body of international law that deals with the resort to force by states as an instrument of their national policy. It consists of both treaty law—principally the United Nations (UN) Charter—and customary international law. Rather than framing *how* hostilities, whether cyber or kinetic in nature, may be conducted, the *jus ad bellum* addresses the question of *when* states may lawfully resort to force. Thus, it deals with such questions as when does a state's act breach the prohibition on the "use of force" found in Article 2(4) of the United Nations Charter, under what circumstances may the United Nations Security Council authorize a state's use of force pursuant to Article 42 of the Charter, and when states may respond to an "armed attack" in individual or collective self-defense under Article 51 of the Charter.

---

[4] GC I–IV, common art. 2.

[5] GC I–IV, common art. 3, Additional Protocol II, art. 1(1).

[6] Additional Protocol II, art. 1(2).

[7] Prosecutor v. Tadić, Case No. IT-94-1-T, Trial Chamber Opinion and Judgment, para. 562 (Int'l Crim. Trib. for the former Yugoslavia May 7, 1997).

[8] Schmitt, "'Attack' as a Term of Art in International Law," 290–93.

[9] Schmitt, "Classification of Cyber Conflict," 241, 248–49.

To illustrate the importance of the distinction between *jus ad bellum* and IHL, consider the frequent, albeit varied, use of the term "act of war"—one no longer recognized in international law—during lay discussions of cyberoperations.[10] The term is sometimes used by nonlawyers to refer to a cyberoperation that international lawyers would label either a prohibited "use of force" or an "armed attack" under *jus ad bellum*. At other times, nonlawyers use the term to indicate the threshold when what an international lawyer would brand an "armed conflict" commences under *jus in bello*. Such terminological imprecision has long hobbled interdisciplinary dialogue between legal and nonlegal communities. As should be apparent, a proper understanding of international law governing cyberoperations, and its likely future evolution, demands terminological fastidiousness.

2.2 Treaty Law

A treaty is an international agreement governed by international law.[11] So long as the parties to such an agreement intended to create legally binding rights and obligations for themselves, the instrument's precise appellation is of no legal significance.[12] Whether styled as a treaty (the most common title), convention, protocol, or any other moniker, the law that applies to formation, application, and interpretation of an international agreement is identical.

The Vienna Convention on the Law of Treaties is generally regarded as an authoritative statement of the law governing treaties. While some states, such as the United States, are not party to the convention, it is, in great part, viewed as reflective of customary international law, a topic examined in section 2.4. Of particular note in this cybercontext is the principle that treaties are governed exclusively by international law, except in cases where the agreement itself refers to domestic law. The fact that a state's domestic law disallows an action required by a treaty—or demands one prohibited by a treaty—does not excuse a state's noncompliance with the terms of the treaty.

Once a treaty has been successfully negotiated, states subsequently consent to be bound by it, which may occur through a number of means. For instance, such consent may be indicated through signature (but not in every case since signature sometimes denotes only adoption), exchange of instruments, ratification, accession,

---

[10] See, e.g., Sanger and Bumiller, "Pentagon to Consider Cyberattacks Acts of War," *New York Times*, May 31, 2011, http://www.nytimes.com/2011/06/01/us/politics/01cyber.html?_r=0; Lin, Allhoff, and Rowe, "Is It Possible to Wage a Just Cyberwar?" *The Atlantic*, June 5, 2012, http://www.theatlantic.com/technology/archive/2012/06/is-it-possible-to-wage-a-just-cyberwar/258106/.

[11] Vienna Convention on the Law of Treaties art. 2(1)(a), May 23, 1969, 1155 U.N.T.S. 331 [hereinafter Vienna Convention on the Law of Treaties].

[12] Vienna Convention on the Law of Treaties, art. 2(1)(a).

or any other means the parties agree upon.[13] Representatives of states often sign treaties subject to ratification. As an example, in the United States, treaty-making power is vested in the president but subject to the "advice and consent" of the Senate.[14] A state "accedes" to a treaty when it did not participate in the negotiations leading to its adoption. Finally, a treaty usually specifies a particular date of its entry into force or includes a provision requiring a particular number of states to ratify the treaty before it comes into effect.[15]

The procedural requirements are important with respect to the application and evolution of legal norms because it is not unusual for a treaty to be adopted and ratified by some states long before it comes into force. For instance, the Rome Statute of the International Criminal Court[16] was adopted in 1998, but it only came into force upon the sixtieth ratification in 2002. Pending a treaty's coming into force, states that have signed it or otherwise expressed an intent to eventually be bound by it may not engage in activities that would defeat the treaty's object and purpose unless they formally provide notification of their decision to not become a party thereto,[17] as was the case with the United States and the International Criminal Court Statute in 2002.[18] Accordingly, the fact that a treaty has not yet come into effect does not preclude it having some normative significance. Similarly in the cybercontext, the eighty-nine states that signed the International Telecommunication Regulations Treaty[19] at the World Conference on International Telecommunications in Dubai, United Arab Emirates, in 2012 had to act in accordance with its object and purpose from the moment they signed the treaty despite the fact that it only came into effect as of January 1, 2015.

States sometimes issue reservations to multilateral treaties when they consent to be bound by them.[20] Reservations act to exclude or modify certain provisions of the treaty with respect to the state concerned.[21] Some treaties prohibit reservations altogether. Even when allowed, reservations cannot be inconsistent with the object and purpose of the treaty. If a state reserves and another state accepts the reservation, the exclusion or modification of the provision in question operates with respect to the obligations of both states. Should a party to the treaty object to the reservation, it will not apply vis-à-vis relations between the objecting and reserving parties. An objecting

---

[13] Ibid., arts. 11–15.

[14] U.S. Constitution, art. II, sect. 2, cl. 2.

[15] See, e.g., Vienna Convention on the Law of Treaties, art. 24.

[16] Rome Statute of the International Criminal Court, July 17, 1998, 2187 U.N.T.S. 90.

[17] Vienna Convention on the Law of Treaties, art. 18.

[18] Press Statement, U.S. Department of State, Richard Boucher, Spokesman, "International Criminal Court: Letter to UN Secretary General Kofi Annan," May 6, 2002, accessed June 1, 2014, http://2001-2009.state.gov/r/pa/prs/ps/2002/9968.htm.

[19] International Telecommunication Regulations, December 14, 2012, deposited with the International Telecommunication Union Secretary-General.

[20] Vienna Convention on the Law of Treaties, art. 19.

[21] Ibid., art. 2(1).

state may also determine that a reservation is so objectionable that the treaty is not in force at all between it and the reserving state. As should be evident, reservations to a multilateral treaty can create an extremely complex maze of legal relationships.

In addition to reservations, states may issue interpretative declarations that clarify their position with regard to a particular provision of the treaty or to how the treaty will be applied by the states concerned. These declarations have no technical legal effect on the state's rights or obligations. However, states sometimes make interpretative declarations that de facto amount to reservations. For example, the United Kingdom has issued a statement concerning the prohibitions on reprisals set forth in Additional Protocol I.[22] The declaration arguably denudes certain of the relevant provisions of their effect. Thus, declarations, like reservations, must always be carefully surveyed when evaluating the actual normative reach of a treaty.

Perhaps the most important aspect of treaty law deals with interpretation. This is so because a treaty's text may be vague or ambiguous. Such ambiguousness is often the only way the parties involved were able to achieve sufficient consensus to adopt the instrument. The Vienna Convention on the Law of Treaties provides that treaties "shall be interpreted in good faith in accordance with the ordinary meaning to be given to the terms of the treaty in their context and in light of its object and purpose."[23] The term "context" refers to the other text of the treaty, as well as to any agreement between the parties made at the conclusion of the treaty.[24] In addition to context, interpretation of a treaty's provision should consider any subsequent express agreement between parties as to its meaning, as well as "subsequent practice in its application that establishes the agreement of the parties regarding its interpretation."[25] If the meaning of a provision remains ambiguous, reference may be made to the "preparatory work of the treaty and the circumstances of its conclusion."[26] In other words, it is appropriate to explore what was in the mind of the parties at the time an agreement was negotiated and adopted.

2.3 Treaties and Cyberconflict

Very few treaties deal directly with cyberactivities. Prominent contemporary examples include the Convention on Cybercrime,[27] its 2006 Additional Protocol,[28] the Shanghai

---

[22] UK Ministry of Defence, *Manual of the Law of Armed Conflict* 422–23.
[23] Vienna Convention on the Law of Treaties, art. 31(1).
[24] Ibid., art. 31(2).
[25] Ibid., art. 31(3).
[26] Ibid., art. 32.
[27] Convention on Cybercrime, November 23, 2001, 2296 U.N.T.S. 167.
[28] Additional Protocol to the Convention on Cybercrime Concerning the Criminalization of Acts of a Racist and Xenophobic Nature Committed through Computer Systems, January 28, 2003, E.T.S. No. 189.

Cooperation Organization's International Information Security Agreement,[29] and the International Telecommunication Union (ITU) Constitution and Convention[30] and International Telecommunication Regulations.[31] No treaties are tailored specifically for cyberconflict.

Despite the absence of cyberconflict-specific treaties, an array of international agreements that govern conflict in a general sense also directly apply to the activities constituting cyberconflict. Central among these, as mentioned, are the UN Charter with respect to *jus ad bellum*, and the 1949 Geneva Protocols and their 1977 Additional Protocols vis-à-vis IHL. Note in this regard that a number of states, including Russia and China, previously expressed some reluctance to acknowledge that existing international agreements extended to cyberactivities.[32] This disinclination seems to have been partially overcome in 2013 with the issuance of the UN Group of Governmental Experts (GGE) report, which found that "[i]nternational law, and in particular the Charter of the United Nations, is applicable and is essential to maintaining peace and stability and promoting an open, secure, peaceful and accessible ICT environment."[33] The report also confirmed the appropriateness of the law of sovereignty and that of state responsibility in the context of cybersecurity.[34] Both Russia and China were represented in the group. Interestingly, and unfortunately, a draft provision endorsing IHL's applicability was removed in order to secure unanimity. However, even beyond the Euro-Atlantic community, many states have publicly confirmed that IHL applies to cyberactivities associated with an armed conflict.[35] There appears to be no serious opposition to the notion in academia.[36]

---

[29] Agreement between the Governments of the Member States of the Shanghai Cooperation Organization on Cooperation in the Field of International Information Security, 61st Plenary Meeting, December 2, 2008.

[30] Constitution and Convention of the International Telecommunication Union, December 22, 1992, 1825 U.N.T.S. 330.

[31] International Telecommunication Regulations, December 9, 1988, deposited with the International Telecommunication Union Secretary-General. The International Telecommunication Regulations, as well as the Radio Regulations, are a legal instrument of the ITU; see Constitution of the International Telecommunication Union, art. 4(3).

[32] As an example, Russia has put forward arguments that instead of regulating armed cyberconflict through IHL, it should be outlawed altogether. On this point, as well as for a comprehensive overview of Russia's views on cyber-conflict, see Giles and Monaghan, "Legality in Cyberspace: An Adversary View," 12, accessed June 1, 2014, http://www.strategicstudiesinstitute.army.mil/pubs/display.cfm?pubID=1193.

[33] "Group of Governmental Experts on Developments in the Field of Information and Telecommunications in the Context of International Security," para. 19, U.N. Doc. A/68/98, June 24, 2013, http://undocs.org/A/68/98.

[34] Ibid., paras. 20–23.

[35] See, e.g., "International Strategy on Cybersecurity Cooperation" (Information Security Policy Council, Japan), 9, accessed June 1, 2014, http://www.nisc.go.jp/active/kihon/pdf/InternationalStrateg yonCybersecurityCooperation_e.pdf; "Defence White Paper 2013" (Australian Department of Defence), 21, accessed June 1, 2014, http://www.defence.gov.au/whitepaper2013/docs/WP_2013_web.pdf.

[36] In addition to the *Tallinn Manual*, referenced below, see, e.g., Dinniss, *Cyber Warfare*; Roscini, *Cyber Operations*. The ICRC has endorsed the same view in ICRC, "International Humanitarian Law and the Challenges of Contemporary Armed Conflicts," 37, Doc. 31IC/11/5.1.2.

Given the general applicability of these instruments to the cyber-conflict, the key issue is how such norms are to be *interpreted* in the cybercontext. This was the subject of inquiry by the International Group of Experts that prepared *The Tallinn Manual on the International Law Applicable to Cyber Warfare*[37] (*Tallinn Manual*) between 2009 and 2013 under the auspices of the NATO Cooperative Cyber Defence Centre of Excellence. Although the *Tallinn Manual* embraces the premise of complete applicability of *jus ad bellum* and IHL norms,[38] it is replete with examples of circumstances in which the Experts could not achieve consensus on their precise interpretation with respect to cyberoperations. Accordingly, the manual often refers to majority and minority views among them. To ensure comprehensiveness, on numerous occasions the manual even acknowledges the existence of reasonable interpretations not supported by any member of the group.[39]

As became clear during the *Tallinn Manual* drafting process, the object and purpose of treaties enjoy particular significance when interpreting existing treaties in the context of new areas of activity such as cyberconflict. Sometimes the activities in question were obviously beyond the contemplation of those drafting the treaties. Indeed, cyberactivities, as we know them today, did not exist when the UN Charter, Geneva Conventions, Additional Protocols, and numerous other relevant treaties were adopted. Therefore, when applying the provisions of these treaties to cyberoperations, it is necessary to examine the foundational rationale underlining both them generally and any particular individual provision concerned.

Four prominent examples illustrate the significance of treaty interpretation techniques with respect to cyberconflict. The first deals with the meaning of the term "use of force" in the UN Charter Article 2(4)'s prohibition thereon. All of the *Tallinn Manual* Experts agreed that a cyberoperation by one state against another that causes injury or death of individuals, or damage or destruction of property, qualifies as a use of force. There was also general consensus that certain cyberoperations lacking such physical consequences likewise qualify.[40] However, no consensus on the exact threshold at which a cyberactivity crosses into the use-of-force realm could be reached. All the International Group of Experts could offer were indicative factors that states are likely to consider when deciding how to legally characterize a cyberoperation in this respect.[41] Delineation of factors should prove useful as states estimate how their activities will be seen by other

---

[37] Schmitt, *Tallinn Manual* [hereinafter Tallinn Manual].

[38] Tallinn Manual, 3, 13.

[39] See, e.g., acknowledgment of a view by which the gap between the thresholds of a "use of force" and an "armed attack" is either so narrow as to be insignificant or nonexistent, but which was not shared by any member of the International Group of Experts. Tallinn Manual, para. 7 of commentary to Rule 11.

[40] Tallinn Manual, para. 7 of commentary to Rule 11.

[41] Tallinn Manual, paras. 8–10 of commentary to Rule 11.

states, as well as when they assess the actions of other states vis-à-vis the norm. But they are not legal criteria as such.

Second, Article 51 of the UN Charter provides that states may use force in response to an armed attack. The interpretation of this article remains a source of uncertainty and controversy. First, there is an ongoing debate over whether the right of self-defense extends to attacks conducted by nonstate actors. For instance, the ICJ appears to have suggested that the article only applies in situations in which the activities concerned reach the level of intensity required for an armed attack and are either conducted "by or on behalf" of a state or with a state's "substantial involvement."[42] However, contemporary state practice, most notably that since the 9/11 terrorist attacks, appears to contradict this position. In particular, the international community unambiguously characterized the Al Qaeda attacks as triggering the United States' inherent right of self-defense.[43] More recently, the Netherlands has taken the position that defensive uses of force in the cybercontext are permissible under Article 51 even if a cyberattack by a nonstate actor cannot be attributed to another state.[44]

It is also unclear when a cyberoperation is severe enough to be regarded as an armed attack in the sense of Article 51. According to the *Tallinn Manual*, significant damage, destruction, injury, or death so qualify, but the International Group of Experts could not articulate a "bright line test" for determining when such harm is "grave."[45] Some experts even took the position that the term should include operations that caused severe nonphysical harm, such as crippling cyberoperations directed at a state's economy.[46] This position seems to have been adopted by the US and Dutch governments, although neither has offered much detail on the issue.

---

[42] Military and Paramilitary Activities in and against Nicaragua (Nicar. v. U.S), 1986 I.C.J. 14, para. 195 (June 27) [hereinafter Nicaragua].

[43] The Security Council adopted numerous resolutions recognizing the applicability of the right of self-defense to attacks by nonstate actors. See, e.g., U.N. Doc. S/RES/1368, September 12, 2001, accessed June 1, 2014, http://www.un.org/en/ga/search/view_doc.asp?symbol=S/RES/1368(2001); U.N. Doc. S/RES/1373, September 28, 2001, accessed June 1, 2014, http://www.un.org/en/ga/search/view_doc.asp?symbol=S/RES/1373(2001). International organizations, including NATO, and many individual states took the same approach. See also Tallinn Manual, 58; U.S. Department of Justice, U.S. Department of Justice White Paper "Lawfulness of a Lethal Operation Directed against a U.S. Citizen Who Is a Senior Operational Leader of Al-Qa'ida or an Associated Force" 2 (Draft, November 8, 2011), accessed June 1, 2014, http://users.polisci.wisc.edu/kmayer/408/020413_DOJ_White_Paper.pdf; Koh, "Obama Administration and International Law" (March 25, 2010), accessed June 1, 2014, http://www.state.gov/s/l/releases/remarks/139119.htm.

[44] "Government Response to the AIV/CAVV Report on Cyber Warfare" (Netherlands), accessed June 1, 2014, http://www.aiv-advies.nl/ContentSuite/template/aiv/adv/collection_single.asp?id=1942&adv_id=3016&page=regeringsreacties&language=UK [hereinafter Dutch Government Response].

[45] Tallinn Manual, para. 6 of commentary to Rule 13, para. 8 of commentary to Rule 11.

[46] Tallinn Manual, para. 9 of commentary to Rule 13.

The third and fourth examples derive from IHL. The paradigmatic interpretive hurdle in IHL deals with the meaning of the word "attack," which is found in various prohibitions set forth in Additional Protocol I. For instance, pursuant to that treaty, it is unlawful to attack civilians, civilian objects, and certain other protected persons and objects.[47] Additionally, states are required to consider expected collateral damage at the "attack" level when assessing the proportionality of their operations,[48] and they must take precautions to minimize such damage whenever they conduct attacks.[49] Therefore, interpretation of the term "attack" in the cybercontext is essential because, to the extent a cyberoperation does not qualify as an attack, these IHL provisions do not apply.

An "attack" is defined in Additional Protocol I as an "act of violence."[50] By one view identified in the *Tallinn Manual*, the notion of attack is strictly limited to cyberoperations that cause physical damage or injury. Another, adopted by the majority of the Experts, looks to the object and purpose of the Protocol and its provisions on point to interpret the term as applying to a situation in which the functionality of an object is affected by a cyberoperation without physical damage having occurred.[51] But even by this approach, there were differences of opinion as to how "functionality" should be interpreted.[52] As this example illustrates there may be layers of interpretation of a treaty's provisions.

Finally, a similar IHL debate is whether the term "civilian object" extends to data. If so interpreted, a cyberoperation designed to destroy civilian data would be prohibited by Article 52 of Additional Protocol I, which bans direct attacks against civilian objects. If not, civilian data would be a lawful object of attack, except in those circumstances where its loss might cause physical damage to objects or injury to persons. The critical, and unresolved, fault line in the debate lies between interpretations that limit the term to entities that are tangible and those based on the argument that in contemporary understanding the ordinary meaning of "object" includes data.[53]

These examples illustrate that the interpretation of certain express *jus ad bellum* and *jus in bello* treaty provisions as applied to cyberconflict remains unsettled. Even strict application of the rules of treaty interpretation set forth in section 2.2 has failed to fully suffice in adding the requisite clarity. Therefore, such interpretive dilemmas concerning treaty law are only likely to be resolved through the recurrent

---

[47] Additional Protocol I, arts. 51–56, 59.
[48] Additional Protocol I, arts. 51(5)(b), 57(2)(a)(iii), 57(2)(b).
[49] Additional Protocol I, art. 57.
[50] Additional Protocol I, art. 49(1).
[51] On the definition of a "cyber attack" under IHL and the "functionality test," see Schmitt, "Rewired Warfare: Rethinking the Law of Cyber Attack," in *International Review of the Red Cross* 96 (2014).
[52] Tallinn Manual, paras. 4, 10–12 of commentary accompanying Rule 30.
[53] Tallinn Manual, para. 5 of commentary accompanying Rule 38.

practice of states in their application. It can include that of states when acting in their capacity as members of international organizations like the United Nations, the European Union, and the North Atlantic Treaty Organization (NATO). Also relevant will be state expressions of opinion as to proper interpretation of the terms and provisions in question. Recent examples include those proffered by former US Department of State legal adviser Harold Koh[54] and by the Dutch government in response to the AIV report, both of which set forth state positions on the meaning of key aspects of relevant treaty law.[55] Judicial interpretation could potentially also shape the meaning of tenebrous treaty norms in the cybercontext, much as the judgments of the International Criminal Tribunal for the Former Yugoslavia have added significant granularity to the understanding of IHL in its noncyber guise. Finally, the work of scholars in the field cannot be understated in light of the stark paucity of overt state practice and interpretive pronouncements. This dynamic is exemplified by the influence the *Tallinn Manual* is having on the formulation of state policies with regard to the respective treaty norms that bind them.

A persistent question is whether new treaties to address the subject of cyberconflict, in particular with respect to arms control, are necessary or likely to materialize. Such treaty law would undoubtedly clear much of the normative fog that presently exists. Yet, new treaties are highly unlikely for the foreseeable future. Historically, treaties are seldom adopted prior to the advent of new technologies. On the contrary, those governing new methods and means of warfare typically emerge only after the technologies have been fielded and employed, thereby revealing lacunae or insufficiencies in the existing law. The paradigmatic examples are the conventions governing antipersonnel landmines and cluster munitions, which were concluded decades after the first employment of the weapons and which are still the subject of much controversy.[56] In the international arena, there is presently little support for proactively addressing cyberweaponry and cyberoperations. As with all other methods and means of warfare, states are hesitant to restrict the use of weapons that may afford them an advantage on the battlefield until they have sufficient experience to allow them to weigh the costs and benefits of prohibitions and limitations on their use.[57]

A further factor rendering cybertreaties unlikely in the near term is the difficulty of verifying compliance with their terms and effectively enforcing them. In the first place, it is sometimes difficult to even ascertain that harm is the result of a cyberoperation. Second, the technical challenges posed by attribution complicate

---

[54] Koh, "International Law in Cyberspace."

[55] Dutch Government Response.

[56] For instance, the United States is not a party to either the Ottawa Convention on antipersonnel mines treaty or the Dublin Treaty on cluster munitions. In both cases it took the position that the instruments run counter to operational needs.

[57] As an example, the 1923 Hague Rules of Air Warfare were never implemented in treaty form, in great part out of the uncertainty of states as to the role of air power in future conflicts.

matters. And third, the law of attribution is complex.[58] In other words, even when the originator of a cyberoperation is known, it may be unclear whether his or her actions can be deemed to be those of a state as a matter of law such that the state is in violation of a treaty obligation.

Perhaps the prospect for evolution of cybertreaty norms was best set forth by the United Kingdom in its 2013 submission to the United Nations Secretary General:

> Experience in concluding these agreements on other subjects shows that they can be meaningful and effective only as the culmination of diplomatic attempts to develop shared understandings and approaches, not as their starting point. The United Kingdom believes that the efforts of the international community should be focused on developing common understandings on international law and norms rather than negotiating binding instruments that would only lead to the partial and premature imposition of an approach to a domain that is currently too immature to support it.[59]

Even if states were to pursue a diplomatic conference, and a cybertreaty was concluded, it would likely be perforated with so many individual reservations that its practical effect would be seriously degraded. While the conclusion of uniform law treaties—those requiring states to adopt measures in their domestic legislation—is usually subject to less intense negotiations than, for example, joint security treaties that impose cybernorms directly, in the cybercontext even the former have proven difficult to conclude. Despite determined international promotion, for example, the 2001 Convention on Cybercrime has been signed by only 53 states; 11 of them have yet to ratify the agreement.[60] Furthermore, 22 reservations and 21 declarations have been attached by the states that have thus far become party thereto. If this track record is illustrative, the prospects for crafting a meaningful legal regime specifically for cyberconflict are grim.

2.4 Customary International Law

The second form of international law recognized in Article 38 of the Statute of the ICJ is "general practice accepted as law," or customary international law.[61] It is a genre of norms unique to international law in the sense that it is unwritten. In the

---

[58] On this topic, see, e.g., Schmitt and Vihul, "Proxy Wars in Cyberspace," 54–73.
[59] "Developments in the Field of Information and Telecommunications in the Context of International Security," 19, U.N. Doc. A/68/156, July 16, 2013, http://undocs.org/A/68/156.
[60] For a list of signatories and ratifications, see *Council of Europe* website, accessed June 1, 2014, http://conventions.coe.int/Treaty/Commun/ChercheSig.asp?NT=185&CM=&DF=&CL=ENG.
[61] ICJ Statute, art 38(1)(b).

field of conflict, customary international law was historically predominant, since it was only in the twentieth century that treaty law on the subject came into its own.[62] Despite the adoption of many treaties addressing conflict in that century, customary law retains its significance. In great part, this is because most conflict law treaty regimes are not universal. As an example, neither the United States nor Israel, two states that have been involved in numerous conflicts since their adoption, are party to the 1977 Additional Protocols. Therefore, to the extent that nonparty states comply with the norms expressed therein, they do so on the basis that their provisions are reflective of customary international law. Additionally, rules expressed in a treaty sometimes crystallize into customary law, even though they did not reflect a customary norm at the time of adoption. The classic case is that of the Regulations annexed to the 1907 Hague Convention IV.[63] And to the extent that a particular point encompassed in the material scope of an agreement is not directly addressed, any existing customary law will govern the matter.[64]

Although unwritten, customary law is as binding on states as treaty law. Such law "crystallizes" upon the confluence of two factors: (1) the objective element of state practice (*usus*), and (2) the subjective element of *opinio juris*.[65] As unwritten law that is developed through an informal process, it is very difficult to both definitively establish when crystallization has occurred and define its precise contours. For reasons that will be explained, this is particularly so with regard to nascent activities such as cyberoperations.

The first prong of the test, state practice, includes both physical and verbal acts of states.[66] To qualify as state practice, the conduct in question must generally occur over an extended period of time. The classic illustration is the 1900 US Supreme Court case, *The Paquete Habana*. There the court looked to the practice of numerous countries over a period measured in centuries to conclude that fishing vessels were exempt from capture by belligerents during an armed conflict.[67] The temporal condition has deteriorated over time. As an example, in the 1969 *North Sea Continental Shelf* judgment dealing with

---

[62] For instance, significant codification in the field occurred during the Hague Conferences of 1899 and 1907. For a list of treaties, see *International Committee of the Red Cross* website, accessed June 1, 2014, http://www.icrc.org/applic/ihl/ihl.nsf/vwTreatiesByDate.xsp.

[63] Regulations Respecting the Laws and Customs of War on Land, annexed to Convention No. IV Respecting the Laws and Customs of War on Land, October 18, 1907, 36 Stat. 2227. This was the finding of the Nuremburg Tribunal. International Military Tribunal at Nuremburg, Case of the Major War Criminals, Judgment, October 1, 1946, I Official Documents 253–54.

[64] See discussion in Dinstein, *Interaction between Customary International Law and Treaties*, 383.

[65] North Sea Continental Shelf (Ger. v. Den.; Ger. v. Neth.), 1969 I.C.J. 3, paras. 71, 77 (Feb. 20) [hereinafter North Sea Continental Shelf]; Continental Shelf case (Libya v. Malta), 1985 I.C.J. 13, para. 27 (June 3); Nicaragua, para. 183.

[66] See, e.g., *Final Report of the Committee on the Formation of Customary (General) International Law*, 13 ff. [hereinafter ILA Report]; Henckaerts and Doswald-Beck, I *Customary International Humanitarian Law*, xxxviii–xxxix.

[67] *The Paquete Habana*, 175 U.S. 686–700 (1900).

the customary law of the sea, the ICJ held that "passage of only a short time is not necessarily a bar . . . [if state practice], including that of states whose interests are specially affected [is] both extensive and virtually uniform."[68] Perhaps the best illustration of the weakening of the requirement of long-term practice is the development of customary space law,[69] an example that suggests that the relative novelty of cyberoperations does not preclude the rapid emergence of cyber-specific customary international law.

The state practice necessary to establish customary law must, even if of limited duration, be consistent. To the extent there are significant deviations from a practice by states, which may include both engaging in an activity and refraining from one, a customary norm cannot crystallize. Although minor infrequent inconsistencies do not constitute a bar to such emergence,[70] repeated inconsistencies generally have to be characterized by other states as violations of the norm in question before a customary norm can be said to exist. For instance, it is clear that the prohibition on the use of force set forth in Article 2(4) of the UN Charter constitutes a customary norm.[71] Obviously, states have historically engaged in uses of force and continue to do so today. Yet, when they do, their conduct is, absent a justification such as self-defense, typically styled by other states as wrongful.[72]

There is no set formula as to the number of states that must engage in a practice before a norm crystallizes, although the greater the density of practice, the more convincing the argument that crystallization has occurred.[73] Of particular importance is the diversity (e.g., geographical, legal system, etc.) of the states involved,[74] and the fact that "specially affected states" have engaged in the practice or expressed their view of such practice by other states.[75] A specially affected state is one upon which the norm will operate with particular resonance. As an example, the International Committee of the Red Cross (ICRC) has opined that "specially affected states" with respect to the legality of weapons include "those identified as having been in the process of developing such weapons."[76] In cyberspace, the United States would qualify as a specially affected state in light of its centrality to cyberactivities and its development of military capacity in the field.

The term *opinio juris* refers to the requirement that states engage in a practice, or refrain from it, out of a sense of legal obligation.[77] In other words, the state must

---

[68] North Sea Continental Shelf, para. 74.

[69] For an early, and classic, treatment of the subject, see McDougal, "Emerging Customary Law of Space," 618–42.

[70] Fisheries Case (U.K. v. Norway), 1951 I.C.J. 116, 131 (December 18).

[71] Nicaragua, paras. 188–90.

[72] See discussion in Nicaragua, para. 186.

[73] Henckaerts and Doswald-Beck, I *Customary International Humanitarian Law*, xlii–xliv.

[74] Ibid., xliv.

[75] North Sea Continental Shelf, para. 74; ILA Report, 25–26.

[76] Henckaerts and Doswald-Beck, eds., I *Customary International Humanitarian Law*, xliv.

[77] S.S. Lotus (Fr. v. Turk.), 1927 P.C.I.J. (ser. A) No. 10, 28 (September 7); Nicaragua, para. 185 (citing North Sea Continental Shelf, para. 77).

believe their actions are required or prohibited by law. It is often the case that state actions are motivated by other factors, such as policy, security, operational, economic, and even moral considerations. For instance, Estonia actively seeks to maintain a clean cyberenvironment. It does so not because it believes that the international legal requirement of due diligence requires such measures, but rather for cybersecurity reasons (e.g., to prevent the use of botnets located in Estonia against the country). Such practices have no bearing on the creation of a customary law norm.

Obviously, it is often very difficult to ascertain the rationale underlying a particular practice; care must be taken in drawing inferences as to *opinio juris* based solely on the existence of state practice.[78] For instance, the ICRC cited many military manuals as evidence of *opinio juris* in its 2005 Customary International Humanitarian Law study.[79] In response, the United States objected that the provisions found in military manuals were often as much the product of operational and policy choice as legal obligation.[80] A similar criticism frequently attends the citation of UN General Assembly resolutions as support for the existence of a customary norm because states can vote in favor of such legally nonbinding instruments for purely political reasons.

Nevertheless, states do engage in conduct or issue statements that clearly indicate their characterization of certain practices as required by customary international law. As an example, although the United States is a party to neither the Law of the Sea Convention nor Additional Protocol I, it often confirms that it views certain of the provisions of those instruments as reflective of customary international law.[81]

Once a customary norm has emerged, it is applicable to all states including those that did not participate in the practice that led to its crystallization. Such norms are even binding on states that are created after the customary norm has developed.[82] However, there are a number of exceptions to this general principle. In particular, a state may "persistently object" to the norm's formation as it is emerging. If the norm nevertheless emerges, the persistent objector is arguably not bound by it.[83]

---

[78] North Sea Continental Shelf, paras. 76–77.

[79] Henckaerts and Doswald-Beck, *I Customary International Humanitarian Law*, xxxviii. See also Prosecutor v. Tadić; Case No. IT-94-1-I, Decision on Defence Motion for Interlocutory Appeal on Jurisdiction, para. 99 (Int'l Crim. Trib. for the former Yugoslavia October 2, 1995).

[80] Bellinger and Haynes, "U.S. Initial Reactions to ICRC Study on Customary International Law," November 3, 2006, accessed June 1, 2014, http://2001-2009.state.gov/s/l/rls/82630.htm.

[81] Department of the Navy and Department of Homeland Security, "The Commander's Handbook on the Law of Naval Operations," NWP 1-14M/MCWP 5-12/COMDTPUB P5800.7A, 2007, paras. 1–2; The United States Army Judge Advocate General's Legal Center and School, "Law of Armed Conflict Documentary Supplement," 232–33, accessed June 1, 2014, http://www.loc.gov/rr/frd/Military_Law/pdf/LOAC-Documentary-Supplement-2013.pdf.

[82] ILA Report, 24–25.

[83] ILA Report, 27–29. The doctrine of persistent objection is not universally accepted. Henckaerts and Doswald-Beck, *I Customary International Humanitarian Law*, xlv.

In this regard, the role of "specially affected states" is paramount.[84] It would be very unlikely that a customary norm could emerge over the objection of such a state. For example, given the military wherewithal of the United States, and its frequent involvement in armed conflicts, it would be difficult for a norm of international humanitarian law to crystallize in the face of a US objection thereto. Assertions of persistent objection are fortunately infrequent. Rather, disagreement regarding customary norms typically surrounds the scope of a rule, not its existence.

In certain limited circumstances, a customary norm may be regional or even local in character. For example, in the Asylum case, the ICJ found that a regional customary norm applies in Latin America,[85] whereas in the Rights of Passage it determined that another existed as between two states with respect to passage across India to Portuguese enclaves in that state.[86]

2.5 Customary International Law and Cyberconflict

There are many obstacles in the path of customary norm emergence vis-à-vis cyberspace. The requirement of practice over time hinders this process to an extent, but is not fatal because, as noted, contemporary customary international law appears to countenance relatively rapid crystallization. A much greater impediment is the visibility of cyberactivities. It is difficult to "see" what goes on in cyberspace. Instead, the effects of cyberoperations are often all that is publicly observed. Therefore, it can sometimes be difficult to point to particular state cyberpractices to support an argument that a norm has emerged. In fact, states, including victim states, may be reticent to reveal their execution, or even knowledge, of a cyberoperation because doing so may disclose capabilities that they deem essential to their security. Undisclosed acts do not amount to state practice contributing to the emergence of customary international law.[87]

Similarly, states will frequently hesitate to proffer opinions regarding the legality of state practice in cyberspace. For instance, a state may be unwilling to definitively articulate a threshold for "armed attack." This could be because it does not want its opponents to discern when it is likely to respond on the basis of the right of self-defense or because it prefers not to clarify the "use of force" threshold as doing so might limit its own options in the future. In other words, it may view strategic

---

[84] North Sea Continental Shelf, para. 74.
[85] Asylum Case (Colom. v. Peru), 1950 I.C.J. 266, p. 276–77 (November 20).
[86] Case Concerning Right of Passage over Indian Territory (Port. v. India), 1960 I.C.J. 6, p. 37 (April 12).
[87] Henckaerts and Doswald-Beck, *I Customary International Humanitarian Law*, xl; ILA Report, 15.

ambiguity as in its national interest. From an international security perspective, normative clarity is not always helpful.

Two recent examples are illustrative. The relative silence of states in reaction to the 2010 Stuxnet attack against Iranian nuclear enrichment centrifuges does not necessarily indicate that states believe that the operation was lawful, assuming for the sake of analysis that it was launched by other states—only states can violate the prohibition on the use of force set forth in Article 2(4) of the UN Charter. On the contrary, they may have concluded that the attack violated the prohibition on the use of force because it was not in response to an Iranian armed attack pursuant to the treaty and customary law of self-defense. Yet, those states may logically have decided that the operation was nevertheless a sensible means of avoiding a preemptive, and destabilizing, kinetic Israeli attack against the facilities. Similarly, the 2012 Shamoon virus attacks against the Saudi Aramco oil company's computers may also have been considered a violation of the use of force prohibition if conducted, as has been speculated, by Iran.[88] Despite this possibility, the relative downplaying of the legal aspects of the attack by states may be attributable to concerns regarding the economic consequences of styling them as a wrongful use of force by Iran.

It is also common for states to support or condemn a cyberactivity in their international rhetoric, but not be specific as to whether the condemnation is based on customary international law or on other considerations, such as moral principles or political concerns. The PRISM surveillance program serves as an example on point. While many states, including Germany and France, criticized the surveillance program, with the first stating that these practices were "completely unacceptable"[89] and the latter that they "cannot accept this kind of behavior from partners and allies,"[90] the comments do not necessarily indicate their position on the legality of the program.

Other requirements that will often be difficult to meet in state cyberpractice are consistency and density. For instance, Brazil argued at the UN General Assembly in 2013 that the interception of communications represents "a case of disrespect to the [country's] national sovereignty,"[91] presumably suggesting that it breaches the international law principle of sovereignty. It is unlikely that a sufficient number of

---

[88] Perlroth, "In Cyberattack on Saudi Firm, U.S. Sees Iran Firing Back," *New York Times*, October 23, 2012, http://www.nytimes.com/2012/10/24/business/global/cyberattack-on-saudi-oil-firm-disquiets-us.html?pagewanted=all&_r=0.

[89] "Merkel Calls Obama about 'US Spying on Her Phone,'" BBC, October 23, 2013, http://www.bbc.com/news/world-us-canada-24647268.

[90] "Hollande: Bugging Allegations Threaten EU-US trade pact," BBC, July 1, 2013, http://www.bbc.com/news/world-us-canada-23125451.

[91] Statement by Brazilian President H. E. Dilma Rousseff on September 24, 2013 at the opening of the general debate of the 68th session of the United Nations General Assembly. Translated reprint of the speech 2, accessed June 1, 2014, http://gadebate.un.org/sites/default/files/gastatements/68/BR_en.pdf.

other states, in particular specially affected states, will embrace the same position such that the criteria will be satisfied.

Indeed, states may be so conflicted regarding their own legal position on customary cyber norms that they take no position on the legality of a particular cyberpractice. To the extent that they wield cybercapabilities that are strategically or operationally useful, states have an incentive to retain the option of employing them. But those same states may be vulnerable to attacks by other states using similar capabilities. Therefore, it may be difficult for a state's political and legal organs to agree on how the state should characterize a particular practice, since they may view the state's national interests from different perspectives. And, of course, states will want to avoid being criticized for adopting a "do as I say, not as I do" approach. The United States, rightly or wrongly, has been the subject of such accusations with regard to its condemnation of Chinese cyberoperations against US businesses.[92]

Finally, state comments regarding their own or other states' activities tend to be crafted by nonlawyers. The legal dimension of the activities is accordingly often neglected. The paradigmatic example was the United States' public statements regarding possible operations against Iraq in late 2002 and early 2003, which focused on alleged Iraqi involvement in transnational terrorism and its development of weapons of mass destruction (WMD) capability.[93] By the time that the United States finally set forth its formal legal justification—a very nuanced interpretation of ceasefire law[94]—it had been rendered inaudible white noise in light of the ongoing geopolitical brouhaha that was underway. As this example illustrates, international security matters generally take on policy and strategic hues, rather than legal ones. The same is likely to be true as states engage in and react to cyberactivities of the future.

Considered in concert, these factors render the immediate crystallization of new customary norms to govern cyberspace somewhat improbable. Therefore, the normative impact of customary law on cyberconflict is most likely to take place in the guise of interpretation of existing customary norms. To the extent this is so, the same interpretive dilemmas belaboring treaty interpretation will surface. In fact, the obstacles will be greater with respect to customary international law because not only are rules themselves not expressly articulated but also there are no explicit rules regarding their interpretation, such as those found in the Vienna Convention on the Law of Treaties.

---

[92] See, e.g., "China Denounces US Cyber-Theft Charges," BBC, May 20, 2014, http://www.bbc.com/news/world-us-canada-27477601.

[93] Address of President George W. Bush, March 19, 2003, accessed June 1, 2014, http://georgewbush-whitehouse.archives.gov/news/releases/2003/03/20030319-17.html.

[94] "Letter dated 20 March 2003 from the Permanent Representative of the United States of America to the United Nations addressed to the President of the Security Council," U.N. Doc. S/2003/351, March 21, 2003.

2.6 Conclusion

Despite the attention that cyberactivities have drawn in the past decade, the conclusion of new treaties or the crystallization of new customary law norms to govern them is doubtful. Opposition from western states is particularly marked on at least the first point.[95] Instead, the application and interpretative evolution of existing international law is the most likely near-term prospect.

Controversy and inexactitude will surely characterize this process, which will be neither linear nor logical. The weakening of the early Russian and Chinese objections to the application of extant international law to cyberspace is a milestone in this regard. Yet, while both states have backed away from their opening bid on the issue, it remains unclear where they stand today. Other states, such as the United States and the Netherlands, are beginning to show a willingness to articulate their positions on how current international law applies in cyberspace. Nonetheless, public pronouncements to date have been vague, probably intentionally so.

As a consequence, the work of scholars such as the International Group of Experts that prepared the *Tallinn Manual*, and those who are engaged in the follow-on "Tallinn 2.0"[96] project, is likely to prove especially influential. This is not necessarily an optimal situation, for states, and only states, enjoy the formal authority to make international law. Unless they wish to surrender their interpretive prerogative to the academy, it is incumbent upon them to engage cyberissues more openly and aggressively.

Bibliography

Additional Protocol to the Convention on Cybercrime Concerning the Criminalization of Acts of a Racist and Xenophobic Nature Committed through Computer Systems. January 28, 2003, E.T.S. No. 189.

Agreement between the Governments of the Member States of the Shanghai Cooperation Organization on Cooperation in the Field of International Information Security. 61st Plenary Meeting, December 2, 2008.

Bellinger, John B. III, and William J. Haynes II. [US Department of State and US Department of Defense, respectively.] "U.S. Initial Reactions to ICRC Study on Customary International Law." November 3, 2006. http://2001-2009.state.gov/s/l/rls/82630.htm. Accessed August 25, 2015.

---

[95] See, e.g., President of the United States, "International Strategy for Cyberspace" 15, May 2011, accessed June 1, 2014, http://www.whitehouse.gov/sites/default/files/rss_viewer/international_strategy_for_cyberspace.pdf; "Cybersecurity Strategy of the European Union: An Open, Safe and Secure Cyberspace," Joint Communication to the European Parliament, the Council, the European Economic and Social Committee and the Committee of the Regions 15, Doc. JOIN (2013) 1 final, February 7, 2013.

[96] On the project, see *NATO Cooperative Cyber Defence Centre of Excellence* website, accessed June 1, 2014, http://ccdcoe.org/477.html.

BBC. "Merkel Calls Obama about 'US Spying on Her Phone.'" October 23, 2013. http://www.bbc.com/news/world-us-canada-24647268.
BBC. "Hollande: Bugging Allegations Threaten EU-US Trade Pact." July 1, 2013. http://www.bbc.com/news/world-us-canada-23125451.
Case Concerning Right of Passage over Indian Territory (Port. v. India). 1960 I.C.J. 6 (April 12).
Constitution and Convention of the International Telecommunication Union. December 22, 1992. 1825 U.N.T.S. 330.
Continental Shelf case (Libya v. Malta). 1985 I.C.J. 13 (June 3).
Convention (I) for the Amelioration of the Condition of the Wounded and Sick in the Armed Forces in the Field. August 12, 1949. 75 U.N.T.S. 31.
Convention (II) for the Amelioration of the Condition of the Wounded, Sick, and Shipwrecked Members of Armed Forces at Sea. August 12, 1949. 75 U.N.T.S. 85.
Convention (III) Relative to the Treatment of Prisoners of War. August 12, 1949. 75 U.N.T.S. 135.
Convention (IV) Relative to the Protection of Civilian Persons in Time of War. August 12, 1949. 75 U.N.T.S. 287.
Convention on Cybercrime. November 23, 2001. 2296 U.N.T.S. 167.
"Defence White Paper 2013" (Australian Department of Defence). http://www.defence.gov.au/whitepaper2013/docs/WP_2013_web.pdf.
Department of the Navy and Department of Homeland Security. "The Commander's Handbook on the Law of Naval Operations." NWP 1-14M/MCWP 5-12/COMDTPUB P5800.7A, 2007.
"Developments in the Field of Information and Telecommunications in the Context of International Security." U.N. Doc. A/68/156, July 16, 2013. http://undocs.org/A/68/156.
Dinstein, Yoram. *The Interaction between Customary International Law and Treaties*. Recueil des Cours 322. N.p.: Martinus Nijhoff, 2007.
Final Report of the Committee on the Formation of Customary (General) International Law, Statement of Principles Applicable to the Formation of General Customary International Law. London: International Law Association, 2000.
Fisheries Case (U.K. v. Norway). 1951 I.C.J. 116 (December 18).
Giles, Keir, and Andrew Monaghan. "Legality in Cyberspace: An Adversary View." Strategic Studies Institute and U.S. Army War College Press. http://www.strategicstudiesinstitute.army.mil/pubs/display.cfm?pubID=1193.
"Government Response to the AIV/CAVV Report on Cyber Warfare" (Netherlands). http://www.aiv-advies.nl/ContentSuite/template/aiv/adv/collection_single.asp?id=1942&adv_id=3016&page=regeringsreacties&language=UK
"Group of Governmental Experts on Developments in the Field of Information and Telecommunications in the Context of International Security." U.N. Doc. A/68/98, June 24, 2013. http://undocs.org/A/68/98.
Harrison Dinniss, Heather. *Cyber Warfare and the Laws of War*. New York: Cambridge University Press, 2012.
Henckaerts, Jean-Marie, and Louise Doswald-Beck, eds. *I Customary International Humanitarian Law*. New York: Cambridge University Press, 2005.
ICRC. "International Humanitarian Law and the Challenges of Contemporary Armed Conflicts." Doc. 31IC/11/5.1.2, 2011.
International Military Tribunal at Nuremburg. Case of the Major War Criminals, Judgment, October 1, 1946. I Official Documents.
"International Strategy on Cybersecurity Cooperation." Information Security Policy Council, Japan. http://www.nisc.go.jp/active/kihon/pdf/InternationalStrategyonCybersecurityCooperation_e.pdf.
International Telecommunication Regulations. December 9, 1988.
Koh, Harold H. "International Law in Cyberspace." Address at the USCYBERCOM Inter-Agency Legal Conference, Ft. Meade, Maryland on September 18, 2012. Reprinted in Harold

Hongju Koh, "International Law in Cyberspace," *Harvard International Law Journal Online* 54 (2012): 1–12. http://www.harvardilj.org/2012/12/online_54_koh/.

Koh, Harold Hongju, Legal Adviser, U.S. Department of State. Address at the Annual Meeting of the American Society of International Law, "The Obama Administration and International Law." March 25, 2010. http://www.state.gov/s/l/releases/remarks/139119.htm.

Lin, Patrick, Fritz Allhoff, and Neil Rowe. "Is It Possible to Wage a Just Cyberwar?" *The Atlantic*, June 5, 2012. http://www.theatlantic.com/technology/archive/2012/06/is-it-possible-to-wage-a-just-cyberwar/258106/.

McDougal, Myres S. "The Emerging Customary Law of Space." *Northwestern University Law Review* 58 (1963–1964): 618–42.

Military and Paramilitary Activities in and against Nicaragua (Nicar. v. U.S). 1986 I.C.J. 14 (June 27).

Perlroth, Nicole. "In Cyberattack on Saudi Firm, U.S. Sees Iran Firing Back." *New York Times*, October 23, 2012. http://www.nytimes.com/2012/10/24/business/global/cyberattack-on-saudi-oil-firm-disquiets-us.html?pagewanted=all&_r=0.

Press Statement, U.S. Department of State. Richard Boucher, Spokesman, "International Criminal Court: Letter to UN Secretary General Kofi Annan." May 6, 2002. http://2001-2009.state.gov/r/pa/prs/ps/2002/9968.htm.

Prosecutor v. Tadić, Case No. IT-94-1-T, Trial Chamber Opinion and Judgment. Int'l Crim. Trib. for the former Yugoslavia, May 7, 1997.

Prosecutor v. Tadić, Case No. IT-94-1-I, Decision on Defence Motion for Interlocutory Appeal on Jurisdiction. Int'l Crim. Trib. for the former Yugoslavia, October 2, 1995.

Protocol Additional to the Geneva Conventions of 12 August 1949, and Relating to the Protection of Victims of International Armed Conflicts, June 8, 1977, 1125 U.N.T.S. 3.

Protocol Additional to the Geneva Conventions of August 12, 1949, and Relating to the Protection of Victims of Non-International Armed Conflicts, June 8, 1977, 1125 U.N.T.S. 609.

Regulations Respecting the Laws and Customs of War on Land, annexed to Convention No. IV Respecting the Laws and Customs of War on Land, October 18, 1907, 36 Stat. 2227.

Rome Statute of the International Criminal Court, July 17, 1998, 2187 U.N.T.S. 90.

Roscini, Marco. *Cyber Operations and the Use of Force in International Law*. New York: Oxford University Press, 2014.

S.S. Lotus (Fr. v. Turk.). 1927 P.C.I.J. (ser. A) No. 10 (September 7).

Sanger, David E., and Elisabeth Bumiller. "Pentagon to Consider Cyberattacks Acts of War." *New York Times*, May 31, 2011. http://www.nytimes.com/2011/06/01/us/politics/01cyber.html?_r=0

Schmitt, Michael N. "'Attack' as a Term of Art in International Law: The Cyber Operations Context." In *Proceedings of the 4th International Conference on Cyber Conflict*, ed. Christian Czosseck et. al., 283-93. Tallinn: NATO Cooperative Cyber Defence Centre of Excellence, 2012.

Schmitt, Michael N. "Classification of Cyber Conflict." *International Law Studies* 89 (2013): 233–51.

Schmitt, Michael N. "Rewired Warfare: Rethinking the Law of Cyber Attack." In *International Review of the Red Cross* 96, no. 893 (2014): 189–206. http://journals.cambridge.org/action/displayFulltext?type=1&fid=9904644&jid=IRC&volumeId=96&issueId=893&aid=9904639.

Schmitt, Michael N., gen. ed. *Tallinn Manual on the International Law Applicable to Cyber Warfare*. New York: Cambridge University Press, 2013.

Schmitt, Michael N., and Liis Vihul. "Proxy Wars in Cyberspace: The Evolving International Law of Attribution." *Fletcher Security Review* 1, no. 2 (2014): 54–73.

Statute of the International Court of Justice. June 26, 1945, 59 Stat. 1055, 33 U.N.T.S. 993.

The Paquete Habana, 175 U.S. 677 (1900).

The United States Army Judge Advocate General's Legal Center and School. "Law of Armed Conflict Documentary Supplement." http://www.loc.gov/rr/frd/Military_Law/pdf/LOAC-Documentary-Supplement-2013.pdf.

U.N. Doc. S/RES/1368, September 12, 2001. http://www.un.org/en/ga/search/view_doc.asp?symbol=S/RES/1368(2001).

U.N. Doc. S/RES/1373, September 28, 2001. http://www.un.org/en/ga/search/view_doc.asp?symbol=S/RES/1373(2001).

U.S. Constitution.

U.S. Department of Justice. U.S. Department of Justice White Paper. "Lawfulness of a Lethal Operation Directed against a U.S. Citizen Who Is a Senior Operational Leader of Al-Qa'ida or an Associated Force." Draft, November 8, 2011. http://users.polisci.wisc.edu/kmayer/408/020413_DOJ_White_Paper.pdf.

UK Ministry of Defence. *The Manual of the Law of Armed Conflict*. New York: Oxford University Press, 2005.

Vienna Convention on the Law of Treaties between States and International Organizations or between International Organizations, opened for signature March 21, 1986, UN Doc A/CONF.129/15 (not yet in force).

Vienna Convention on the Law of Treaties. May 23, 1969, 1155 U.N.T.S. 331.

3

Distinctive Ethical Issues of Cyberwarfare

RANDALL R. DIPERT

3.1. Toward a Full Ethics of Cyberwarfare

A number of articles have now appeared dealing with the strictly ethical issues of nation-on-nation intentional acts of cyberharm, that is, cyberwarfare.[1] The ethical discussion started among nonphilosophers (and nonethicists),[2] and more recently these issues have been addressed by professional philosophers.[3] Discussions of the status of acts of cyberwarfare in international law predated the ethical literature by more than a decade; but with the publication of the *Tallinn Manual*, these discussions have reached a more systematic and settled state than have debates on the ethics of cyberwarfare.[4] I will say something later about the difficult issue of the difference and relationship between ethical and legal considerations.

Most discussions of the ethics of cyberwarfare have begun with traditional just war theory and argued or assumed that it is valid or the best available framework for approaching the ethical issues of cyberwarfare. They have also argued or assumed that cyberwarfare fits squarely within the concepts of traditional warfare to which just war

---

[1] Earlier versions of this paper were given at the conferences "Ahead of the Curve: Anticipating Ethical, Legal, and Societal Issues Posed by Emerging Weapons Technologies," John J. Reilly Center for Science, Technology, and Values, University of Notre Dame, April 22–23, 2014; Ninth International Conference on Cyber Warfare and Security ICCWS-2014, Purdue University, West Lafayette, Indiana, March 24–25, 2014; and the First Workshop on Ethics of Cyber Conflict, Rome, Italy, November 2013, organized by the NATO Center of Excellence for Cooperative Self-Defense (CCD COE).

[2] Arquilla, "Ethics and Information Warfare," 379–401; Rowe, "Ethics of Cyberwar Attacks"; Rowe, "Ethics of Cyberweapons," 20–31; Owens, Dam, and Lin, *Technology, Policy, Law, and Ethics Regarding US Acquisition of Cyberattack Capabilities*.

[3] Dipert, "Ethics of Cyberwar," *Journal of Military Ethics* 9, no. 4 (2010): 384–410; Strawser, "Special Issue on Cyberwar and Ethics."

[4] Schmitt, "Bellum Americanum," 1051–90; Schmitt, *Tallinn Manual on the International Law Applicable to Cyber Warfare*.

theory applies. By "just war theory" I mean a criteria-based approach for when it is morally permissible to resort to acts of war, *jus ad bellum*—traditionally understood as a use of lethal force that is also typically massively destructive, is likely to have undesired consequences (such as famine, environmental damage, and economic decline affecting noncombatants), and is accompanied by invasions or intrusions from humans or weapons. Just war theory has a second criteria-based approach for morally permissible actions within a war, *jus in bello*. Historically, just war theory has foundations in natural law or natural rights theory, such as in its two loci classici, Aquinas and Grotius.[5] However, the philosophical assumptions and principles for these foundations or logical derivations are not widely shared among contemporary philosophers. This nevertheless leaves other possible sources of support, such as through widely shared intuitions, or in modern ethical-political theories for which there is more widespread sympathy, such as consequentialism or various contract theories. These meta-ethical or foundational issues in the theory of the morality of war have not been dealt with in any extensive way. Instead, there is such widespread agreement within most theories about the contours of what is moral in war that discussion has settled far more on "intermediate" principles, such as that some wars in self-defense are justified, that deaths are generally to be avoided, and that especially noncombatants (including prisoners of war and wounded soldiers) should be spared as much as possible.[6]

This lack of agreement about foundations is, I would argue, especially problematic for the ethical issues involved in cyberwarfare. It appears, for example, that many forms of current and future cyberwarfare will often *not* involve deaths and widespread permanent physical destruction. Consequently, even if there is widespread agreement across diverse meta-ethical frameworks about the outlines of a morality in war as traditionally understood, we are in need of much more fine-grained moral theories for international relations that include lesser forms of intentional harm, such as economic damage and espionage. These lesser forms of intentional harm deserve much more critical attention in ethics than they have received. The damage that may be done by sanctions, or through trade, diplomatic, and communication embargoes, is enormous; yet they have been treated as if they are ethically and legally unproblematic compared with war.[7] The moral status of espionage is

---

[5] The historical foundations are discussed in Christopher, *Ethics of War and Peace*, 68–74; and appear at many points in Reichberg et al., *Ethics of War*.

[6] They are intermediate in the sense that they are not basic, foundational principles (such as "minimize harm") and they are not highly specific, applying to particular circumstances, or even to a single instance. They may also have a different moral force, being more rules-of-thumb to guide ethical behavior than exceptionless statements. The hypothesis that even the most basic ethical principles have this character is found in W.D. Ross's arguments for prima facie ethical principles and in what may be called "Wittgensteinian ethics" along the lines of A. Stroll's "Ethics without Principles."

[7] This is surprising given the widespread sympathy for various forms of consequentialism. See Brunstetter and Braun, "From Jus ad Bellum to Jus ad Vim," 87–106; and Ford, "Jus Ad Vim and the Just Use of Force-Short-of-War."

especially problematic, I will argue, but there is widespread agreement that this is one of the most widespread forms of cyberwarfare.

A further and much more controversial claim is that just war theory and most modern alternatives actually yield mistaken judgments. For example, one tenet of just war theory is to return to the original status quo as soon as possible. This does not permit "punishment," and certainly not disproportional punishment of a defeated, unjust enemy. But such immoderate measures may be necessary to prevent an enemy from attacking again, or necessary to dissuade other parties from attacking in the future. Notoriously the threat of disproportionate and indiscriminate counterattack using nuclear weapons is widely regarded as having kept the relative peace of the period in Europe from 1950 to the present. Respected foreign policy and international relations experts believe this. But according to just war theory as normally interpreted, such actions, even the threat of them, are morally impermissible. Perhaps we should rather take this as good evidence that just war theory is simply mistaken. Its weakness involves a number of policies, such as possibly wrong moral judgments about deterrence and preemption (prevention). For reasons detailed below (especially the cost of defensive measures) cyberwarfare is an area where deterrence is likely to be regard as the best of bad options. Just war theory seems to prohibit some policies that game theory would regard as the most likely to result in the least total harm over the long run.

A full "ethics of cyberwarfare" would involve much more than simply applying widely accepted intermediate moral principles for war and warfare. It would involve a robust ethics for all of international relations—of what any state, or political organization, morally may do, or not do, to another state, or that affects another state, even if short of killing and permanent destruction.[8] In cyberwarfare, consideration of incidental harm to third-party or neutral countries acquires major ethical significance. This phenomenon occurs because cyberwarfare typically involves such forms of attack that use botnets or pivots: the use of third-party information systems and servers that disguise the origins of an attack. These may be in highly networked nations where numerous users do not update their software and malware protection, as well as in nations that are, from the perspective of cybersecurity, lawless—not having or enforcing laws against cyber misdeeds or lacking extradition treaties for cyber misdeeds or that do not criminalize or regulate Internet activity outside of their borders.[9]

---

[8] In the earliest criticism of my view that cyberwarfare is morally distinctive and that just war theory is insufficient to deal with many forms it, James Cook, in "'Cyberation' and Just War Doctrine: A Response to Randall Dipert," 411–23, argued that just war theory cannot, as a matter of logic, be criticized for failing to deal with phenomena that are not war: that is not its aim. However, the point remains that surely there is *some* ethical theory that *should* deal with behavior of one nation to another even if these behaviors do not literally constitute warfare.

[9] Pano Yannakogeorgos in a number of presentations and papers has argued that the special category of these relatively lawless states is one of the main obstacles to regulating international cyberbehavior. Panayotis Yannakogeorgos, "Internet Governance and National Security."

The full ethics of cyberwarfare would address not just the behavior of nations to each other, but also what a nation may do to its own citizens or to the citizens of other nations as part of cyberwarfare operations. Traditional theorizing about justice consists mainly in addressing the relationship of a state to its citizens. Further extensions would include what a citizen or corporation may do to its own or a foreign nation.[10] This is something of a "hot" topic since it is clear that some major US companies have taken, or considered taking, cyberactions against foreign governments. Foremost among the constrained behaviors affecting its own citizens are those that might invade privacy. For example, may a state examine a citizen's communications or information system without explicit permission; may it even involuntarily "patch" a citizen's computer if it is found to be the source of malware or DDoS attacks because of failure of firewalls or software updates? Even more problematic (and to date little discussed) are what actions a state may take in manipulating the communications or the information system of a citizen. May it interrupt or alter communications, especially when these are taking place without the user's knowledge (such as its communications to a botmaster), as a legitimate part of national defense?

3.2 The Ethical Distinctness of Cyberwarfare

One key issue in ethics is whether cyberwarfare raises distinctive and new issues that cannot be addressed by traditional just war theory and other theories of the morality of war and warfare. This has been contested. I have argued elsewhere that there are several distinctive ethical issues in cyberwarfare. One difference is that the *quantity* of harm being inflicted by a cyberattack will often not rise to the level of traditional "kinetic" weapons.[11] It will often be more like harm that has been called "measures short-of-war," or the application of "soft" force, such as sanctions or boycotts, and so on. Secondly, the *quality* of the harm is often not going to be like harm caused by traditional weapons. Killing or injuring human beings may be absent, as well as permanent physical destruction of other entities. These harms will instead often be to the *functioning* of information systems in financial, energy, communication, and other sectors of an economy. A third way in which cyberwarfare

---

[10] Namely, states might issues Letters of Marque, or tacitly permit individuals to take cybermeasures against state actors. Rabkin and Rabkin, "To Confront Cyber Threats, We Must Rethink the Law of Armed Conflict," essay by the Koret-Taube Task Force on National Security and Law, Hoover Institute, accessible at: http://media.hoover.org/sites/default/files/documents/EmergingThreats_Rabkin.pdf. I first heard of this possibility in conversation with George Lucas. Elsewhere I have called this a matter of "internal" justice, while the obligations and moral limits of a state to other states, and to the citizens of other states, is a matter of "external" justice.

[11] This simple thesis has been recently drawn into a book-length treatment in Rid, *Cyber War Will Not Take Place*.

is unique is epistemic: we will often not know immediately who attacked us, a situation that is rare in the history of warfare. This is the well-known attribution problem—although I believe it is on the way to being technically solved and in any case need not lead to complete and indefinite inaction. I have argued elsewhere (Dipert 2010) that this makes defending oneself against cyberattacks much like the problem of preemptive or preventive war (Dipert 2006). James Cook and others have argued that these differences are matters of degree, and do not require truly new ethical paradigms. However, there are at least three threshold differences between cyber- and traditional warfare that surely are ethically important: lethality, permanent physical destruction, and collateral damage. It is the irreversible nature of these effects that justifies the "last resort" condition of just war theory as well as the narrow and stringent requirement for "just cause."[12] There are reasons to hope that cyberweapons can have the precision that had been claimed for earlier weapon systems (e.g., Stuxnet) and that the use of some cyberweapons does not require the strict and narrow fulfillment of last resort and the usual conditions for just cause.

If the intentional or foreseeable effects of cyberwarfare *do* rise to the level and kind of harm typically seen with traditional forms of warfare, then there is every reason to think that traditional international law and ethics (e.g., just war theory) do straightforwardly apply. A careful calculation of these various parameters (aggregate harm, intensity, duration, etc.) has been a strong component of the thinking of Michael Schmitt[13] and of the legal thinking in US Cyber Command: this gives content to the "armed attack" as it occurs in the self-defense clause of the UN Charter (Article 51) and would also be a guide when a traditional kinetic attack might be an ethically permissible response to a cyberattack.

Traditional theories of morality in war arise from ethical values that are widely shared across cultures: namely, it is generally wrong to kill human beings and permanently to destroy the physical entities that are necessary for human well-being. It is still more wrong to kill large numbers of people or cause widespread, permanent destruction. Some careful definitions of war (Orend 2005) have stipulated that deaths and damage in war (as opposed to warfare) strictly understood must be "widespread."

However, many instances of cyberwarfare arise in far more slippery moral terrain: intrusion into information systems, exfiltration of data (cyberespionage), and "theft" of intellectual property, as well as the placing of software entities in an enemy's information systems that could eventually be used to cause harm but needn't have that purpose or might remain inert until activated. Although many users are not informed about the extent to which their information systems can be "read" by

---

[12] Neil Rowe had earlier argued for the development of cyberweapons that could be neutralized or their damage reversed: "Towards reversible cyberattacks." In *Proceedings of the 9th European Conference on Information Warfare and Security*, 261–67.

[13] Schmitt, "Cyber Operations and the Jud Ad Bellum Revisited," 569.

others using the Internet (open ports, OS version, etc.), and would not want that, they *could* know, and they are, after all, voluntarily connecting themselves to the Internet. Consequently there is an argument that they have consented to allow their information to travel along pathways where others can access it. They are connecting themselves to the informatics-analog of a pipe delivering untested and unguaranteed water, which should be more clearly labeled, "possibly not drinkable."[14]

The defenders of US government and military infrastructure have overdramatized (especially in congressional testimony) the supposedly precarious position of technologically advanced states as being subject to thousands or hundreds of thousands of "attacks" per week or even per hour. Very few of these so-called attacks do any damage at all, and so the word "attack" is misused. They fall into different categories, with different ethical implications, many being merely probes or guesses at passwords. A probe of a system (detecting open ports, for example) is the weakest such "attack." Slightly more aggressive is an attempt to "hack into" a website or system; this is discussed extensively below by the use of a moral analogy. One bit of damage even the least aggressive "attacks" do is to raise the cost to a defender in order to protect against them. The cost of enhanced cybersecurity to governments, corporations, and individuals in the future is likely to be staggering. Now estimated in the tens of billions in the United States (the revenue of Kaspersky, Macafee, and Symantec, plus many smaller businesses and dedicated cybersecurity professionals in government and industry), it is likely to be hundreds of billions in the near future.[15]

Unlike traditional warfare, many forms of cyberwarfare do not involve the intrusion of physical objects or human agents into a state's territory. This fact is striking and marks yet another difference with traditional warfare. Intended harm is accomplished by the conveyance of information entities from one information system to another. As I have argued elsewhere, there are important forms of cyberwarfare, broadly defined as nation-on-nation intended harm to information systems, that may be accomplished by means other than the Internet.[16] Far too little attention has been

---

[14] Of course, even this would have to assume (or somehow ensure) that most people are literate enough to read such warning words, and have a decent comprehension of what they mean.

[15] Reliable estimates of the total costs of cybersecurity are difficult to determine and to defend against cyberwarfare intrusions, but also cybercrime and vandalism. The costs would be borne by national governments (defense and nondefense), state and local governments, businesses and private individuals. Yadron, "Companies Wrestle with the Cost of Cybersecurity," *Wall Street Journal*, February 25, 2014: "Global cybersecurity spending by critical infrastructure industries was expected to hit $46 billion in 2013, up 10% from a year earlier"; Corrin, "Defense Budget Routes at Least $5B to Cyber," *Federal Times*, March 5, 2014; Messmer, "Gartner Security Report: McAfee Up, Trend Micro Down" *NetworkWorld*, May 30, 2013: "Symantec, with modest growth, still at top with 19.6% of overall share in $19.1 billion worldwide market."

[16] Dipert, "Other-Than-Internet (OTI) Warfare," 34–53. This paper was based on public, published remarks of M. Chertoff, M. McConnell, and M. Hayden, as well as on speculation about what

given to this other-than-Internet information warfare, whose extent has only recently become public (Sanger and Shanker 2014).

The properties of most forms of cyberwarfare that do not involve human or projectile violation of territorial integrity are significant because the just war principles and virtually all of international law have been interpreted in modern times through the lens of what may be called "Westphalian intermediate principles." By "Westphalian principles," I mean the linked notions of sovereignty and territory and permitted and nonpermitted activities in that territory, described in the Westphalian (Münster) Treaty of 1648 that ended the Thirty Years' War. So the modern interpretation of the just war criterion of "just cause" paradigmatically involves invasion of armies, that is of organized, armed human beings traveling into another state's sovereign territory, or physical destruction caused by physical objects such as arrows or cannon shells entering into that territory. As is the case with artillery shells and missiles, drones are intentional destruction in another state's territory without the need for human agents that might be captured and held responsible.

International law has frequently used the notion of "armed" aggression, attack, invasion, and so on. What constitutes being "arms" or being "armed" has been bent by those eager to apply existing international law to cyberwarfare. Past uses of "arms" have almost certainly meant exactly and only artifacts (and hence material objects) specifically crafted to kill human beings or destroy or render useless material entities[17] used by human beings.[18] However, as we have already seen, cyberwarfare may not aim for deaths, incapacitation of human beings, or permanent destruction of material entities. Furthermore, the information systems' hardware is typically general purpose. Malware could better be said to be weaponized, and it is often software (algorithms, or the specific implementation of an algorithm in a high-level computer language).[19] The hampered functioning, namely of information processing, or corrupted data are still different kinds of entities; neither are material objects. Software is of the nature of an idea or thought, but we do not generally consider ideas to be, literally, weapons, even if they too can be used to do harm.

---

was technologically feasible based on my knowledge of electronics and conversations with electronics and computer experts who also lacked access to classified sources. We now know something more about the level of this activity because of charges and countercharges between the United States and China.

[17] One destroys weapons, munitions, buildings, etc. and renders useless, say, the water behind a dam. In a strict sense I develop in *Artifacts, Art Works and Agents* (1993), such arms or weapons can be tools or artifacts (but would not be nontool instruments such as David's use of a stone against Goliath).

[18] In its discussion of "armed attack" (Rule 13, 54f) the *Talinn Manual* does not consider the precise meaning of "arms."

[19] Information-theoretic entities and what it means to "process" them has proven notoriously difficult to analyze. In the formal ontology (a Semantic Web technology) I have contributed to (Dipert 2013b), the Basic Formal Ontology (BFO) developed at the University at Buffalo, the relevant entities are handled in the Information Artifact Ontology (IAO).

3.3 The Legal and Ethical Issues Contrasted

The legal and moral questions of war differ in interesting and complicated ways. The details of what counts as a law in international law is similar to that of most domestic legal systems. There are kinds of *statutes*, namely treaties, including especially the UN Charter, as well as precedents and often unwritten customary prohibitions. The fragility of judgments about legal aspects of cyberwarfare is immediately demonstrated by what appears to be the inherent vagueness of the meaning of "force," "threat of force," "armed," and "attack" in the UN Charter, and whether and how they apply to cyberattacks.[20] Likewise the term "[civilian and joint-use] objects" in the widely accepted Addition Protocol I to the Geneva Conventions of 1977 gives examples that are all physical objects.

The ethical account of cyberwarfare would grant no privileged status to the Security Council, except as a matter of procedural, but not substantive, justice. The Security Council having passed a resolution is neither a necessary nor a sufficient condition for its being ethical. The international courts and the UN do have a kind of moral force that weakly derives from the status of treaties as promises of signatory states. But the treatment of states as having binding promises over decades and through dramatic changes in their forms of government derives from a very complicated theory of agency, of the identity and endurance of states, and is again Westphalian.

As mentioned earlier, ethical judgments are most often understood by ethical theorists as being based on one single (or a small number of) basic, foundational principle(s) of widespread application. Examples of such proposed basic principles are maximizing the aggregate well-being of present and future humans (utilitarianism), maximizing one's own rationally considered well-being (enlightened egoism), or permitting universalizability: what if everyone acted like that (Kantianism)? In certain areas of application—such as the "Hippocratic Oath" background of medical ethics,[21] as well as in the ethics of warfare—there had already arisen widely agreed-upon ethical principles before modern ethical theories applied their axiomatic approach to ethics. This gives such principles in applied ethics, such as the just war conditions, a problematic status. They are not derived, in any obvious way, from truly basic ethical principles. In practice, this has led modern ethicists to mix and match various foundational principles with traditionally accepted intermediate criteria like just war theory, and with intuitions or with what seem to be conclusions of historical study, as elegantly practiced in Michael Walzer's *Just and Unjust Wars*.[22]

---

[20] The lack of any clear definition for "force," etc., are discussed in the *Talinn Manual* (Rule 11, 45f). Also noted is that, as a legislative history of intent of the UN Charter, economic and political coercion do not in themselves constitute impermissible uses of force.

[21] For one example of such principalism, see Beauchamp and Childress, *Principles of Biomedical Ethics*.

[22] Walzer, *Just and Unjust Wars*.

The use of intuitions is particularly problematic. While we might have intuitions—our ethical consciences—about ways individuals should behave if they were in a moral dilemma, there is far less reason to think that we have decision-making skills, and developed intuitions, as leaders of states responsible for many, many lives. Surely this is a difficult, practiced skill.

Ethical theories of war have oddly ignored certain game-theoretic results,[23] and so they have generally rejected "realist" approaches that form one of the major schools of international relations.[24] In particular, philosophers have rejected deterrent strategies, including Mutually Assured (nuclear) Destruction, while most geopolitical thinkers and leaders have endorsed them. Although deterrence is always a complicated phenomenon, with many preconditions for success,[25] it would seem that it would be exceptionally usable in cyberwarfare, at least between rational cyberpowers. Deterrent strategies in cyberwarfare do not have the serious failures in just war theory that they have in the nuclear case, where they fail the probability of success and proportionality conditions. Furthermore it appears to be the case that in cyberwarfare (as much as we can say about it from relatively few cases) we are heading toward a game-theoretic equilibrium, in which certain limited cyberattacks and extensive cyberespionage are tolerated.

The lack of an objection by Iran to Stuxnet in forums such as the Security Council and international courts, as well as the silence of the other major cyberpowers, probably indicate a tacit international acceptance of Stuxnet, perhaps as a limiting case of the most severe such cyberattack that would be tolerated. George Lucas, in a set of wise and perceptive essays on permissible cyberattacks,[26] formulates criteria that more or less conform to the Stuxnet case and its apparent ethical acceptance. Avoiding an unstable escalation nevertheless remains a difficulty for all deterrent strategies.

---

[23] Dipert, "Preventive War and the Epistemological Dimension of the Morality of War," 32–54.

[24] Waltz, *Theory of International Politics*. Waltz distinguishes classical realists in international relations from neoclassical realists, among other distinctions. Unlike what philosophers such as P. Christopher, M. Walzer, and B. Orend call (military) realists, the term in international relations does not necessarily mean that ethical principles are ignored. As I have argued elsewhere, the philosophical realist in military or international affairs (e.g., Sherman's "War is hell" remark) is a strawman and is a position no major figure in modern times has held. See Dipert, "Defense of the Tactics of William T. Sherman."

[25] Libicki, *Cyberdeterrence and Cyberwar*.

[26] George Lucas, "Permissible Preventive Cyberwar: Restricting Cyber Conflict to Justified Military Targets," presented at Oxford University's Ethics, Law, and Armed Conflict Center, 2012, accessible at: http://www.elac.ox.ac.uk/downloads/Permissible%20Preventive%20Cyberwar%20UNESCO%202011.pdf; and Lucas, "Can There Be an 'Ethical' Cyber War?" Presented to the US Naval Academy, 2012, accessible at: http://www.usna.edu/Ethics/_files/documents/Just%20War%20and%20Cyber%20War%20GR%20Lucas.pdf.

3.4 Applying Just War Theory and Other Moral Principles to Cyberwarfare

Just war theory and its variants can only be taken as intermediate guiding principles or rules of thumb, since they lack a clear, widely accepted derivation from foundational principles. Another source of legitimacy is that they have some status as conventions that have come to be widely accepted, and that might limit the damage of war, if everyone abides by them.

Of the four core principles of just war theory for going to war—initiating the use of force—two are especially problematic for some forms of cyberwarfare. The four are just cause, last resort, probability of success, and proportionality. The widely accepted "high" barrier for just cause—namely armed invasion by an enemy with an intention to use lethal force—does not seem to apply to many forms of cyberwarfare. Likewise, cyberwarfare is not necessarily a *last* resort. That would continue to be the use of lethal force or force that brings extensive permanent destruction. Some forms of cyberwarfare would fall in the next-to-the-last-resort category, such as threats and ultimatums, sanctions, unilateral breaking of diplomatic and economic ties, and so on. Modern ethical and legal theory has largely ignored these smaller acts of intentional nation-on-nation harm.[27]

A just cause for war has never included another state's distribution of misleading or faulty information, conveyed in human-to-human communications. This would simply be "disinformation." However, Internet-based injection of malware can be described in terms that involve the unwitting and undesired transference of information entities. This develops the useful insights of Floridi (2008) and Taddeo (2012) that place cyberwarfare in a wider landscape of information warfare.

No person familiar with the technology could think that information coming through the Internet is protected by diplomatic conventions or international principles. By connecting oneself to the Internet, one knowingly opens one's own information systems to all manner of inspection, information, disinformation, and simply noise. A convention might arise in which certain, ideally encrypted, messages are protected by a special status from examination and manipulation. However the history of espionage, and especially of the morality and legality of espionage, seems broadly to permit examining and even manipulating another state's messages without incurring a justified armed attack.[28] Note that this might not be so for messages within a state, which should enforce high standards of privacy.

---

[27] There is an emerging discussion on this issue of "force short of war"; see Brunstetter and Braun, "From *Jus ad Bellum* to *Jus ad Vim*"; and Ford, "*Jus Ad Vim* and the Just Use of Force-Short-of-War."

[28] Goldman, *Ethics of Spying*.

Especially instructive is a careful moral examination of "hacking into" a website. The maker or owner of the website might not desire nonauthorized users even to access the public webpage, although this is inconsistent with using the Internet to make it visible. If access and ability to alter information on a website is password-protected, we have a scenario that raises clearer ethical issues. Note that it is actually fairly rare to encounter government, corporate, or individual public webpages that have warnings against their improper use by unauthorized personnel. Partly this would betray a naïveté and even inconsistent thinking about the Internet. Without enforced statutes, or ways of pursuing or even correctly identifying noncitizen violators, there have not arisen even nominal attempts clearly to separate permissible from nonpermissible activities.[29]

Various kinds of access to information via the Internet can be described. It is indeed clear that many of these forms of access are *undesired* by the owner; the owner might also have good reason to believe that no hacker will break through these protective barriers. The owner might declare that unauthorized users may not attempt to hack into the system. But with what moral and legal force? That is, when is a hacker doing something unethical, and why is it unethical?

One approach to a difficult ethical question is to find relevantly similar actions about which we do have intuitions or theoretical resources. What is morally like "hacking into" a website or system? A search for useful moral analogs is difficult. Consider this extended analogy: A business hangs an "Open for Business" sign or otherwise gives indications of its entry conditions—with windows displaying goods, a description of the goods to be obtained there, and perhaps an "Enter Here" sign. If the door is locked, then by convention one may not try to obtain entry, despite the "Open for Business" sign. If the door is unlocked, then one may reasonably enter the store and look around. If the goods are openly displayed, and without a sign to the contrary, one may pick them up. Though we must recognize that there are societal conventions, perhaps even laws, that differ across cultures. In most American grocery stores one may pick up fruit. In many small European ones, the grocer must handle them for you. The store will likely have a declared "public" area and a private one, where the general public may not go. The rules of information gathering are not strict. If I can stand in the public area and see a sheet of the store's accounts on a desk, then perhaps I would be rude to attempt to scrutinize it, but it is unlikely that it constitutes a punishable illegal or even an unethical deed to read the numbers. The grocer does not expect or want me to see this

---

[29] The existence of the Deep Web raises additional ethical issues, since here the owner of a webpage may take measures to block access, such as by avoiding links to it, not listing it with common search engines or with any metadata in its HTML code, and most effectively by using dynamic (and possibly encrypted) generation of the webpage address. However even here it is not clear if it is unethical to attempt to access the webpage and succeed.

information, but that fact alone does not constitute a strong case that it is unlawful or unethical.[30]

A store that is open to customers' literal, physical entry is in some respects like a website. It has a dual public-private nature. Members of the public may perform certain acts with or on this private property, but not others. Certain information is clearly directed at anyone, such as the price of items, the signs for a bathroom, and so on. Other information that the owner may not want any customer to access nevertheless might be such that it can be accessed or inferred with little effort, such as the stores' markup percentage, inventory, list of expenses, and so forth.

Unfortunately, although this is the best one can do in the way of moral analogs, there are some failures of the analogy. For one thing, being private property, this is a case in which there is physical space governed by rules and conventions, much as in the case of the application of Westphalian principles. This physical space is owned, and there are conventions governing what I may do in it. Secondly, there are elaborate conventions governing these various permitted and unpermitted activities—where I may go, how far I can reach my arm, what I may do with the merchandise, and so on. It has proven beastly difficult to build artificial intelligence systems that understand these transactions and the background information governing behavior. Normally, for example, I may not eat the products in a grocery store. But if there are small pieces of say, pastries, displayed to allow easy access by customers, then they are probably samples that may be eaten.

The concepts of an owned space and what others are permitted to do in them—for both private land like a store and for a state's territory—evolved over centuries if not millennia. They are based on a shared, literal notion of space and of boundaries in that space. By comparison, the ethics of cybersecurity has developed very recently, with almost laughably undeveloped concepts. There are few clear customs and very few clear, enforceable, extraditable laws.[31] It uses a metaphor of space, "cyberspace," but without key structural features of space. What counts as "movement" and thus intrusion into another agent's cyberspace? How many dimensions does it have? What counts as distance in cyberspace? And most troublesome of all, what are the acknowledged or declared boundaries of one person's, or one state's, area (or volume) of cyberspace?

In order to determine what is and what is not ethical regarding access to another's data, either to exfiltrate information or to alter it, we would need generally acknowledged principles of where the "borders" are. In order for such "border" notions to be

---

[30] Of course, our intuitions on this case can become murkier if we spell out just what we mean by "scrutinize." Is it beyond merely rude if, for example, rather than merely looking at the store's accounts, one took a photo of it? Thanks to Adam Henschke for this point.

[31] Even without regard to widely acknowledged criminal acts, extradition is most often a complicated affair, relying on numerous bilateral treaties (with 120 nations, one needs 7, 140 separate bilateral treaties).

useful, one would need well-developed techniques of determining what counts as violating them. This is unproblematic in the literal notion of owned space, but more difficult in the cyberrealm. There are some ways of starting to make some progress, however. The devices that support the Internet, and the devices that constitute the hardware "component" of information systems, are all owned—or at least there is a more usual way of tracing ownership and boundaries. Some information is stored in, or resident on, parts of these devices. Likewise, information entities do travel through devices where they might, or might not, be stored.[32]

However, any attempt to ground useful notions of cyberspace, and ownership of space, on owned material devices quickly breaks down. The public webpage of a website is resident on certain sectors of a hard drive; the password-protected data is elsewhere; and the operating system is located still elsewhere. But it is in the nature of property and territorial boundaries that they occupy fixed places and, as much as possible, are contiguous and not fragmented. Yet this encroachment on another's territory will generally only occur with the permission of its owners, and hence, by Westphalian principles, such an approach will be inherently compromised. So it does not advance our analysis to locate public and less public amounts of information by the material parts of storage devices in which they reside. We cannot say precisely where these boundaries are, and they can shift second-to-second.

3.5 Perfidy and Deception

In a series of papers the computer scientist Neil Rowe has argued that many forms of cyberwarfare involve perfidy, in the sense it is used in international law. [33,34] He is almost certainly mistaken to use the term "perfidy" here. "Perfidy" is used in a very narrow sense in international law to describe the exploitation of explicitly protected behaviors in the laws and customs of war to further an attack (*Talinn Manual 2013*).[35]

A correct term would be deception. It is much easier to see that when a Ukrainian, for example, hacks into a US Department of Defense computer, he is pretending to be someone else, namely the user with a certain username who has permitted access. But there is no blanket prohibition, either ethical or legal, on deception.

---

[32] For an interesting discussion of the metaphysical boundaries of the cyberrealm, see Jenkins, "Is Stuxnet Physical, Does It Matter?" 68–79.

[33] See also Heather Roff's essay in this book.

[34] Rowe, "Cyber Perfidy."

[35] Paradigms of perfidy include the deceptive use of the "white flag" or surrender or truce, or of using signs for protected vehicles, such as the Red Cross, to disguise weapons or healthy soldiers, with the intent to use that deception to cause death or damage. Again, see Heather Roff's chapter for more on this.

Even in the ethical theory most hostile to deception, namely Kantian ethics, there would be exceptions.[36] In games of cards (bridge, poker) and many board games one tries to deceive other players as to one's cards and intentions. This is one of the most important, essential features of what it is to try to play the game. If one is aware of a high probability that another party could be bluffing, and there are no rules to the contrary, then deceiving is not "deceptive" with any moral force suggesting wrongdoing. It is one of the design features of the Internet, and of software such as browsers interacting with it, that they provide varying degrees of anonymity that one may choose. The concept of deception only makes sense if there is a "reasonable expectation" that one can expect an honest representation. Agents using the Internet, at least in many uses, are virtually carrying a sign saying, "I may not be who I claim I am and what I say may not always be what I believe." In this situation meaningful deception cannot logically arise.

Furthermore, even if some Internet practices and intrusions are sometimes morally wrong, they are not wrong in the sense required for anything substantive to follow in the high-stakes arena of military ethics. In that arena we are concerned with, such consequences are what morally justifies a military counterattack, or war (including deaths and permanent destruction), or what justifies punitive use of force by an international body on a perpetrator. No mere Internet deception is likely to rise to that level, unless it intentionally or negligently results in deaths and permanent destruction. And in those cases it is covered by a plausible "effects-based" assessment according to traditional laws and standards of the use of force.

3.6 The Future of Cyber Conflict

As I have argued, there are key differences between traditional warfare and what we are most likely to see in the near future in cyberwarfare. The differences are ethically significant. Just war theory arose in contexts where warfare was assumed to always cause widespread death and destruction. The majority of forms of cyberwarfare we are likely to see do not rise to that level. Internet communication has not solidified around customs, practices, and a legal environment in which Westphalian principles of territory and sovereignty can be usefully applied.

Major cyberpowers have engaged in cyberespionage on each other, and they have also committed acts of disruption of information operations or degrading of stored information. They have harassed each others' information operations, including degrading operations of large commercial interests (one thinks especially

---

[36] In what seems to be a perfectly inconsistent position, Rowe seems to endorse the use of honeypots, which seem equally if not more deceptive. See Rowe et al., "Defending Cyberspace with Fake Honeypots," *Journal of Computers* 2, no. 2 (2007): 25–36.

of South Korean banks). They have apparently avoided causing deaths or extensive and permanent physical destruction that would extensively impact citizens' lives. By not crossing these lines they have conformed to the traditionally permitted (if undesired) behavior of states toward other states in international relations. In fact, as others have noted, we have witnessed the development of an increasingly firm norm of international relations that such death and destruction is impermissible. Crossing that line risks a cybercounterattack by a much superior cyberpower.[37] Attackers are deterred in other, much more powerful ways. China and Russia would risk grave harm to the banking and economic system of Europe and North America, and thus to their own economic well-being.

As I suggested in "The Ethics of Cyberwarfare,"[38] it is likely we will enter a long period of a kind of cold war, a game-theoretic equilibrium in which the worst weapons are not used and in which cyberpowers and others test the limits of what will be tolerated by the international community. Pursuing only cyberdefense, as opposed to having, demonstrating, and threatening cyberoffensive capability, will become enormously expensive and disadvantage the purely defensive strategy.

Bibliography

Arquilla, John. 1999. "Ethics and Information Warfare." In *Strategic Appraisal: The Changing Role of Information in Warfare*, ed. Z. Khalilzad, J. White, and A. Marsall. Santa Monica, CA: Rand Corporation, 379–401.

Christopher, Paul. 2004. *The Ethics of War and Peace: An Introduction to Legal and Moral Issues*. Rev. ed. Upper Saddle River, NJ: Prentice Hall.

Cook, James. 2010. "'Cyberation' and Just War Doctrine: A Response to Randall Dipert." *Journal of Military Ethics* 9, no. 4: 411–42.

Dipert, R. R. 1993. *Artifacts, Art Works, and Agency*. Philadelphia: Temple University Press.

Dipert, R. R. 2006. "Preventive War and the Epistemological Dimension of the Morality of War." *Journal of Military Ethics* 5, no. 1: 32–54.

Dipert, R. R. 2010. "The Ethics of Cyberwarfare." *Journal of Military Ethics* 9, no. 4: 384–410.

Dipert, R. R. 2013a. "Other-than-Internet (OTI) Warfare: Challenges for Ethics, Law, and Policy." *Journal of Military Ethics* 12, no. 1: 34–53.

Dipert, R. R. 2013b. "The Essential Features of an Ontology for Cyberwarfare." In *Conflict and Cooperation in Cyberspace*, ed. Panayotis Yannakogeorgos and Adam Lowther. New York: Taylor & Francis, 35–48.

Dipert, R. R. 2013c. "A Defense of the Tactics of William T. Sherman: Rehabilitating Moderate Military Realism." International Society for Military Ethics (ISME), University of Notre Dame, October 13–15.

Floridi, Luciano. 2008. "Information Ethics, Its Nature and Scope." In *Information Technology and Moral Philosophy*. Vols. 40–65. Cambridge: Cambridge University Press.

Goldman, Jan, ed. 2010. *Ethics of Spying: A Reader for the Intelligence Professional*. Lanham, MD: Scarecrow Press.

Jenkins, Ryan. 2013. "Is Stuxnet Physical, Does It Matter?" *Journal of Military Ethics* 12, no. 1: 68–79.

---

[37] I note here that in deterrence theory one should not say precisely where that line is.

[38] Dipert, "Ethics of Cyberwarfare."

Kaspersky. 2013. "Kaspersky Security Bulletin 2013. Overall Statistics for 2013." http://www.securelist.com/en/analysis/204792318/Kaspersky_Security_Bulletin_2013_Overall_statistics_for_2013

Kaspersky. 2014. http://cybermap.kaspersky.com/.

Libicki, M. 2009. *Cyberdeterrence and Cyberwar.* Santa Monica, CA: RAND Corporation.

Libicki, M. 2012. "Panel on Response to Cyberattacks: The Attribution Problem." The McCain Conference, organized by the Stockdale Center for Ethical Leadership. U.S. Naval Academy, Annapolis, MD, April 2012: http://www.youtube.com/watch?v=bI7TLqTt0H0. Accessed November 3, 2013.

Lucas, George. 2012a. "Permissible Preventive Cyberwar: Restricting Cyber Conflict to Justified Military Targets." http://www.elac.ox.ac.uk/downloads/Permissible%20Preventive%20Cyberwar%20UNESCO%202011.pdf. Accessed August 25, 2015.

Lucas, George. 2012b. "Can There Be an 'Ethical' Cyber War?" http://www.usna.edu/...%20and%20Cyber%20War%20GR%20Lucas.pdf. Accessed November 3, 2013.

Orend, B. 2005. War. In *The Stanford Internet Encyclopedia of Philosophy.* http://plato.stanford.edu/entries/war/. Accessed September 13, 2013.

Owens, W., K. Dam, and H. Lin. 2009. *Technology, Policy, Law, and Ethics Regarding US Acquisition of Cyberattack Capabilities.* Washington, DC: National Research Council of the National Academies of Science.

Reichberg, G., H. Syse, and E. Begby, eds. *The Ethics of War: Classic and Contemporary Readings.* Malden, MA: Blackwell, 2006.

Rid, Thomas. 2013. *Cyber War Will Not Take Place.* New York: Oxford University Press.

Rowe, Neil C. 2006. "A Taxonomy of Deception in Cyberspace." International Conference on Information Warfare and Security. Princess Anne, MD.

Rowe, Neil C. 2007. "The Ethics of Cyberwar Attacks." In *Cyber War and Cyber Terrorism,* ed. A. Colarik and L. Janczewski. Hershey, PA: Idea Group.

Rowe, Neil C. 2009. "The Ethics of Cyberweapons." *International Journal of Cyberethics* 1, no. 1: 20–31.

Rowe, Neil C. 2010. "Towards Reversible Cyberattacks." In *Proceedings of the 9th European Conference on Information Warfare and Security,* 261–67. N.p.: Academic Conferences Limited, 2010.

Rowe, Neil C. 2013. "Cyber Perfidy." In *Routledge Handbook of War and Ethics,* ed. N. Evans, chap. 29: at http://faculty.nps.edu/ncrowe/cyberperfidy.htm. Accessed November 16, 2013.

Rowe, Neil C., and Han Ho. 2007. "Thwarting Cyber-Attack Reconnaissance with Inconsistency and Deception." Proceedings of the Eighth IEEE Workshop on Information Assurance, West Point, NY, June 2007: http://faculty.nps.edu/ncrowe/iaw07_reconnaissance.htm. Accessed November 13, 2013.

Rowe, Neil C., and E. John Custy. 2010. "Deception in Cyber Attacks." In *Warfare and Cyber Terrorism,* ed. L. Janczewski and A. Colarik. Hershey, PA: Information Science Reference.

Rowe, Neil C., E. John Custy, and Binh T. Duong. 2007. "Defending Cyberspace with Fake Honeypots," *Journal of Computers* 2, no. 2: 25–36.

Sanger, David, and Thom Shanker. 2014. "N.S.A. Devises Radio Pathway into Computers." *New York Times,* 14 January 2014.

Schmitt, M. 1998. "Bellum Americanum: The U.S. View of Twenty-first Century War and Its Possible Implications for the Law of Armed Conflict." *Michigan Journal of International Law* 19, no. 4: 1051–90.

Schmitt, Michael N. 2011. "Cyber Operations and the Jud Ad Bellum Revisited." *Villanova Law Review* 56, no. 3: 569.

Schmitt, M. Gen. ed. 2013. *Tallinn Manual on the International Law Applicable to Cyber Warfare.* New York: Cambridge University Press.

Strawser, Bradley J. 2013. "Issue on Cyberwar and Ethics." *Journal of Military Ethics*: 1.

Stroll, Avrum. "Ethics without Principles." In *Wittgenstein and the Philosophy of Culture,* ed. K. S. Johannessen and T. Nordenstam. Vienna: Holder-Pichter-Tempsky, 1996, 310–20.

Taddeo, Mariarosario. 2012. "An Analysis for a Just Cyber Warfare." Fourth International Conference on Cyber Conflict. Ed C. Czosseck, R. Ottis, and K. Ziolkowski. NATO CCD

COE Publications: Talinn Estonia. http://www.ccdcoe.org/publications/2012proceedings/3_5_Taddeo_AnAnalysisForAJustCyberWarfare.pdf. Accessed November 15, 2013.
Verizon. "2014 Data Breach Investigations Report": http://www.verizonenterprise.com/DBIR/2014/
Waltz, Kenneth. 1979. *Theory of International Politics*. New York: McGraw-Hill.
Walzer, Michael. 2006. *Just and Unjust Wars*. 4th ed. New York: Basic Books.
Yannakogeorgos, Panayotis A. 2012. "Internet Governance and National Security." *Strategic Studies Quarterly* (Fall 2012): 102–125.

Part II

CYBERWARFARE AND THE JUST WAR TRADITION

4

Cyber *Chevauchées*

Cyberwar Can Happen

DAVID WHETHAM

4.1. Virtual War or Real War?

It is not clear whether cyberwarfare deserves to be called "warfare" or if it should instead be considered as something short of war, like espionage, or even a discrete form of criminal activity. For example, it is difficult to conceptualize how a denial of service attack against civilian social media, even on a massive scale, can overthrow an adversary, so how can such an event be thought of as war? Some, like Thomas Rid, argue that "cyberwar" is a misnomer. It has never happened, and it is unlikely to ever happen. Cyberwar is not real war because it does not involve violence or the threat of violence; thus it is not instrumental in the same way that military force can be, nor is it attributable to a specific actor. We are therefore simply confusing one kind of activity for another when we call a cyberattack an "act of war."

To consider this matter, it may be helpful to hark back to medieval history, even though that period appears to have little relevance to the computer age. During the Hundred Years' War, *chevauchées* were common, involving mounted soldiers deliberately spreading out over an enemy's territory to plunder and destroy everything in their path. Many military historians saw these actions as a diversion from the real focus of war, which was, surely, to find and defeat the enemy. Therefore, for many years *chevauchées* were wrongly dismissed as peripheral activities—viewed simply as mounted plundering expeditions rather than actual war. However, their real strategic purpose was to weaken the moral and economic base of rivals, undermining their political legitimacy by raising the cost of conflict to a point where negotiation and compromise would follow, often without the need to fight at all.

If some forms of cyberattacks are seen in this context, it becomes clearer how they can be conceptualized within a broader notion of war as a social phenomenon. If the intent is to demonstrate that a state cannot protect the day-to-day lives of its citizens, the cost in moral and economic terms, as well as political legitimacy, may be raised to a point that it is easier to offer concessions than submit to further prolonged attacks. This chapter will examine the arguments put forward by Rid and, by drawing parallels with historical warfare examples, will conclude that Rid's position is both misguided and unrealistic: cyberwar can happen.

4.2 There Is No Such Thing as Cyberwar

In 2012, Thomas Rid challenged many assumptions about cyberwar when he published a provocative and influential piece in the *Journal of Strategic Studies* titled "Cyber War Will Not Take Place."[1] With governments all around the world striving to show that they were meeting the challenges posed by these new threats,[2] and apocalyptic claims about "cyber Hiroshimas,"[3] Rid challenged some of the hype that has grown up in this area of domestic and international security. He argued that the "war" in "cyberwar" has more in common with the use of "war" in terms such as the "war on obesity" than with anything we might consider real war.[4] Real war, such as that experienced during World War II, is a completely different phenomenon; according to Rid, it is simply incorrect to think of cyberthreats in the same way. Put simply, he says, "Cyber war has never happened in the past. Cyber war does not take place in the present. And it is highly unlikely that cyber war will occur in the future."[5]

Rid built his argument around a definition of war derived from the writings of Carl von Clausewitz, the famous Prussian philosopher of war. Three elements must all be satisfied if an action, whether defensive or offensive, is to be considered as a stand-alone act of war.[6] Clausewitz captures the first of these elements in the phrase, "War is an act of force to compel the enemy to do our will."[7] Rid takes this, quite sensibly it appears,

---

[1] Rid, "Cyber War Will Not Take Place," 5–32. This argument was subsequently expanded in his 2013 book, also titled *Cyber War Will Not Take Place*.

[2] For example, the UK government followed the United States when it placed "hostile attacks upon UK cyber space" in Tier One of the National Security Strategy priority risks. https://www.gov.uk/government/publications/the-national-security-strategy-a-strong-britain-in-an-age-of-uncertainty, accessed June 6, 2014.

[3] Gross, "Declaration of Cyber-War."

[4] Rid, *Cyber War*, 15.

[5] Ibid., 6.

[6] Ibid., 7. It is worth noting that a serious cyberattack is likely to be employed in conjunction with other forms of influence operations, be they diplomatic or kinetic (i.e., physical) in character. The cyberattacks that preceded military operations in South Ossetia in 2008, combining online graffiti, denial-of-service attacks, and the distribution of malicious software, were limited in scope. They were not apparently coordinated with or used to facilitate the military operation, but perhaps give an idea of the way that cyberactions will be used in conjunction with other available tools in future conflicts. See Nazario, "Georgia DDoS Attacks."

[7] Clausewitz, *Vom Kriege*, 27.

to say that "if an act is not potentially violent, it is not an act of war."[8] The second essential element is that an act of war needs to be directed toward a purpose—if it is to be considered an act of war, that act must be instrumental in some way toward achieving an end. This, of course is captured by Clausewitz's most famous phrase: "War is a mere continuation of politics by other means."[9] An act of war is not an end in its own right, but physical violence (or even the threat of physical violence) is merely a means to achieving the objective of forcing the other party to accept your terms. This is achieved when one opponent renders the other side defenseless, or when they place their opponent in such an undesirable position that they are forced to accept that their situation will only get worse if defeat is not accepted now.[10] Finally, Rid argues that to satisfy his definition of war, the violent act, carried out with a political goal in mind, also "has to be attributed to one side at some point during the confrontation. History does not know acts of war without eventual attribution."[11]

As other chapters in this volume will attest, problems arise in trying to apply these essential elements to the cyberrealm. Cyberattacks are not easy to equate with acts of force—they are simply different from an airstrike or bayonet charge because they are very unlikely to directly lead to someone being harmed in the same way. Pulling the trigger on a gun shoots a bullet and, if it is aimed accurately, this either causes a wound or kills the target. Any connection between a cyberattack and physical injury, violence, or loss of life will be indirect or even incidental. For example, the cooling system for a nuclear reactor might be switched off, leading to a radiation leak and contamination for inhabitants nearby; or a commuter train might be derailed by changing its safety signals, leading to a crash and the death of passengers and crew. Rid concedes that such an intentional attack, if launched with a political goal, *could be* considered an act of war even if the means used were not violent, only the ends. However, Rid then argues, quite rightly, that while large-scale cyberattack scenarios—such as those involving infrastructure damage, disruption to essential services (such as electricity and water, leading to potential loss of life), or massive economic harm—have been predicted many times, they still remain an entirely hypothetical threat owing more to science fiction than reality.[12] Anything short of this level of destruction and/or harm, namely the very limited cyberactivity that we have actually seen to date, fails the threshold test. Simply put, Rid makes two claims: we have seen nothing in the cyberrealm that qualifies as war so far; and we are unlikely ever to do so.

The examples he provides bear this out. For example, in 2007 the Baltic state of Estonia decided to move the Bronze Soldier of Tallinn, a Russian World War II

---

[8] Rid, *Cyber War*, 7.
[9] Clausewitz, *Vom Kriege*, 44.
[10] Rid, *Cyber War*, 8.
[11] Ibid.
[12] Ibid., 10.

memorial, from the center of the capital to a military cemetery on the outskirts of the city. Being an ex-Soviet state, Estonia has a significant Russian-speaking minority, and this decision was met with shock and anger by both this group and neighboring Russians across the border. There were riots in Tallinn as well as in the towns of Johvi and Kohtla-Jarve, leading to over 100 injuries, 800 arrests, and a fatality.[13] Cyberattacks began on the second night of rioting, starting with relatively "inept, low-technology methods" such as flooding official websites with negative comments and bombarding various banks and government services with Internet traffic in order to overload their capacity.[14] As the attacks continued over subsequent weeks, they became much more sophisticated. The volume of distributed denial-of-service (DDoS) attacks increased and they also became more coordinated, coming from an estimated 85,000 hijacked computers at its peak.[15] These attacks were certainly large-scale and sustained, but the result was hardly comparable to an invasion or even a very-low-level armed attack. As Rid notes, "the effect of these coordinated online protests on business, government, and society was noticeable, but ultimately it remained minor."[16] The Estonian case also fails each of the criteria that Rid has set out above:

> [U]nlike a naval blockade, the mere "blockade" of websites is not violent, not even potentially; unlike a naval blockade, the DDoS attack was not instrumentally tied to a tactical objective, but an act of undirected protest; and unlike ships blocking the way, the pings remained anonymous, without political backing.[17]

Rid argues that due to the lack of the essential features war requires, political (as opposed to merely criminal) cyberoffenses should be considered as neither crimes nor warfare but instead placed in the same category as subversion, spying, or sabotage, existing somewhere on the spectrum between apolitical crime at one end and genuine war at the other.[18]

Rid places the most sophisticated known cyberattack to date—the Stuxnet worm used against Iran's nuclear enrichment program—firmly in the category of sabotage.[19] The Stuxnet worm was capable of transmitting and replicating itself through a variety of ways, including the Internet and USB drives, and was

---

[13] "Tallinn Tense after Deadly Riots," http://news.bbc.co.uk/1/hi/world/europe/6602171.stm, accessed June 6, 2014.

[14] Rid, *Cyber War*, 11.

[15] Tikk, Kaska, and Vihul, *International Cyber Incidents*, 17.

[16] Rid, *Cyber War*, 12.

[17] Ibid., 13.

[18] Ibid., 7.

[19] Ibid., 17.

programmed to specifically target the logic controllers of uranium-enrichment plants. The infection resulted in random changes in the speed of centrifuges, with the worm disabling their warning systems to prevent operators from being notified of the problem. The random changes in speed, often beyond the permissible limits of technical equipment, resulted in permanent damage to certain parts of the centrifuges. The net result was that Iran's nuclear enrichment program was set back by an estimated period of approximately five years.[20] Iran blamed Israel and the United States for the attacks, but both countries denied any hand in it. This attack clearly went beyond a mere criminal act, but for Rid, the lack of violence and the lack of acknowledged attribution means that Stuxnet fell short of a genuine act of war. It was computer sabotage on a whole new level, but it was not on a sufficient scale to be considered an act of war.[21]

Of course, sabotage along with espionage and subversion can be used in conjunction with military operations as part of real war, just as they have been "since time immemorial."[22] But these acts are not classified as war by themselves, just as their cyber equivalents cannot be considered acts of war. One can expect that cyberelements will be incorporated into any sophisticated military operation in the future. An example suggested by Rid that illustrates this prospect is the cybersabotage that preceded a 2007 Israeli airstrike on a nuclear reactor site at Dayr ez-Zor in northern Syria. To facilitate the daring strike (codenamed "Operation Orchard") deep into Syrian territory, the highly capable Syrian air-defense system was rendered temporarily "blind" through an electronic attack on a radar site at Tall al-Abuad near the Turkish border.[23] There is speculation that this involved some kind of "kill switch" embedded in the system by a contractor to disable the equipment.[24] Photographs later made public by the US government demonstrate that the suspected reactor building was reduced to rubble in the successful air attack. The cyberelement of the attack was clearly a highly effective enabler for the bombing raid; however, Rid argues that on its own it "would not have constituted an act of war."[25]

Cyberattacks carried out to date, including the neutralization of the Syrian air-defense system described above, fail to satisfy the strict definition of war that Rid suggests. Moreover, Rid believes, they are exceedingly unlikely ever to do so. Indeed, it would appear that by accepting such a narrowly demarcated definition of war, cyberattacks are always going to be excluded from this taxonomy. But is this a

---

[20] "How Stuxnet Works," http://www.telegraph.co.uk/technology/8274488/How-Stuxnet-works-what-the-forensic-evidence-reveals.html, accessed June 7, 2014.

[21] Rid, *Cyber War*, 20.

[22] Ibid., 16.

[23] Fulghum, Wall, and Butler, "Israel Shows Electronic Prowess"; Fulghum, Wall, and Butler, "Cyber-Combat's First Shot," 28–31.

[24] Adee, "The Hunt for the Kill Switch."

[25] Rid, *Cyber War*, 17.

valid position to take? Given the inherent risk of escalation in any conflict, having a clear idea about what *we think* war actually is and also what *other people think* it is is very important, making this something far from merely an academic debate.

4.3 *Chevauchées* and Medieval Warfare

The history of warfare can provide insight into this new realm of conflict despite the separation of many hundreds of years and radically different ways of conducting military operations. Until the 1950s, medieval warfare was viewed in fairly straightforward terms. Influential historians such as Hans Delbrück regarded knights (obviously the most important actors on the medieval battlefield) as not particularly sophisticated individuals, who from the start of battle were "moved only by the instincts of the mass itself."[26] Courage, rather than guile or wit, was what was required for success in such mass encounters of armored fighting men. Another influential historian, Charles Oman, argued that a young Frankish noble "deemed his military education complete when he could sit on his charger firmly and handle a lance and shield with skill."[27] *Chevauchées* were mounted plundering expeditions—perks of the trade—carried out in-between the real business of conducting war and fighting battles. Writing around the turn of the twentieth century, military historians such as Oman and Delbrück captured the common assumptions underpinning how medieval warfare was understood at the time. They themselves were influenced by what they understood the philosopher of war, Clausewitz, to mean when he articulated the meaning of "war." Therefore, they saw the primary objective of any military campaign as the destruction or overthrow of the enemy's forces in order to impose one's will on him. This meant an almost exclusive focus on battles, because it was obviously through this means that the political objective was supposedly satisfied. The idea that warfare could be dominated by a subtly different conception of strategy was simply not entertained. The result was that other core aspects of warfare in the Middle Ages, such as sieges or raids, were acknowledged as present but were then practically dismissed because they did not appear to directly contribute to the key objective. Effectively, they were dismissed as distractions rather than being considered real "war" at all.

These views were challenged in the 1950s by a new wave of scholars such as J. F. Verbruggen and R. C. Smail, who had a much more nuanced appreciation of medieval strategy and military affairs.[28] Verbruggen forced a thorough revaluation of

---

[26] Delbrück, *History of the Art of War*, 158.
[27] Oman, *Art of War in the Middle Ages*, 172.
[28] Verbruggen, *Art of War*. Originally published as *De Krijgskunst in West-Europa in de Middeleeuwen, IXe tot begin XIVe eeuw* (Brussels, 1954). A new 1996 edition in English includes more of the annotation that was heavily abridged in the 1977 translation. Also, Smail, *Crusading Warfare*.

medieval strategy by demonstrating the sophistication of medieval military thinking. He showed, for example, that the heroic knight was but one of many actors on the medieval battlefield, even if chivalric accounts tended to concentrate on individual acts of heroism with "the fighting of entire formations . . . represented as a duel fought out by two champions."[29] By carefully rereading the sources, Verbruggen was able to show that political and social organization was actually very well developed rather than rudimentary.

Smail, on the other hand, was the first among a growing number of historians who saw raids and sieges in their true context as part of an attritional strategy aimed at undermining the economic base of the enemy and gaining control of the means of producing wealth. Although obviously battles could still be of vital importance, "the primary military objective was normally the control of fortifications because they were the key to controlling the land. It was not to destroy or overthrow the enemy in order to impose one's will on him."[30] Battles themselves were extremely risky affairs because of what was at stake—apart from the high risk of death or capture for the main protagonists on each side, the verdict was likely to be seen as a clear judgment by God on the merits of their respective claims. It is hardly surprising that there were so few real battles in the Middle Ages, but this was *because* the framework of war was so well understood and appreciated by those conducting it.[31]

The repeated *chevauchées* that were so common during the Hundred Years' War in the fourteenth and fifteenth centuries, for example, could now be seen not merely as mounted plundering expeditions but as having a clear strategic purpose. Obviously, the acquisition of plunder was going to be an element of any captain of war's plan if he intended to keep his followers contented. However, the *chevauchée* was so much more than this. The literal translation of *chevauchée* is not raid, as it is often rendered, but rather *procession*.[32] By being able to move unchallenged through an opponent's territory, destroying and plundering, one could demonstrate their inability to defend both their lands and their people. "In a period that had a very positivist conception of rights, if one was unable to defend something, one did not have a right to it."[33] For example, the contemporary chronicler Chandos Herald records the actions of the English army between the Seine and the Somme in 1346: "[T]he English, to disport themselves, put everything to fire and flame. There they made many a widowed lady and many a poor child orphan."[34]

---

[29] Verbruggen, *Art of War*, 19.

[30] Whetham, *Just Wars*, 11.

[31] Ibid.

[32] Burne, *Crecy War*, 245.

[33] Whetham, *Just Wars*, 12. The theory and practice of the medieval *chevauchée* is explored in Whetham, *Just Wars*, chap. 3.

[34] Herald, *Life of the Black Prince*, lines 236–39.

Rather than being merely a distraction from what was really important, or even an enabler for other lines of military operation, the *chevauchée* was an act of war in its own right. It was a means to, very publicly, undermine the legitimacy of an impotent lord while demonstrating the justice of one's own cause. While a battle *might* be the result of a *chevauchée* if the mounted force was intercepted, this was seen as merely another means of achieving the actual military aim, which was in this case to weaken the moral and economic base of rivals, forcing them to come to the negotiating table and accept terms.

Can this understanding of medieval warfare help us to frame our analysis of attacks in the cyberrealm? Is it possible to conceive of a cyberattack that does not involve physical harm, death, or destruction but is carried out with a clear intent to achieve a political purpose, even if ownership of the attack is not admitted by any party? If such a situation could be easily imagined, based on documented historical events without straying into the realms of science fiction or cyber Hiroshimas, it would appear to challenge both Rid's deliberate exclusion of cyberattacks from the category of "acts of war" and also his confident prediction about its unlikeliness in the future.

4.4 A Cyber *Chevauchée*

To give an example, consider that two states are participating in a protracted diplomatic disagreement over territory and resources. They are evenly matched in military capabilities. Despite clear warnings to the contrary, State A moves into the disputed territory and begins drilling for petrochemicals. Rather than initiating kinetic military operations in response to the actions of State A, State B responds with an unattributed (and due to its nature, initially un-attributable) cyberattack that demonstrates an ability to turn off State A's air defense system in different areas at will. The cyberattacks are always preceded with a clear and very public (and yet unattributed) announcement of what area in State A will be targeted next, demonstrating to the whole world State A's vulnerability and inability to protect itself. No one is harmed, no one is physically attacked, and State B very publicly pulls its military forces back to ensure that it is not seen as a military threat. At the same time, State B pursues normal diplomatic activity, robustly condemning the intervention and illegal drilling operations of State A. It also seeks international arbitration and sanctions on the offending party, while denying it has anything to do with the very embarrassing technical problems that State A appears to be having with its supposedly very capable (and very expensive) defense systems.

The government of State A finds itself under increasing pressure from an irate public to "do something" but finds itself uncertain how to act. Its people have not been harmed, nor have they even been obviously threatened. And yet, they feel fearful, vulnerable, and unprotected. Not wishing to risk being seen as the aggressor by mobilizing its military (and not willing to take the risks associated with direct confrontation), State A offers a series of concessions to State B over the sharing of proceeds from mineral rights in the disputed territory, while the decision regarding ownership of the actual territory is lodged with an international arbitration panel, unlikely to reach a verdict for many years. This leaves both State A and State B able to claim they have moved forward. The cyberattacks mysteriously cease.

Was this hypothetical cyberattack an act of war? Following Rid's reasoning, clearly not, and yet it still seems worth returning to the three criteria set out above—an act of war must be (1) violent (or potentially violent), (2) instrumental, and (3) attributed. Disarming a protective air defense system using electronic means can have the same effect as bombing the radar stations. One would be called an act of force (and therefore an act of war); the other, according to Rid, would not as there has been no physical violence, thereby failing the first of the three criteria tests. At best it would be considered sabotage, but as nothing has been destroyed or even permanently damaged, the act of turning something off temporarily is clearly not violent or even potentially violent. Therefore, pace Rid, it cannot be an act of war.

Perhaps it is unsurprising, but contra Rid, that a number of cyber, legal, and ethical experts do consider the Stuxnet attack as an "act of force" consistent with an act of war. According to the *Tallinn Manual on the International Law Applicable to Cyber Warfare*, while specifically excluding acts that merely generate inconvenience or irritation, "acts that kill or injure persons or destroy or damage objects are unambiguously uses of force."[35] The international group of researchers who wrote the *Manual* were unanimous in their view that Stuxnet constituted an "armed attack," meaning that Iran would have been entitled to respond in self-defense.[36] While the researchers do not speak for the United Nations or even NATO, they do represent a broad international consensus on the way existing law applies in cyberspace.

However, some types of sabotage do not even involve something akin to the physical destruction of the centrifuge caused by Stuxnet. In the cyber *chevauchée* example above, nothing is destroyed at all—merely temporarily switched off. Surely, that cannot be considered violent? Now admittedly, even if most people erroneously did not regard *chevauchées* as "proper war" until fairly recently, the *chevauchée* did still involve physical violence. Returning to each of the definitional elements of war proposed by Rid, just as there is more to medieval war than battles, there

---

[35] Schmitt, *Tallinn Manual*, 48.
[36] Waterman, "U.S.-Israeli Cyber Attack on Iran,"
http://www.washingtontimes.com/news/2013/mar/24/us-israeli-cyberattack-on-iran-was-act-of-force-na/?page=all, accessed June 7, 2014.

is surely more to some kinds of cyberattacks than physical violence (or the threat of it).[37] Demanding that physical violence is required does seem to be an overly restrictive interpretation of an act of war—causing an inability to defend oneself is certainly not a *nonviolent* action, even if it does not fit easily into our normal notions of violence as necessarily physical.

However, we understand this concept in other areas of human activity. For example, a rape that does not use physical force, but instead employs fear, a power relationship, perceived authority, or something else interfering with informed consent, is still an assault and deserves to be regarded as violence as a result. It is instructive to note that in the United Kingdom and other common law jurisdictions, the legally accepted definition of assault does not require physical harm, or even the threat of physical harm, to be satisfied. The Crown Prosecution Services guidance sets out the offense of Assault Occasioning Actual Bodily Harm (contrary to section 47 of the Offences against the Person Act 1861):

> The offence is committed when a person assaults another, thereby causing Actual Bodily Harm (ABH). Bodily harm has its ordinary meaning and includes any hurt calculated to interfere with the health or comfort of the victim: such hurt need not be permanent, but must be more than transient and trifling. (R v Donovan 25 Cr. App. Rep. 1, CCA)

The guidance also makes clear that psychological harm, including fear, distress, or panic, can amount to actual bodily harm.[38] An assault need not be physical to cause injury, and there are clearly and demonstrably many injuries that are not physical. Given that the aim of the protagonists in a struggle is to overthrow their opponent, then it is often the moral injury that causes the most profound kind of damage on an adversary. Indeed, it was the moral injury that was so damaging in the medieval conception of the *chevauchée*—the lord was supposed to be able to protect the people, and the lord's authority was sapped by a demonstrable inability to do so.[39]

Disabling a target's air defenses could clearly be instrumental in the same way that the medieval *chevauchée* was—both are demonstrating a freedom to act in another's territory and an inability of the other to do anything about it. The attacker can, if they desire, act with impunity at a time and place of their choosing, thereby causing political and, most of all, moral assault on the target who is seen as, and perceives itself to be, helpless in the face of the enemy.

---

[37] Clausewitz notes that a threat of action can have the same effect as the actual action. Clausewitz, *On War*, 181.

[38] http://www.cps.gov.uk/legal/l_to_o/offences_against_the_person/, accessed June 7, 2014.

[39] Rid himself acknowledges that trust itself was a target of the Stuxnet attack, something clearly outside of the physical realm.

Finally, according to Rid, the attack must apparently be attributed before it can be regarded as war. But surely one can be under attack and not know from where? This is not a situation unique to the cyberrealm, and in actuality, "attribution is often challenging even in circumstances of kinetic warfare, especially at sea."[40] I fully accept that in accordance with international law and the just war tradition, some kind of positive attribution is required before an appropriate response can be applied, but to say that a war does not commence until intention has been advertised would seem to be overly artificial and even slightly ludicrous. One would hardly claim that a genocidal nuclear strike launched without warning from concealed submarines was not an act of war, just because no one "owned up" to it at the time. If a country has experienced a cyberattack that demonstrates an ability to turn off its air defense systems at will, I find it difficult to accept that there is any government on the planet that would not see that as an act of war. To pretend it is something else (*merely* espionage or sabotage?) for six months while forensic investigations into server origins go on, or until a state "owns up," would be ridiculous. It would be more honest (and more likely) for a state to say, "we are under attack, but we cannot yet prove by whom." If deploying cyberattacks, any competent attacking state is likely to employ "correlation is not causation" diplomatic language as tensions rise, to obscure and confuse, while perhaps signaling through unofficial (and of course un-attributable) channels that the attack is very likely to cease once x, y, or z has been ceded.

Of course, in the hypothetical situation set out earlier, if State A is also in possession of a developed cybercapability, it may be that it chooses to authorize its own cybersabotage attacks against specific targets in order to raise the cost for State B. They might choose to do this without attribution and would probably not wish to directly cause death and destruction (at least at first). Would such a situation not qualify as a cyberwar? No one would be killed, but a state of hostilities would pertain with the threat of this escalating into other more kinetic activity due to the inherent dynamic of war that Clausewitz identifies so pertinently.

4.5 Cyberwar Is War

Looking back at the historical understandings of warfare in the Middle Ages, much confusion was caused by an understanding of war that was focused almost exclusively on battles. This artificially constrained analysis applied to a very specific, but also very rare, activity in the endemic conflicts of the period relegated all other forms of fighting to being at best peripheral or otherwise simply irrelevant. This was a mistake. The hypothetical but hardly fantastical scenario above involving State A versus State B represents a form of cyber *chevauchée* undertaken in order to

---

[40] Dinstein, "Cyber War and International Law," 281.

achieve a clear political goal without the necessity of the contemporary equivalent of a pitched battle. Taking a definition of war that is so narrow (or rather interpreting Clausewitz in such a restrictive way), so that it excludes the cyber equivalent of the medieval *chevauchée*, has similar problems and can dangerously skew perceptions in such a way that it obscures what is really happening.

The fault, therefore, lies not with Clausewitz's definition of war but rather with Rid's interpretation of it. At its heart, Clausewitz explains, the essence of war is "a complete, untrammeled, absolute manifestation of violence."[41] However, it is also important to understand that as a philosophical tool, Clausewitz distinguished between "absolute" and "real" war. While the former represents the theoretical realization of war's unrestricted nature, this only exists in a philosophical sense. Clausewitz fully understood that real war would be restricted by a range of factors, including friction and the influence of policy. The distinction that Clausewitz makes between the nature of war (immutable but therefore purely theoretical) and its character (what war actually looks like when translated into a particular time and place) also acknowledges that real war is a social phenomenon and cannot simply be divorced from this context.[42] Violence will indeed normally involve bloodshed, but violence, in some contexts, need not equate to *physical* harm—violence can still involve hurt and injury without physical harm in the real world.

One can obviously trade Clausewitz quotes back-and-forth in order to prove a point either way, but it is useful to refer to this section in Book I of his work:

> When we attack the enemy, it is one thing if we mean our first operation to be followed by others until all resistance has been broken; it is quite another if our aim is only to obtain a single victory, in order to make the enemy insecure, to impress our greater strength upon him, and give him doubts about his future. If that is the extent of our aim, we will employ no more strength than is absolutely necessary.[43]

Real war takes place on a spectrum, framed by the unattainable concept of absolute war at one end and the absence of war at the other. That spectrum should, rightly, include cyberattacks that do no obvious physical harm.

That is not to say that all cyberattacks are acts of war, just as not all physical attacks are acts of war. However, if it involves "an act of force to compel the enemy to do our will" and is employed as a means to achieve a political goal, it does appear to satisfy the first two of the criteria (as long as violence may be rightly interpreted in a less restricted sense). The third of Rid's criteria, attribution, appears to be a practical

---

[41] Clausewitz, *On War*, 87.
[42] Whetham, "Just War Tradition," 67.
[43] Clausewitz, *On War*, 92.

consideration rather than an essentially definitional one. It is at least theoretically possible to be subject to an act of war without knowing from where the attack originates. That makes formulating an appropriate response exceedingly difficult, and the attribution problem is therefore one of the most difficult areas of defense thinking and, indeed, of military ethics.[44] It is possible, as demonstrated above, that the other two essential criteria may be satisfied without attribution being possible, and that the act could still be legitimately regarded as an act of war. Therefore, this last element cannot be an essential part of the required definition of war.

To deliberately exclude all cyberattacks from the definition of an "act of war" seems to go too far, making the resulting classification decidedly not useful and unrealistic as a result.[45] Unfortunately, war does not stop being war just because one has chosen to define it as something else.

Bibliography

Adee, Sally. "The Hunt for the Kill Switch." *IEEE Spectrum*, May 2008.
Allhoff, Fritz, Nicholas G. Evans, and Adam Henschke, eds. *The Routledge Handbook of War and Ethics*. Oxford: Routledge, 2013.
Burne, Alfred H. *The Crecy War: A Military History of the Hundred Years War from 1337 to the Peace of Bretigny, 1360*. London: Eyre and Spottiswoode, 1955.
Chandos Herald. *Life of the Black Prince, by the Herald of Sir John Chandos*. Ed. Mildred K. Pope and Eleanor C. Lodge. Oxford: Clarendon, 1910, http://trove.nla.gov.au/work/11076500?selectedversion=NBD339261.
Clausewitz, Carl von. *Vom Kriege*. Berlin: Ullstein, 1832.
Clausewitz, Carl von. *On War*. Ed. Michael Howard, Peter Paret, and Bernard Brodie. Princeton, NJ: Princeton University Press, 1989.
Delbrück, Hans. *History of the Art of War within the Framework of Political History*. Vol. 3, *The Middle Ages*. Trans. Walter J. Renfroe Jr. Westport, CT: Greenwood, 1982.
Dinstein, Yoram. "Cyber War and International Law: Concluding Remarks at the 2012 Naval War College International Law Conference." *International Law Studies* 89 (2013): 276–287.
Fulghum, David A., Robert Wall, and Amy Butler. "Cyber-Combat's First Shot." *Aviation Week & Space Technology* 167 (16 November 2007).
Fulghum, David A., Robert Wall, and Amy Butler. "Israel Shows Electronic Prowess." *Aviation Week & Space Technology* 168 (25 November 2007).
Gross, Michael Joseph. "A Declaration of Cyber-War." *Vanity Fair*, April 2011.
"How Stuxnet Works." *The Telegraph*, 21 January 2011.
Lucas, George R. "Can There Be an Ethical Cyberwar?" In *Conflict and Cooperation in Cyberspace*, ed. A. Panayotis Yannakogeorgos and Adam B. Lowther. (Boca Raton, FL: Taylor and Francis, 2013, 195–210.
Lucas, George R. "*Jus in Silico*: Moral Restrictions on the Use of Cyber Warfare." In *The Routledge Handbook of War and Ethics*, ed. Fritz Allhoff, Nicholas G. Evans, and Adam Henschke. Oxford: Routledge, 2013, 367–80.
Nazario, Jose. "Georgia DDoS Attacks—A Quick Summary of Observations." *Security to the Core* (Arbor Networks), 12 August, https://asert.arbornetworks.com/georgia-ddos-attacks-a-quick-summary-of-observations/. Accessed August 25, 2015.

---

[44] For example, Strawser, *Journal of Military Ethics Special Issue*.

[45] A point also made by Lucas, "Can There Be an Ethical Cyberwar?," 195–210, and also see Lucas, "*Jus in Silico*," 367–80.

Oman, Charles. *The Art of War in the Middle Ages*. 2 vols. London: Methuen, 1924.

Rid, Thomas. "Cyber War Will Not Take Place." *Journal of Strategic Studies* 35, no. 1 (February 2012): 5–32.

Rid, Thomas. *Cyber War Will Not Take Place*. Oxford: Oxford University Press, 2013.

Schmitt, Michael N., ed. *Tallinn Manual on the International Law Applicable to Cyber Warfare*. Cambridge: Cambridge University Press, 2013,

Smail, R. C. *Crusading Warfare, 1097–1193*. 2nd ed. Cambridge: Cambridge University Press, 1995.

Strawser, Bradley J., ed. *Journal of Military Ethics Special Issue: Cyberwar and Ethics* 12, no. 1 (April 2013).

"Tallinn Tense after Deadly Riots." BBC News, http://news.bbc.co.uk/1/hi/world/europe/6602171.stm, 28 April 2007.

Tikk, Eneken, Kadri Kaska, and Liis Vihul. *International Cyber Incidents*. Tallinn: CCDCOE, 2010.

Verbruggen, J. F. *The Art of War in the Middle Ages*. London: North Holland, 1977.

Waterman, Shaun. "U.S.-Israeli Cyber Attack on Iran Was 'Act of Force,' NATO Study Found." *Washington Times*, 24 March 2013.

Whetham, David. "The Just War Tradition: A Pragmatic Compromise." In *Ethics, Law and Military Operations*, ed. D. Whetham Basingstoke, UK: Palgrave Macmillan, 2010, 65–89.

Whetham, David. *Just Wars and Moral Victories: Surprise, Deception and the Normative Framework of European War in the Later Middle Ages*. Leiden: Brill, 2009.

Whetham, David, ed. *Ethics, Law and Military Operations*. (Basingstoke, UK: Palgrave Macmillan, 2010.

Yannakogeorgos, A. Panayotis, and Adam B. Lowther, eds. *Conflict and Cooperation in Cyberspace*. Boca Raton, FL: Taylor and Francis, 2013.

5

Cyberwarfare as Ideal War

RYAN JENKINS

Is there an ideal war, a best possible war? Is there a war greater than which no war can be conceived?[1] What would such a war be like, and are there any means of waging war that satisfy this description? I will suggest that cyberwarfare offers the possibility of just such an ideal war.

The notion of an ideal war will strike some as incredible. War's unspeakable brutality informs its mythical status as a paradigm of evil: war as one of the Four Horsemen of the Apocalypse; or war identified with Hell. But wars can clearly be morally better or worse.[2] This implies that some type of war could be as morally good as possible. As long as the concept of an *ideal war* is coherent—as I argue in this essay—we should answer the opening question like this: An ideal war would be a war wherein civilian casualties were minimal or nonexistent and where acts of violence perfectly discriminated between combatants and noncombatants (discussed in section 5.1). Cyberwarfare has made possible this kind of ideal warfare for the first time by profoundly improving a state's ability to direct its force discriminately and to ensure that force is proportional (section 5.2).[3] Since cyberwarfare does not raise any moral concerns serious enough to countervail its clear benefits, we are obligated to prefer cybermeans where practical (section 5.3). These benefits

---

[1] See, for parity of argument, Anselm's ontological argument. There, Anselm seeks to infer God's existence merely by examining the concept of "something than which nothing greater can be thought." See Anselm, *St. Anselm*, 82.

[2] The contrary view, that war is *necessarily* bad, is certainly defended only at the fringes of the philosophical literature. As theists occasionally, in the interest of time, begin by assuming that God is conceptually possible, so we must assume that war can be better or worse *at all*.

[3] It is worth pointing out that, while cyberwarfare may be the first technology to make this kind of ideal warfare possible, it is not obviously the only technology to do so. For example, a war fought between fully autonomous robots, without harming any humans, could also count as an ideal war. That examination must be outside the scope of this paper, which focuses solely on cyberwarfare.

of cyberwarfare undermine the moral stringency of the proportionality and probability of success criteria of *jus ad bellum* (section 5.4).[4,5]

5.1 *In Bello* Moral Continua

By *war*, I will mean the organized, intentional, large-scale application of force for political ends. A full defense of this definition is outside the scope of this paper.[6] This definition is meant to be broadly consonant with our colloquial understanding of war and with points of scholarly agreement.[7] It is meant to be plausible and uncontroversial without being vacuous, and without ruling out the possibility of cyberwarfare, which remains at least controversial.[8] This definition is meant to count

---

[4] Whether a war's moral status *in bello* is independent of its status *ad bellum* is a central point of contention between traditional and revisionist just war theorists. This essay does not argue for the superiority of either traditional or revisionist just war theory. The main points will be relevant to both schools.

[5] I am bracketing concerns about *jus post bellum*, that is, justice in ending war and securing the peace. For what it is worth, in the cyberattacks I contemplate below most or all of the damage wrought is reversible. This is a reminder that cyberattacks, more than any other means of waging warfare heretofore used, offer the real possibility of restoring the status quo ante bellum.

[6] For a fuller explication of *war*, including a more comprehensive survey of the definitions on offer, see Steinhoff, "What Is War—And Can a Lone Individual Wage One?" Steinhoff agrees that war is essentially about politics, i.e., who has the proper authority to rule a citizenry or area, through the use of force that tries to "kill or at least wound or incapacitate" the enemy, using "lethal or physically destructive means" as part of an ongoing action. For his full definition, see his "What Is War," 145.

[7] Grotius defines war as "the state of contending parties, considered as such," implying that war must be organized, though not necessarily between states. Grotius, *Rights of War and Peace*, 18. Steinhoff adds rightly that these parties must be contending *by force* ("What Is War," 146). Walzer tells us only, in the early pages of his magnum opus, that war "has a recognizable and relatively stable shape, that its parts are connected and disconnected in the recognizable and relatively stable ways ... [that] states and soldiers [are] the protagonists of war," and that combat is its "central experience." Walzer, *Just and Unjust Wars*, 22. Orend says that war should be understood as the "actual, intentional, and widespread armed conflict" between "entities which either are states or intend to become states," or, at least, "associations of people with a political purpose." Orend, *Morality of War*, 3. Hobbes is perhaps the most permissive of these when he acknowledges the possibility of a cold war, as for him *war* is the "known disposition" to fight, even if no battles occur. Hobbes, *Leviathan*, 76. See also UN Declaration 3314, which defines *aggression* as '[the use of armed force by a State against the sovereignty, territorial integrity or political independence of another State," though war is at least an ongoing aggression, and with a political purpose. Definition of Aggression, United Nations General Assembly Resolution 3314 (1974).

[8] Some cyberoperations surely rise to the level of war. See United States, "International Strategy for Cyberspace" (2011). Moreover, though perhaps surprisingly, all cyberoperations are plausibly applications of *physical* force. See on this point Jenkins, "Is Stuxnet Physical? Does It Matter?"; Yannakogeorgos, "Internet Governance and National Security," 102–25; and, especially, Landauer, "Information Is Physical," 23–29. Would this definition count *mere* changes in code as warfare? To deny this would be to rule out the possibility of cyberwar by definition, but that would be too quick. This is

out the application of mere economic force, such as in the case of sanctions or other methods of *jus ad vim*, the use of force short of war or coercion short of combat.⁹ It is meant to be agnostic on whether all warfare must be conducted, sanctioned, supported, or enabled by states as this question is immaterial to my discussion. My definition perhaps bears most resemblance to Carl von Clausewitz's suggestion that war is "the act of force to compel our enemy to do our will."¹⁰ Finally, what I say below is compatible with many addenda, for example, that war must exist against a background of a publicly acknowledged conflict between belligerents.

We judge that some forms of waging warfare are morally better than others, and these judgments reveal a moral continuum along which any particular war falls. At one end of that continuum lies a morally perfect war: a war that is the *least morally bad* as is possible.¹¹,¹² It will be useful at this point to distinguish what I mean by "ideal war" from what others have meant. Clausewitz in particular means something utterly different when he discusses "ideal war."¹³ For Clausewitz, war is the pursuit of politics by other means. Various ways of waging warfare fall along a continuum, from pure politics at one end to pure war at the other.¹⁴ What defines this continuum is the level of violence that a state employs in pursuit of its goals. The more violent a conflict, the more it resembles an "ideal state of war."¹⁵ The more peaceful the

---

especially true given the colloquial familiarity of "cyberwarfare," because cyberwarfare could conceivably rise to the level of harmfulness of kinetic warfare, and because some actions we might describe as cyberwarfare have already taken place. We should take seriously cyberwarfare as a new instrument of international policymaking on a par with warfare.

⁹ For the original discussion of *jus ad vim*, see Walzer, *Just and Unjust Wars*, particularly the preface to the 4th ed. For discussion since, see especially Ford, "Just Use of Lethal Force-Short-Of-War," 63; also see Walzer, "Regime Change and Just War," 103–8; and Brunstetter and Braun, "From *Jus ad Bellum* to *Jus ad Vim*," 87–106. The phrase "coercion short of combat" is care of Heather Roff in conversation. The phrase appears to have been coined in Feaver, "Blowback: Information Warfare and the Dynamics of Coercion," 88–120. For a discussion of whether cyberattack could rise to the level of a casus belli, see Allhoff and Jenkins, "Facebook War."

¹⁰ See Clausewitz, *On War*, 75. Steinhoff's complaint with Clausewitz's definition is that it allows a legally authorized duel to count as a war, which is unsatisfying. Steinhoff for this reason includes in his definition the criterion that there is no other authority with a credible and respected monopoly on force that sanctions the ongoing conflict.

¹¹ As an exercise in imagination we can consider the other end of the continuum, a war that is *as bad as possible*. This war would plausibly contain gruesome crimes against humanity, directed entirely against those not liable to harm, inflicting unconscionable suffering on the most vulnerable segments of the population, and achieving some unjustifiable purpose. The Holocaust, as one example, falls far to this side of this continuum.

¹² I am treating the locutions "as good as possible" and "as least bad as possible" as synonymous. Making war less bad *just is* making it better; making it as good as possible *just is* making sure it contains as little badness as possible. Thus the "morally best" war is the least morally bad war, and vice versa.

¹³ See Clausewitz, *On War*, I.I.25.

¹⁴ See Darley, "Clausewitz's Theory of War and Information Operations," 75.

¹⁵ At the risk of complicating matters, we might also say that: the more violent a conflict, the more it participates in the Form of War.

conflict—such as at the opposite extreme, "elections in stable democratic societies"[16]—the more the conflict resembles pure politics, that is, the pursuit of political goals in a manner totally devoid of violence.[17] Clausewitz takes war to be inherently violent, in which case a more violent war is a closer approximation of war's "Platonic abstraction."[18] The continuum I consider is orthogonal to Clausewitz's: it instead evaluates warfare as falling along a *moral* continuum. On my analysis, the reduction of the violence of warfare moves a war toward its ideal state, rather than away from warfare and toward the purely political. Following this, one could read my account as an account of a *"morally* ideal war" as opposed to a *"conceptually* idealized war." In the interest of easier reading, I keep to "ideal war."

Prior to asking if a war could be *maximally* morally good, we need to investigate whether this maximum is conceptually possible. Some properties have no intrinsic maxima.[19] Mass is one such property: an object cannot be *as massive as possible*, since for any given mass, we can always add an additional gram.[20] If the *in bello* moral continua have no *intrinsic* maxima, then the concept of an ideal war would be incoherent in the same way that the concept of a largest number is incoherent: there could always be a war that was morally better.[21] But the *in bello* moral continua of discrimination and proportionality do have intrinsic maxima. An act of violence is more discriminate as the extent to which it harms particular individuals more closely approximates their individual liability. An act of violence is *maximally discriminate* if the harm that it causes befalls *only* those liable to attack. Few acts of modern warfare discriminate successfully, much less perfectly

---

[16] Darley, "Clausewitz's Theory of War and Information Operations," 76.

[17] It seems not to occur to Darley that information operations—whose "five pillars" are "operations security, psychological operations (PSYOP), deception, computer network operations, and electronic warfare" ("Clausewitz's Theory of War and Information Operations," 74)—can be violent. More precisely, computer network operations such as cyberattacks can be properly understood as the application of force for political goals. This is less obvious with regard to the other four pillars of information operations, which call for the separation of computer network operations from its purported conceptual brethren. Instead, Darley groups information operations at one end of Clausewitz's continuum of nearer to "ideal politics," i.e., activities totally devoid of violence.

[18] Darley, "Clausewitz's Theory of War and Information Operations," 75.

[19] Some object to Anselm's ontological argument along these lines: since moral goodness has no intrinsic maximum, there could be no being who is as morally good as is conceivable; we could always conceive of a morally better being. See Broad, "Arguments for the Existence of God"; and Wierenga, "Intrinsic Maxima and Omnibenevolence," 41–50.

[20] There may be a nomological limit to the mass of an object given by the total mass and energy in this universe, but this maximum is extrinsic, not intrinsic, since it depends on the contents of the actual world and is not internal to the concept of mass.

[21] Clausewitz, again concerned with the violence of war, rather than its moral goodness, suggests there is no logical upper bound on violence. See his *On War*, I.I.3; also quoted in Darley, "Clausewitz's Theory of War and Information Operations," 75. The only limit is practical: the adversary's ability to resist with reciprocal action.

successfully. But that does not show that perfectly discriminatory violence is impossible, only that it is rare.

We can show similarly that proportionality has an intrinsic maximum: an act of violence in warfare is maximally proportionate if the harm it inflicts is the minimum necessary for securing its strategic goal.[22] Note that proportionality is a two-place relation between the harm to be inflicted and the strategic importance (or moral good) of the goal to be achieved. Thus, the value of the goal of an action determines the internal maximum for a calculation of proportionality: it sets the limit for what violence is proportional in pursuing that end.

In traditional just war theory, the requirement of proportionality is not typically construed as a requirement that violence be minimal, only that it be proportionate to the strategic goal that is sought. This is less demanding than requiring that the harm be the minimum that is strictly necessary, since the harm that is proportionate to the value of securing some good end may be more than is strictly necessary to secure that end. There is a plausible historical explanation for why proportionality is the common requirement in this respect rather than minimal violence. Previously, there has rarely been a way to make sure that warfare is minimally violent, given the chaotic nature of warfare and the fact that engagements unfurl unpredictably. Inflicting *too little* violence risks failing to secure some goal entirely, which could be strategically (and perhaps morally) disastrous. Accordingly, states have acted understandably when inflicting more violence than might be strictly necessary to accomplish their strategic goals. They have erred on the side of violence. Still, it should be clear that the plausibility of a requirement of proportionality depends on our more basic desire to minimize the overall harm inflicted in warfare. Thus, a maximally proportionate war would be one in which as little violence as possible was inflicted. Again, while no war in history has plausibly satisfied this criterion, this is no evidence against its coherence. Thus, some method of waging war could minimize harm, in the limiting case by perpetrating no harms at all. Would such a conflict—in which no harm was perpetrated—count as war? We will see below that the organized, intentional application of force for political motives could in fact inflict no harm, and could even prevent harms from being inflicted. Yet the organized, intentional application of force for political motives is war.

Proportionality implies a fit between the harm caused by an action and the value of that action's end. At least conceptually, then, the harm caused by an action can be disproportionate in two ways: it can be too great or too little, given the value of the action's goal.[23] We should pause to clarify a potential misunderstanding; that is,

---

[22] This strategic goal must of course be morally justified, otherwise we would not be permitted to inflict any amount of harm in its pursuit.

[23] Recall that proportionality is a two-place relation. There are at least two ways of understanding this relation. It could mean "roughly equivalent," in which case it would be disproportionate to inflict too great *or too little* harm. Or, it could mean "not *excessively* more than." The first interpretation is unreasonable for the reasons I give in this paragraph: we have not failed morally if we inflict *less*

we might worry that some harms in warfare could fail to be proportionate by being *less* than could be justified, given their strategic goal. Suppose I am being attacked and I may either break my attacker's arm or sprain his wrist in self-defense, either of which would prevent my attacker from harming me. If I successfully defend myself by merely spraining his wrist, we might say that the harm I inflicted was disproportionate because I would have been justified in inflicting a greater harm. This criticism of my action rests on a confusion, and calls for two responses. First, it mistakes the difference between liability to harm and desert of harm.[24] If someone deserves to be harmed, there is a noninstrumental reason to harm him. Some will claim it is bad if he is not harmed. If someone is liable to be harmed, there is only an instrumental reason to harm him. It is not bad if he is not harmed. In warfare, we are typically concerned only with liability to harm, not deservingness to be harmed—we do not think of ourselves, at least in our clearer moods, as *punishing* our enemies for their wrongs, or of meting out karmic retribution.[25,26] Thus it is not regrettable if a harm fails to be proportionate by being less than could have been justified.

---

harm than *would* have been justified. Rather, the proportionality constraint ought to be understood as forbidding gross excesses of harm. In this way, we can see that the proportionality constraint is but an application of our general—and hopefully uncontroversial—moral duty to minimize harm. Honoring this duty to the fullest would mean inflicting the minimum necessary amount of harm to secure our strategic goal.

[24] See McMahan, *Killing in War*, 8.

[25] See Walzer, for example, on reprisals, which are the closest *in bello* analogue of punishment (*Just and Unjust Wars*, 209). Walzer is understated when he says that the willingness to repay offenses by punishing innocents is "an unattractive principle," and never mind that it may simply be incoherent to punish a person who is not guilty of the crime in question. Walzer ultimately argues that reprisals are more properly thought of as deterrence against future violations rather than retributive harms in the sense I have in mind above. Likewise, Orend cites McMahan approvingly when he points out that repelling aggression may be one and the same action as punishing aggression, but is properly thought of as defensive rather than punitive action (*Morality of War*, 81).

Also note that this behavior is importantly different from punishing war criminals *after the fact* for crimes against peace or crimes against humanity. The view I argue against here is that the normal prosecution of war is best thought of as meting out punishment, both because it is repugnant and because it would be epistemically impossible to ensure.

[26] Consider an analogy from criminal justice: we may think it is unjust to let a murderer off easy, i.e., to give them less punishment than they deserve. If this is right, and if this is properly analogous to war, then I may be giving this objection short shrift in this paragraph. However, it could be dangerous to apply this thinking to wartime: it is difficult enough in wartime to ensure that our harms befall all and only those who are liable to be harmed. It would multiply our difficulties to attempt to ensure in addition that our harms were roughly equal to the individual liabilities of those who were harmed. All of this is in addition to the general metaphysical difficulties in specifying an individual's desert. This is a practice more likely to be achieved through individual criminal trials lasting weeks rather than dropping the markedly less discriminate and less discerning bombs. It would be practically impossible, even if we could determine individual desert, to ensure that individuals were punished *to the proper extent* in wartime. See on this point Orend (*Morality of War*, 120). I am grateful to Keith Abney for this objection.

Moreover, this objection overlooks the fact that necessity is internal to liability. Even those who are liable to harm are liable only to the minimum amount of harm that is *necessary* to achieve some justifiable goal. If they are harmed more than is strictly necessary, they are harmed unjustly. In practice, the uncertainty surrounding scenarios of self-defense or war can excuse the infliction of harm that is greater than is necessary, such as if I had broken my attacker's arm rather than merely sprained his wrist. Of course, we should never *knowingly* inflict such disproportionate harms.[27]

A war could maximally appropriately distribute any harms that did occur by ensuring that only those liable to attack were actually harmed. A war could be maximally proportionate by inflicting the minimum necessary amount of harm or, ideally, no harm at all. Thus, since war can be morally better or worse, and since these properties have intrinsic maxima, we must accept that some war could be morally best. This is to say we believe in the possibility of ideal war. Below I will suggest that cyberwarfare offers the possibility of an ideal war with regard to *jus in bello*.

Finally: if ideal war is possible, it is obligatory.[28] Ideal war is the morally best war. Of course, we have more moral reason to choose one option over another if it is morally better than the other, and we have the *most* moral reason to pursue the morally *best* option. Note, though, we are not always *morally required* to do what we have *most moral reason* to do. Many people, for example, accept that we have most moral reason to donate our money to charities that work to reduce preventable deaths around the world, but deny that this is what we are *morally required* to do. If war were like giving to charity in this way, then improving warfare would be a good option, but not required; it would be supererogatory. But this seems implausible: certainly we are obligated to humanize the conduct of warfare as much as possible. War has historically been one of the most ruinous phenomena in human history, which gives us reason to think that future wars will be similarly destructive. Inspired by our general motivation to minimize harm, we ought to focus especially on reforming those practices that tend to wreak the most spectacular harms. Accordingly, humanizing warfare has been the desperate plea of millions and the laudable work of generations of moral philosophers. It does not seem like a project that is merely optional.[29]

---

[27] We might also think that, in order for our action to be just, we have a further *epistemic* obligation to be *confident* that our action does not inflict more harm than is necessary.

[28] This is, of course, if it is practically feasible and does not sacrifice or seriously endanger any competing values. If this were the case, an ideal war would only be morally best in a respect, and not be morally best all things considered. I discuss and reject several objections of this kind in section 5.4 below.

[29] Much of this discussion on the relationship between moral reasons and moral requirements is thanks to Duncan Purves.

5.2 Cyberwar and Ideal *Jus in Bello* Warfare

In the previous section I showed that ideal warfare is conceptually possible, such that a war could be morally perfect as far as our *jus in bello* judgments are concerned. In this section, I will argue that the advent of cyberwarfare is already rendering this kind of ideal war attainable, and that cyberwarfare is thus obligatory.

A case will help to advance this discussion:

> A militarily weak state, R, shares a border with its much more powerful neighbor S. In a moment of geopolitical avarice, S invades R. Suppose that repelling S's invasion of R would be a just cause for war, but because S has a powerful conventional military, the disvalue caused by the disproportionate harms R could expect a conventional war with S to cause would easily eclipse the value to be won by repelling the invasion. Now, T is a state with sophisticated cybercapabilities and a strong commitment to international humanitarian law. T appreciates that it could leverage its cybercapabilities against S in order to coerce S to withdraw from R. Accordingly, T offers S an ultimatum to withdraw from R or else suffer serious cyberconsequences.[30] S refuses, and T responds with a cyberattack. S's conventional military capabilities are disabled—its command and control centers become inoperable, its logistics databases become scrambled, its lists of targets in R (and T) are encrypted and rendered inaccessible, and so on. Furthermore, T's cyberattack is so discriminatory that only those people and services that are legitimate targets are impacted by the attack. S sues for peace, and T lifts the veil, restoring S's conventional military capabilities, on the condition that S withdraw from R.

The war that T wages against S is ideal: its harms are perfectly discriminate and totally reversible. We have good reason to think that cybermeans are already making this kind of warfare possible. Take, for instance, the discrimination capabilities of Stuxnet, the computer worm uncovered in June 2010, which had been planted in an apparent attempt to sabotage Iran's nuclear enrichment facilities at Natanz: "Each instance of Stuxnet attempts to propagate for only 21 days, only to three other computers, and only if those other computers are running particular software manufactured by Siemens to interface with specific programmable logic

---

[30] Suppose that T's ultimatum is backed by whatever authority that the criterion of right authority requires. Suppose, for example, that T's ultimatum has been sanctioned by the UN General Assembly and Security Council, as well as a relevant proportion of S's population.

controllers."[31] Ralph Langner, the consultant who discovered Stuxnet, notes further restrictions on Stuxnet's propagation routines in his seminal analysis:

> [P]ropagation can only occur between computers that are attached to the same logical network or that exchange files via USB sticks. The propagation routines never make an attempt to spread to random targets for example by generating random IP addresses. Everything happens within the confined boundaries of a trusted network.[32]

In fact, Stuxnet was written to search out and conform to precise environments, and there is no known case of Stuxnet harming civilian assets.[33] This should give us confidence about the capacity for cyberweapons to inflict discriminate harm.[34] While Stuxnet propagated to tens of thousands of computers that were not its primary target, it sat inert on those computers. It would be implausible to object that having an inert copy of Stuxnet on one's computer constitutes a harm, given that it would do no lasting damage and would occupy a negligible amount of disk space.[35] In fact, taking this one step further, we can imagine a kind of cyberweapon that would propagate between computers and, once done, would destroy itself without leaving a trace in a game of cyberhopscotch. In this way, a cyberweapon could kick away the ladder once it had found its next host, not harming any of its intermediate hosts, and any rights violations would be very limited in their scope.

The war that T wages against S is maximally discriminate. Showing that cyberweapons herald great advances in proportionality is more difficult, but there are two observations that are relevant. The first is that cyberweapons could be programmed to communicate constantly with command and control servers elsewhere. Stuxnet in fact had this capability.[36] Thus, a cyberweapon could be kept on

---

[31] Jenkins, "Is Stuxnet Physical?" 74. For further discussion of the features of Stuxnet, see Chen, "Stuxnet, the Real Start of Cyberwarfare?" 2–3; and Farwell and Rohozinski, "Stuxnet and the Future of Cyber War," 23–40.

[32] Langner, "To Kill a Centrifuge: A Technical Analysis of What Stuxnet's Creators Tried to Achieve," 18.

[33] See Lucas, "Permissible Preventive Cyberwar: Restricting Cyber Conflict to Justified Military Targets," 16.

[34] For a conflicting opinion, see Rowe, "Ethics of Cyberweapons in Warfare," 20–31.

[35] Stuxnet was roughly half a megabyte in size. See Falliere, Murchu, and Chien, "W32. Stuxnet Dossier." While this is enormous for a piece of malware, it is no more than the disk space occupied by a family photo.

[36] See, again, Langner's analysis: "Given the fact that Stuxnet reported IP addresses and hostnames of infected systems back to its command-and-control servers, along with basic configuration data, it appears that the attackers were clearly anticipating (and accepting) a spread to non-combatant systems, and quite eager to monitor it closely" ("To Kill a Centrifuge," 18).

a short leash, and its harms could be redirected, terminated, or reversed if need be.[37] Suppose, in a modification of the above case, that T mistakenly attacks state U instead of S. If this happened, T could send a command that would recall its cyberassets, erasing any trace left on U's systems and restoring them to their ante bellum state. In this case, U would have been inconvenienced, but not seriously harmed.[38] The second observation is that, if cyberweapons are capable of achieving some strategic purpose while inflicting totally reversible harms, then they are maximally proportionate in the way contemplated in section 5.1, namely, that the harms are minimal. It is hard to imagine a case where inflicting totally reversible harms would be disproportionate to securing a justified goal.[39]

If the kind of cyberwarfare I have discussed should obtain, would it even count as a war? How could it, if it inflicts no harms on anyone? This prompts two clarifications: one about the consequences of cyberwarfare and the other about the conception of harm I have in mind. These clarifications together entail that cyberwarfare is necessarily harmful.

I cannot give a full defense to an account of harm here. The most popular accounts of harm on offer hold that A harms B if A makes B worse off than B otherwise would have been,[40] if A brings it about that B's life as a whole contains less overall well-being than it otherwise would have,[41] or if A puts B in some state that is objectively bad and that B otherwise would not have been in.[42] I will assume that one of these accounts of harm, or something similar, is correct. If I cause you stress, anxiety, or inconvenience that you would not otherwise have endured, then

---

[37] See the capabilities of Flame, discussed below, which executed a "kill" command to remove itself from any infected computers.

[38] I return to this modified example below in the context of the *ad bellum* probability of success criterion.

[39] One concern is that this appraisal rests on an overly optimistic view of policymakers. To accomplish all I suggest in this paragraph, it may seem that a state makes themselves overly vulnerable to counterattack by making it easier to attribute an attack to them. (Imagine, for example, that R publicly complains to T, and the attack suddenly shifts immediately thereafter to target S. This would leave little mystery as to who was responsible.) This charge is difficult to answer, except to point out that it is an instance of the general problems associated with balancing misdirection and ruse against probability of success, i.e., making sure your enemy knows who you are to ensure your strategic goals are met.

[40] This is the counterfactual comparative account of harm. See Feinberg, *Harm to Others*; Klocksiem, "Defense of the Counterfactual Comparative Account of Harm," 285–300; Thomson, "More on the Metaphysics of Harm," 436–58; Boonin, "How to Solve the Non-Identity Problem," 127–57; Bradley, *Well-Being and Death*; and Purves, "Accounting for the Harm of Death."

[41] This is the global account of harm and is usually taken to be a subspecies of the counterfactual comparative account. They differ most notably with regard to their explanation of the badness of death. See Feldman, "Some Puzzles about the Harm of Death," 205–27.

[42] This is the objective state account of harm. See Harman, "Can We Harm and Benefit in Creating?" 89–113; and Shiffrin, "Wrongful Life, Procreative Responsibility, and the Significance of Harm," 117–48.

I harm you on all of these accounts.[43] Cyberwarfare, in order to be successful, must cause stress, anxiety, or inconvenience.[44,45] Therefore, in order to be successful, cyberwarfare must inflict some harm. This suggests that truly ideal warfare may be unattainable, but cyberwarfare may make possible the least harmful kind of war that *is* attainable. And cyberwarfare clearly does the best of any technology on offer of approximating ideal warfare for the foreseeable future.

Cyberwarfare makes possible the realization of ideal warfare, a war waged in a way that is maximally discriminate and proportionate. Because of this, cyberwarfare is obligatory on any occasion when a similar act of conventional warfare is justified. Consider the following principle of ideal warfare:

(I1) If some means of waging warfare A promises to better satisfy the requirements of discrimination and proportionality than some other means of waging warfare B, it is obligatory to deploy A rather than B, other things being equal.

This principle is eminently plausible and should be acceptable to traditionalists and revisionists alike.[46] If we deny with revisionists that warfare is a special moral

---

[43] Of course, this is consistent with our use of the term *harm*: an account of harm had better agree that I harm you if I cause you stress, anxiety, or inconvenience that you would not otherwise have endured.

[44] It should be clear that cyberwarfare, as an act of war, can only be successful if it puts an enemy in a state that they would better off *not* to be in. This may be through spectacular means that are on par with kinetic warfare, such as causing physical damage to important equipment. But even more mundane or "low boil" forms of cyberwarfare cause harms by temporarily disabling equipment or destroying information. These harms may be strictly psychological, and so they may differ in kind from the physical violence that is the essence of conventional warfare. But they remain harms that, in the broadest sense, make people worse off than they otherwise would have been.

[45] Suppose that I cause you stress, anxiety, or inconvenience but that, in the relevant alternative world, I kill you instead. On some accounts, I make you much better off, even though I cause you stress, anxiety, or inconvenience that you otherwise would not have endured (because you otherwise would have been dead). These are exceptional cases where the relevant alternative world is one in which you are *even* worse off if I had *not* acted the way I did. We can bracket these cases since, if I wage cyberwarfare on you, it seems that the nearest relevant alternative world is one in which I do not harm you at all, i.e., there is a state of peace. For more on this, see Norcross, "Harming in Context," 149–73; and Feit, "Plural Harm."

[46] Traditional just war theorists typically hold that the ethics of war is divisible into at least two categories: justice in the resort to war and justice in the prosecution of wars. Michael Walzer is the preeminent traditionalist, whose *Just and Unjust Wars* has been foundational to the study of the morality of war since its publication in 1977. (See also Brian Orend's notable contribution to traditionalism in his *The Morality of War*.) Traditionalists further hold that the ethics of killing in war is importantly different from the ethics of killing outside of the context of war, while revisionists deny this. Walzer's "naked soldier" case makes this disagreement clear: the traditional ethics of war make it sometimes permissible to kill a defenseless soldier, even one who fights for a just cause, because the combatants on both sides of a conflict have allegedly surrendered their rights by taking up arms against the other

context, we could still assent to a corollary principle concerning violence in general, since there exists a strong presumption in favor of minimizing and appropriately directing the more quotidian harms that we inflict. Of course, more proportionate and discriminate harm is morally better, other things being equal.

Consider this stronger version of the principle that is similarly plausible:

(I2) If some means of waging warfare A *likely* better satisfies the requirements of discrimination and proportionality than some other means of waging warfare B, it is obligatory to deploy A rather than B, other things being equal.

It is clear that if I can minimize and more discriminately direct my harm, I have a strong moral reason to do it. If I am not certain, but instead have *good reason to believe* that I can minimize and more appropriately direct my harm, I still have a strong moral reason to do it.[47] Notice (I2) contains a conjunct: it must be likely that A better satisfies *both* discrimination *and* proportionality than B. Consider (I3), which involves a disjunct instead, and so is weaker:

(I3) If some means of waging warfare A likely better satisfies the requirements of discrimination *or* proportionality than some other means of waging warfare B, it is obligatory to deploy A rather than B, other things being equal.

Notice (I3) requires that we meet a lower epistemic burden than (I2). Consider a cyberweapon that propagated uncontrollably but caused very little harm to each individual affected. (For example, this cyberweapon could appropriate a user's computers for use in a botnet, in the process gaining access to her address book and occasionally slowing her Internet connection noticeably.)[48] The harms inflicted by this weapon could be more proportional but less discriminate than those inflicted by some other means of waging war. According to (I3), we are required to deploy this cyberweapon rather than some other means of waging war. But, given the lack

---

and thus share a *moral equality* (*Just and Unjust Wars*, 143). Revisionists deny that a state of hostilities between two parties renders everyone in each army liable to be killed. McMahan's *Killing in War*, a lucid statement of revisionism, and Rodin's earlier *War and Self-Defense* are the founding documents of the movement. For a recent reply to some of these criticisms, see Benbaji, "Defense of the Traditional War Convention," 464–95.

[47] Perhaps I must be 51 percent certain that A better satisfies these requirements than B, or I must be all-things-considered justified in believing it, or I must have undefeated reasons for believing it. Whichever we choose from among these various options is not significant for the discussion here.

[48] This access to a user's address book could be construed as a minor rights violation, i.e., an invasion of privacy. It is more plausible to construe this as minor if we suppose this cyberweapon propagates automatically and no human ever has access to any victim's address book.

of discrimination, we would *not* be required to deploy this cyberweapon according to (I2). Whether we think (I3) or (I2) is more plausible will likely depend on whether we think discrimination and proportionality are competing values that can be traded off against one another, such that a lack of one can be *made up for* or *compensated by* an increase in the other. We may believe, instead, that in order for there to be a moral presumption in favor of some means of waging warfare, we must expect that weapon to inflict harms that are *both* more discriminate and more proportional than the relevant alternative.

Are discrimination and proportionality commensurable values that can be traded off against one another? Supposing that discrimination and proportionality cannot be reduced to one another—that is, they are not simply facets of the same value. Then there are three possibilities: either (i) they are incommensurable and cannot be compared at all; or, if they can be compared, then (ii) one must always be more valuable than the other, that is, one is *lexically prior*[49] to the other; or (iii) there exists a relative weighting between the two.[50]

It is implausible to hold that one of these values is *lexically prior* to the other, such that one is always more valuable than the other, and *any* increase in the one could offset *any* decrease in the other. (It is implausible to hold that one weapon is better than another weapon because it inflicts harms that are *very slightly more* discriminate while being *wildly less* proportionate. The inverse is similarly implausible.) When faced with a choice between two weapons, we must know how successful each weapon will be in inflicting harms that are discriminate and proportionate. If one of these values were lexically prior to the other, then we would need to compare *only* the discriminatory capabilities *or* the likelihood of proportionate harm, and nothing more, to make this decision. It should be enough to show this is true to reflect that we do, in fact, care about *both* values when we make a choice between weapons.

It is even stronger to hold that the values of discrimination and proportionality are incommensurable. This is not just to say that one always trumps the other—in fact, it is to deny that it is *possible* for one to trump the other, since it is to deny that

---

[49] Rawls introduces the term "lexical priority" to characterize the superiority of political liberty over economic well-being in his *Theory of Justice*, 37. According to Rawls, no reduction in political liberty is justified by any increase in economic wellbeing.

[50] Consider possibility (iv): our weapons must meet some minimum threshold of both discrimination and proportionality and failing to meet either would render that weapon immoral. Another way of putting this would be to say that, within some range, discrimination and proportionality are competing values that can be traded off against one another, until either one falls too low. (For example, we might think that no increase in discrimination capabilities could justify the use of a weapon that was likely to inflict hugely disproportionate harms.) This possibility seems to specify baseline requirements for the use of any weapon, rather than revealing the nature of the competing values involved. Therefore, I will simply note this possibility and assume that any weapons we are considering meet at least these minimal standards for discrimination and proportionality. Possibilities (i) through (iii) explore the ways alternative weapons might be further evaluated.

any coherent comparison between the two values could be made. But it is not incoherent to say that one weapon is better than another *all things considered* because it inflicts harms that are more discriminate but less proportional. We can make valid judgments of this kind.[51]

If the values of discrimination and proportionality are not incommensurable, and if neither is lexically prior to the other, then there must be some relative weight to them.[52] One possible avenue for determining the respective weights of the values of discrimination and proportionality is to reduce either to a monist basis, such as *liability-adjusted harm*.[53] Liability-adjusted harm is the sum of the harms inflicted by our actions, adjusted with regard to the liability of each victim. For example, liability-adjusted harm would treat harm to innocent civilians as worse than a similar amount of harm to a combatant who is culpable for an unjust threat. (In fact, the moral badness of harm to such a combatant may be negligible or nil.) Then, the best weapon would be the one that we could expect to minimize total harm, where each harm is adjusted by the liability of the individual victim it befalls. If our concern for both discrimination and proportionality can be shown to be parasitic on some more basic desire, for example, to reduce the liability-adjusted harms of war, then we could derive a principle like:

(I4) If some means of waging warfare A likely *increases the liability-adjusted sum of the values* of discrimination and proportionality than some other means of waging warfare B, it is obligatory to deploy A rather than B, other things being equal.

---

[51] Note that I am not denying that there are any values that are lexically prior to any others. For all I know, there may be some such pair of values. Nor do I mean to deny that there are genuine moral dilemmas involving incommensurable goods. There may be such dilemmas, though I doubt it. All I mean to say is that the moral values of discrimination and proportionality are not incommensurable nor lexically prior to one another.

[52] Some will suspect that talk of the relative weighting of moral values is a way of smuggling consequentialism into the discussion, but this would be wrong. Deontologists are also often concerned with the relative weighting of competing values. W. D. Ross is the most prominent example of this. See his *Right and the Good*. Ross's prima facie duties have relative weights whose weighting is allegedly inscrutable, or at least not codifiable. See also Shelley Kagan's suggestion that the badness of actions that violate rights is determined by the number and stringency of the rights being violated in his *Limits of Morality*.

[53] For a discussion of a similar basis, "morally weighted harm," see Lazar, "Necessity in Self-Defense and War," 6–7. For Lazar, the harms to innocent bystanders weigh much more heavily against an action than similar amounts of harm to those who are liable to be harmed, for example, those responsible for an unjust threat. Thus the moral badness of some level of harm depends on the liability of the individual it befalls. For a much earlier use of a similar strategy elsewhere in moral philosophy, see Feldman, *Pleasure and the Good Life*, 108 ff. Feldman introduces "desert adjusted hedonism" as a measure of how well someone's life is going, where the pleasure one experiences is more or less valuable depending on whether one deserves to experience it.

Determining the proper weighting between these two values would enable us to adjudicate difficult cases like the cyberweapon mentioned above whose harms are less discriminate but much more proportionate than those inflicted by other means of waging war. Determining this weighting would also enable us to make finer-grained moral appraisals of means of waging war than (I2) or (I3) allow.

But any method of quantifying harms or wrongs would prove controversial. Even with a system for quantifying harms in hand, reliably anticipating the harms that would result from any attack would be epistemically impossible. Determining this comparative weighting is what stands in the way of adopting (I4), which is more plausible than (I2) and more informative than (I3). But the correct weighting is far from obvious, and is going to be contingent on the specific cases at hand. While such an empirical discussion of a whole *cyberwar* is beyond the scope of this paper, I consider specific examples of cyberweapons below to show how such weightings might be conducted. Finally, note that these sorts of difficulties threaten any comparison of means of waging war—which is at least essential to any discussion of proportionality, whether *in bello* or *ad bellum*.

Examination of Stuxnet and more recent, more sophisticated cyberweapons gives us good reason to think that cyberweapons satisfy at least (I3) above and perhaps (I2), the more restrictive principle, in relation to conventional warfare. If this is true, then a strong moral presumption exists in favor of deploying cyberwarfare in place of conventional warfare whenever it is practicable.

5.3 Other Things Being Equal

Recall that the principles (I1) through (I4) contain *ceteris paribus* clauses: they only generate moral presumptions that may wither in the face of countervailing concerns. I will now consider several ways in which cyberweapons may compromise other competing goods, concluding that these concerns are unjustified.

There are two suggestions we can dispatch quickly. The first is that cyberweapons could increase the risk borne by just soldiers. If this were true, it would be problematic. But it seems clearly false.[54] If cyberweapons increased the risk to noncombatants or violated their immunity, then they would be seriously problematic. But we have already seen above that, on the contrary, cyberweapons make possible a war of perfectly discriminate and totally reversible harms.[55] Here, I will consider some ways in which cyberweapons could be indiscriminate or destructive.

---

[54] Thus, under a Principle of Unnecessary Risk (PUR) there exists a strong moral presumption in favor of cyberweapons. This constitutes an independent route to the moral requirement to deploy cyberweapons in place of conventional weapons. See Strawser, "Moral Predators," 342–68; and Denning and Strawser, "Moral Cyber Weapons," 85–103.

[55] It is worth pointing out that cyberweapons are only as good as their creators. Cyberweapons also threaten new kinds of havoc. Below I discuss the objection that cyberweapons could usher in a new era of terribly destructive warfare.

First, cyberweapons are only as discriminate as their programming guarantees. It is conceivable that a weapon like Stuxnet could be *mistaken* about where it finds itself, and end up wreaking havoc on a programmable logic controller that controls a water treatment plant, power plant, or other essential civilian infrastructure. It is extremely unlikely that Stuxnet could have made such a mistake, however, once we consider the specificity of its directives: there are likely very few Siemens programmable logic controllers, connected to a particular model of uranium centrifuge, with Iranian IP addresses, which are in fact operating a power plant, water treatment facility, or other civilian infrastructure. The ultimate question is how likely it is that a state could furnish and deploy a cyberweapon such that its code picked out a target with a definite referent, where it is confident that referent is liable to attack. It is conceivable that a state could conduct an exhaustive investigation of the civilian infrastructure in the enemy state to ensure this, though this may be a greater demand than any existing intelligence agency could meet.[56]

A cyberweapon may inflict indiscriminate harms by damaging or interfering with civilian infrastructure. This is an especial concern with regard to *dual use* infrastructure. It may seem impossible to disable, say, only that portion of a power grid that services a military base without also disabling power to essential civilian infrastructure. However, note that for any particular strategic goal, there are plausibly multiple combinations of attacks that are coextensive in their effects. Consider two possible routes of attack: (1) A cyberattack disables some portion of a power grid, which disables military machinery at the target but also risks enveloping essential civilian infrastructure in darkness. (2) A collection of tailored cyberattacks *individually* disable all of the machinery at a military base. Each of these methods would have the same practical effects on those targets liable to harm, but these attacks vary with regard to their success in discriminating. The effectiveness of these attacks depends crucially on two factors: the intelligence capabilities and technical capabilities of the creators of the cyberweapon. Perhaps the most useful intelligence available to cyberattackers would be the serial numbers of particular computers. This knowledge, in concert with technical expertise, would enable cyberattackers to reliably target individual computers. Of course, this requires both outstanding intelligence and technical skill, and this combination may be rare even among the world's foremost intelligence capabilities and cyberpowers.[57]

---

[56] Even if this is within the realm of possibility for the states with the most sophisticated cybercapabilities, it is a separate question whether those states would be *willing* to expend the necessary resources to tailor their cyberweapons so carefully. Note an additional cost that comes with such custom tailoring: the most narrowly tailored cyberweapons might only be useful against a single target. For this reason, the most expensive bespoke cyberweapons could also be the least useful in terms of reuse. Thomas Rid discusses this in conjunction with what he terms *the problem of generics*. See his *Cyber War Will Not Take Place*, 50–52.

[57] Thanks to Herb Lin for raising these points in correspondence. Note, further, that targeting a specific computer is not the same thing as targeting its owner. In this way, talk of "military computers"

In the absence of such perfect intelligence and programming capabilities, cyberoperations should be aimed as far *causally downstream* as possible, that is, they should be focused on machines as causally close to their desired effects as can be achieved. Otherwise, cyberoperations focused too far causally upstream risk propagating downstream in ways that are unpredictable or indiscriminate. By targeting assets further downstream, a state could better control the effects of its cyberoperations. I suspect that for most justified cyberoperations, there will be some combination of cyberoperations that may perfectly discriminate between military and civilian assets, even if those attacks must take place causally downstream from where they would otherwise be easiest to deploy. Here, as in many other instances during warfare, combatants may be required to shoulder increased burdens (e.g., in terms of risk, resource expenditure, or inconvenience) in order to shield civilian assets from unjustified harm.[58]

While cyberwarfare offers the possibility of greatly morally improved warfare, it also raises the specter of more harmful, less discriminate warfare. Of particular concern is the possibility that cyberweapons, however well designed, could be reverse engineered and deployed for nefarious purposes. In this way, cyberwarfare could also represent the moral nadir of war.[59]

This worry rests on a mistake: modern conventional warfare clearly *makes possible* greater evils than the warfare of the nineteenth century, but we ought not count it against a means of waging war that it enhances the ability of evildoers to do evil, since it also enhances the ability of good-doers to do good. We could say the same about the Internet, which through its profound anonymity facilitates all manner of unseemly behavior.

How should we morally judge the advent of new weapons technologies? Should we judge them by their most encouraging potential consequences, or their most

---

and "civilian computers" is admittedly coarse-grained, since one and the same computer may be shared by both military personnel and civilians.

[58] Note, again, that in order to fight justly, a state might have to shoulder significant burdens to ensure just execution of their cyberattacks. These issues are covered, amongst other places, in the *Tallinn Manual*: Schmitt, *Tallinn Manual on the International Law Applicable to Cyber Warfare*. For instance, see Rule 39: "An object used for both civilian and military purposes—including computers, computer networks, and cyber infrastructure—is a military objective"; Rule 50: "A cyber attack that treats as a single target a number of clearly discrete cyber military objectives in cyber infrastructure primarily used for civilian purposes is prohibited if to do so would harm protected persons or objects"; and Rule 59: "The Parties to an armed conflict shall, to the maximum extent feasible, take necessary precautions to protect the civilian population, individual civilians, and civilian objects under their control against the dangers resulting from cyber attacks."

[59] Note that a cyberwar could conceivably be worse than *any* conventional war. This is because a cyberattack could conceivably be used to launch all of the nonconventional weapons in a country's arsenal—their nuclear, biological, or chemical weapons—and then continue to wreak havoc in other ways. Thus, a cyberattack could be *at least as bad* as any conventional attack, and even any nonconventional attack.

upsetting? Rather, we tend to appraise weapons technologies by the effects they *actually* have. Thus, while the advent of precisely targeted means of killing, such as GPS-guided bombs, made possible new and terrible means of waging war, they also made possible new moral refinements in warfare. At least those states acting with a good-faith commitment to the principles of *jus in bello* are now able to fight in ways that are remarkably morally improved. Consider, for example, the enhanced discrimination capabilities of unmanned aerial vehicles, even compared to some means of waging war with "boots on the ground."[60] Intense moral pressure from the international community and new supranational structures of political authority have provided additional legal pressure to ensure restraint in warfare. The covenants underlying these structures urgently need to be clarified, amended, and extended to make clear that cyberwarfare falls under the purview of international legal regimes. Cyberwarfare presents new challenges to policymakers in this regard, perhaps most notably through the attribution problem.[61] While these problems are daunting, they do not seem in principle insoluble.

It is conceivable that a cyberweapon, even one that is meticulously crafted to be discriminate and proportional, could be captured, quarantined, and reengineered to attack targets not liable to harm, such as civilian targets. As with the other moral problems introduced by cyberwarfare, this is actually a new facet of an old problem: any weapons technology "in the wrong hands" could be used against

---

[60] The claim that unmanned aerial vehicles (UAVs, commonly called "drones") are better at discriminating between combatants and noncombatants than other means of waging war plays a crucial role in Strawser's argument in his "Moral Predators." There, he cites data compiled in Avery Plaw's "Sudden Justice": "This data shows that UAV strikes were far better at noncombatant discrimination than all other methods used for engaging Taliban fighters in the region. For example, the UAV strikes resulted in a ratio of over 17 to 1 of intended militant targets to civilian deaths compared with a 4 to 1 ratio for Pakistan Special Weapons and Tactics Teams team offensives or a nearly 3 to 1 for Pakistan Army operations in the same region during the same time period. Or, compare the 17 to 1 ratio for the UAV employment to the shocking 0.125 to 1 militant to civilian casualty ratio estimate for all armed conflict worldwide for the year 2000." See Plaw, "Paper presented at 7th Annual Global Conference on War and Peace," Prague, (2010).

A metastudy conducted by the Council on Foreign Relations compiling independent reports of UAV strikes between 2002 and 2012 estimated that civilians accounted for about 12 percent of those killed by UAV strikes, i.e., in a ratio of roughly seven combatants for every one civilian killed. See Zenko, "Reforming U.S. Drone Strike Policies," 13. Notice, finally, that we should be concerned about conflicting definitions of "noncombatant" or "civilian" being used in these studies, or our ability to determine the identity and liability of any particular person killed by a strike. Still, data from various sources continue to provide support for the view that the harms inflicted by unmanned aerial vehicles are typically more discriminate than those inflicted by other means.

[61] There is reason to think that the attribution problem may be mitigated in coming years by the adoption of technologies such as IPv6 combined with reliable packet forwarding. For a recent discussion of attribution, including the strategic interests an attacker might have in making their identity known and the suggestion that states mandate the use of IPv6 to enable more reliable methods of attribution, see Rowe et al., "Challenges in Monitoring Cyberarms Compliance," 1–14.

civilian targets. Cyberwarfare, of course, could make this easier for nonstate actors to accomplish.[62]

How could this reengineering problem be solved? The first response that suggests itself is simply to patch those vulnerabilities that allow a cyberweapon to compromise its targets. This would harden a state's assets against any future attacks that make use of a reverse engineered cyberweapon. But this is impracticable and could be self-defeating, since it is unclear how a state could propagate a patch to all and only its allied civilian resources without letting on crucial strategic intelligence about the particular vulnerability at issue. And even with diligent state-sponsored assistance, civilians as a whole remain notoriously lackadaisical about computer security.[63]

A second response would be to obfuscate a cyberweapon's code beyond any hope of reverse engineering. While this may deter amateurs from reverse engineering a cyberweapon, which would provide some benefit, it is unlikely to be a successful defense against the highest echelons of computer scientists and engineers comfortable coding in assembly languages, especially those supported and funded by powerful states.

A third and final hope is that a cyberweapon may be crafted to execute a "kill" command upon discovery or tampering, destroying itself like an ouroboros.[64] This would prevent reengineering, though it is unclear how reliably this could be implemented. In the end, the threat of reverse engineering of cyberweapons for nefarious purposes remains a real one. The summed effects of the three suggestions given here will go some way toward neutralizing this threat, but for now it is unclear if the threat can be totally avoided.

If (I2) is true, if cyberwarfare really does promise an increase in the discrimination and proportionality of the harms of war, and there exist no sufficient countervailing moral considerations, then cyberweapons are obligatory any time a relevantly similar act of conventional war is justified. In fact (I2) does seem true, and there is no clear morally sufficient countervailing consideration.

---

[62] The level of technical expertise necessary to wage full-fledged cyberwarfare will likely remain high for the foreseeable future. But the spread of computer technology and knowhow is democratizing the ability to carry out cyberespionage and other "low boil" means of cyberattack to less powerful groups.

[63] Note, for example, that the four most common passwords for 2013 were "123456," "password," "12345678," and "qwerty." SplashData, "'Password' Unseated by '123456' on SplashData's Annual 'Worst Passwords' list." More worrisome is that 44 percent of consumers do not even require a password to log into their personal computers. For more harrowing statistics like these, see NetSecurity.org's "Consumers' Bad Data Security Habits Should Worry Employers," February 26, 2014.

[64] Stuxnet's progeny, the Flame virus, executed just such a "kill" command when its discovery was made public, but this command was initiated from a third party, not by the weapon itself. See Goyal et al., "Obfuscation of Stuxnet and Flame Malware," *Latest Trends in Applied Informatics and Computing* (2012); and Symantec, "Flamer: Urgent Suicide" (2012).

5.4 Cyberwarfare's *in Bello* Success and the *ad Bellum* Requirements

The preceding sections argued that the advent of cyberwarfare makes possible an ideal war by profoundly increasing a state's ability to abide by the *in bello* requirements of discrimination and proportionality. In closing, I will consider how the availability of cybercapabilities dramatically alters the *ad bellum* moral calculus.

The *ad bellum* requirements of just war theory derive much of their moral force from the evil of warfare. Because war has historically been a grave evil, we ought to satisfy stringent criteria of just cause, right intention, probability of success, and so on before resorting to it. Conversely, the force of many of the traditional *ad bellum* requirements melts into air when we consider the kind of ideal warfare outlined above. This is because we would be considerably less worried about the harms that would accompany an ideal war, especially if we take seriously the claim that the harms of cyberwarfare could be maximally discriminate and perfectly reversible. The less likely a war is to inflict indiscriminate harms, the lower the moral burden of proof to demonstrate probability of success.[65]

What would it mean for a war to have a maximally just cause? In fact, it seems that the justice of a cause could have no intrinsic maximum. Consider a resort to war that is a response to unjust aggression, the least controversial—some would say the only legitimate—just cause for war.[66] Whatever unjust aggression we imagine, it could always be *more unjust*: the threat more ghastly, the impending massacre or enslavement more brutal. One potential maximum in the injustice of an aggression is genocide, the total denial of the rights of the citizenry, a rejection of their basic humanity, an aim at the utter extirpation of its culture and tradition. Perhaps an aggression that foreshadows those injustices could be maximally unjust, thus a response to such an aggression could constitute a maximally just cause for war. But this seems false: equivalent rights violations may still differ with regard to the suffering or anguish they inflict, and so even maximal rights violations involve some additional morally important consideration, such as suffering, that itself does not obviously have an intrinsic maximum. The oppression of a people *prior* to their genocide seems to have no upper limit, for example.

While this is interesting, it is ultimately immaterial to the discussion, since the justice of a cause is independent of the means of the response. That is, the availability

---

[65] This claim may seem controversial, but its contrapositive is undeniable: the greater the risk of disproportionate or indiscriminate harms, the more stringent the requirement to demonstrate probability of success. To deny this would be to hold that the requirement to demonstrate probability of success is identical whether the possible harms are minuscule or massive, and that is plainly implausible.

[66] See, for example, Walzer's characterization of the war convention: "Nothing but aggression can justify war" (*Just and Unjust Wars*, 62).

of cybercapabilities as a means of responding to aggression does nothing to ensure that the cause of any particular war is just.[67] Cyberwarfare's effect on the criteria of right intention, proper authority, and last resort are similarly minimal (or nonexistent). Cyberwarfare can be wielded with any intention, by any authority—proper or improper—and can be used as other than a last resort.[68]

The advent of cyberwarfare does have an important impact on the *ad bellum* consideration of the proportionality of warfare, however. Resort to war is justified only in cases where the foreseen harms are proportionate to the value of the war's aim. If these foreseen harms are reduced drastically, then so is the moral burden we are required to meet before a resort to war is justified. Consider the case of the wholly cyberwar that achieves maximum discrimination and maximum proportionality—a war waged with very little harm, as contemplated above in the case of S and T. In this case it seems that the moral threshold for resort to war falls significantly. This is to say that if we are contemplating launching a cyberoffensive that (we will suppose) satisfies all the other *ad bellum* criteria and targets only those who are legitimate targets, and if we can be confident this offensive will wreak negligible harm, then the minimum value of the aim we are justified in pursuing falls as well.

Under the definition of war offered earlier, T's actions constitute a war, since its cyberoperations are an organized, intentional, large-scale application of force for political aims. T's actions were justified, moreover, since we are supposing that S's ongoing rights violations constitute a just cause for war. What is remarkable is that the advent of sophisticated cybercapabilities like T's makes proportional warfare easier to achieve, and thus raises the likelihood that T's resort to war is just.

---

[67] Note that some means of war can allegedly do just the opposite: means *mala in se* can guarantee that a war is *unjust*. At any rate, cyberwarfare does not seem like a means *mala in se*.

[68] On the last resort criterion, consider that cyberweapons may lower both the psychological and/or political thresholds to waging war. There is significant worry that, as a result, states will find it easier to resort to cyberwarfare at times that are unjustified. This is the *threshold problem* for cyberwarfare, though notice that it generally affects any novel means of waging war that is less costly than existing means warfare, e.g., killing by remotely piloted aircraft. For discussion of the threshold problem in the context of cyber, see Barrett, "New Warriors/New Weapons, Executive Summary of the 2010 McCain Conference of Service Academies and War Colleges," 416–23; Lucas, "Ethics and Cyber Conflict," 20–31; Lucas, "Postmodern War," 289–98; and Denning and Strawser, "Moral Cyber Weapons."

The threshold problem arises in many other debates over military technology. For the threshold problem's appearance in the debate over the morality of unmanned aerial vehicles, see Sparrow, "Killer Robots," 62–77; Sparrow, "Predators or Plowshares? Arms Control of Robotic Weapons," 25–29; and Strawser, "Moral Predators." For the threshold problem's relevance with regard to military robotics generally, see Lucas, "Industrial Challenges of Military Robotics," 274–95; and Lin, Bekey, and Abney, *Autonomous Military Robotics: Risk, Ethics, and Design*.

Notice that the general form of the threshold problem arises in still other areas of applied ethics, for example, some policymakers oppose needle exchanges: this policy's goal is to lower the risk of harm to those who are presently using drugs, though some fear that it lowers the overall cost of being addicted to drugs, and thus makes it less likely that addicts will quit.

T's actions are an example of an ideal war: a war where harms discriminated perfectly between those liable and not liable to harm, and where the harm inflicted was maximally matched to the justified goal. Given the care taken in discrimination, we could even argue that no relevant rights were violated, as the legitimate targets had already forfeited the relevant rights through their unjust behavior.[69] Recall that T's aggression is an ideal case, since the harms it inflicted were perfectly discriminate and totally reversible. Given these considerations, while T would not have been justified in pursuing its just aims by means of conventional war, they were justified in pursuing those same aims by cyberwarfare. Hence the potential of cyberwarfare to inflict harms that are perfectly discriminate and totally reversible lowers the moral justification required for resorting to war.[70] Finally, the widespread availability and use of cybermeans could increase the number of wars that are just by increasing their proportionality and discrimination.

Notice that there remains a real burden on T to demonstrate that it is acting in the interest of a just cause. Irrespective of whether T's actions actually harm S, T surely interferes with S's sovereign affairs, and this requires justification. The arguments in this chapter are not meant to provide a blank check for cyberoperations or to suggest that, because cyberoperations could conceivably inflict no harm at all, they are justified in any case. Because launching cyberoperations against a state infringes on its sovereignty, cyberoperations stand in need of justification. T's actions may be harmless, but they are undoubtedly a rights violation, and even harmless rights violations require justification. It is only part of a state's justification for going to war to demonstrate that the harms foreseeably caused by the war will be proportionate to the value of the end. I am suggesting here that this requirement becomes much easier to meet in the case of ideal warfare.

The same could be said about the criterion of reasonable probability of success: if we anticipate inflicting very little harm on a state, and if those harms are reversible as in the case of the harms T inflicted on S above, then the moral burden to demonstrate reasonable probability of success falls as well. This is because the outcome would be less disastrous if we are mistaken (one way in which the probability of success could be lowered), thus our requirement to be cautious is less urgent. Suppose T's intelligence had been grossly mistaken and their cyberoperations were instead directed against state U rather than S, as first contemplated above. This would have been a serious error, perhaps a grave breach of international norms. But if all of the harm was perfectly discriminate and totally reversible, in the final calculus it seems that T has not seriously wronged U. Of

---

[69] See, for example, McMahan, *Killing in War*, 159; and Leverick, *Killing in Self-Defence*, 54–68.

[70] For a technical discussion of the reversibility of the effects of cyberweapons, see especially Rowe, "Towards Reversible Cyberattacks," 20–31. See also Rowe et al., "Challenges in Monitoring Cyberarms Compliance," §4.3.

course, T will have other powerful reasons to be profoundly careful before launching such an attack: at least a political reason and a strategic reason. The political reason is that T's relations with U would be significantly worsened if this modified version of the scenario obtained. The strategic reason is that T would have shown its cards to S, at least to some degree, and so would sacrifice some measure of its tactical advantage. I suspect these reasons are serious enough that states will not simply throw caution to the wind when planning their cyberoperations: of course states will still have powerful *prudential* reasons to maximize the probability of success of their cyberoperations. In a case of cyberoperations that inflict perfectly discriminatory and totally reversible harm, these prudential reasons seem to provide as much a foundation for the reasonable probability of success criterion as do the moral reasons.

I suspect this is correct, and that if the kind of ideal war discussed above becomes normalized, the demands of the *ad bellum* requirements of proportionality and reasonable probability of success will be reduced. What would a maximally justified resort to war look like? Its cause would be maximally just and supported by unimpeachable public evidence. For example, it must be a response to an utterly unjustified territorial encroachment, an upraised sword with certain slavery and widespread dehumanization following from it. An ideal war would be maximally likely (i.e., certain) to be won, and the harms it threatened would be maximally proportionate to the value of the end being sought (i.e., minimal). Of course, it is implausible any resort to war has boasted such a clear and compelling case, but this is a separate question from what such an ideally justified war would look like. Cyberwarfare makes it possible to satisfy the *ad bellum* criteria of probability of success and proportionality with a higher degree of confidence than conventional warfare.

All of this is dependent, of course, on the willingness and ability of states with powerful cybercapabilities to deploy cyberweapons in ways that are proportionate and discriminatory. Answering these questions requires the application of game theory and is outside the scope of this narrow argument.[71] This chapter has set out instead to elucidate the concept of ideal warfare and to convey the remarkable potential that cyberwarfare holds to promote moral progress in the prosecution of wars. As humanity continues its inexorable march toward greater computerization and connectedness, cyberwarfare shows promise for humanizing and pacifying the landscape of war, of ushering in an age of ideal warfare. Of course, what is said here only underscores the urgency of developing robust international covenants for governing cyberattacks and the technological means of attributing them reliably.

---

[71] See Dipert, "Ethics of Cyberwarfare," 384–410. This paper is a promising attempt to bring the methods of rational choice theory to bear on the question of cyberwar.

Bibliography

Allhoff, Fritz, and Ryan Jenkins. "The Facebook War: Would Taking Down the Social Network Justify a Real-world Attack?" *Slate*. June 11, 2014: http://www.slate.com/articles/technology/future_tense/2014/06/cyberwar_ethics_when_is_a_real_world_response_to_a_cyberattack_justifiable.html.
Anselm of Cantebury. "Proslogion." In *St. Anselm: Basic Writings*. Trans. Thomas Williams. Indianapolis: Hackett, 2007.
Barrett, Edward. "New Warriors/New Weapons, Executive Summary of the 2010 McCain Conference of Service Academies and War Colleges." In *Special Issue on Ethics and Emerging Military Technologies*, ed. George R. Lucas Jr. *Journal of Military Ethics* 9 (2010): 416–23.
Benbaji, Yitzhak. "A Defense of the Traditional War Convention." *Ethics* 118, no. 3 (2008): 464–95.
Boonin, David. "How to Solve the Non-Identity Problem." *Public Affairs Quarterly* 22 (2008): 127–57.
Bradley, Ben. *Well-Being and Death*. Oxford: Oxford University Press, 2009.
Broad, Charlie Dunbar. "Arguments for the Existence of God." In *Religion, Philosophy and Psychical Research: Selected Essays*. New York: Routledge, 2000.
Brunstetter, Daniel, and Megan Braun. "From *Jus ad Bellum* to *Jus ad Vim*: Recalibrating Our Understanding of the Moral Use of Force." *Ethics & International Affairs* 27 (2013): 87–106.
Chen, Thomas M. "Stuxnet, the Real Start of Cyberwarfare?" *IEEE Network* 24 (2010): 2–3.
Clausewitz, Carl von. *On War*. Ed. and trans. Michael Howard and Peter Paret. Princeton: Princeton University Press, 1976.
Darley, William M. "Clausewitz's Theory of War and Information Operations." *Joint Forces Quarterly* 40 (2006): 73–79.
Denning, Dorothy E., and Bradley J. Strawser. "Moral Cyber Weapons." In *The Ethics of Information Warfare*. New York: Springer International, 2014, 85–103.
Dipert, Randall R. "The Ethics of Cyberwarfare." *Journal of Military Ethics* 9 (2010): 384–410.
Falliere, Nicolas, Liam O. Murchu, and Eric Chien. "W32. Stuxnet Dossier." 2010. Symantec publication, available at: https://www.symantec.com/content/en/us/enterprise/media/security_response/whitepapers/w32_stuxnet_dossier.pdf.
Farwell, James P., and Rafal Rohozinski. "Stuxnet and the Future of Cyber War." *Survival: Global Politics and Strategy* 53 (2011): 23–40.
Feaver, Peter D. "Blowback: Information Warfare and the Dynamics of Coercion." *Security Studies* 7, no. 4 (1998): 88–120.
Feinberg, Joel. *Harm to Others*. Oxford: Oxford University Press, 1984.
Feldman, Fred. "Some Puzzles about the Harm of Death." *Philosophical Review* 100 (1991): 205–27.
Feldman, Fred. *Pleasure and the Good Life: Concerning the Nature, Varieties, and Plausibility of Hedonism*. Oxford: Clarendon Press, 2004.
Ford, S. Brandt. "Just Use of Lethal Force-Short-Of-War." In *Routledge Handbook of Ethics and War: Just War Theory in the 21st Century*, ed. Fritz Allhoff, Nicholas G. Evans, and Adam Henschke. New York: Routledge, 2013, 63.
Goyal, Ravish, Suren Sharma, Savitri Bevinakoppa, and Paul Watters. "Obfuscation of Stuxnet and Flame Malware." *Latest Trends in Applied Informatics and Computing* (2012): 150–54.
Grotius, Hugo. *The Rights of War and Peace*. Trans. Archibald Colin Campbell. Westport, CT: Hyperion, 1990.
Harman, Elizabeth. "Can We Harm and Benefit in Creating?" *Philosophical Perspectives* 18 (2004): 89–113.
Hobbes, Thomas. *Leviathan*. Ed. Edwin Curley. Cambridge and Indianapolis: Hackett, 2001.
Jenkins, Ryan. "Is Stuxnet Physical? Does It Matter?" *Journal of Military Ethics* 12 (2013): 68–79.
Kagan, Shelly. *The Limits of Morality*. Oxford: Clarendon Press, 1991.
Klocksiem, Justin. "A Defense of the Counterfactual Comparative Account of Harm." *American Philosophical Quarterly* 49 (2012): 285–300.
Landauer, Rolf. "Information Is Physical." *Physics Today* 44 (1991): 23–29.

Langner, Ralph. "To Kill a Centrifuge: A Technical Analysis of What Stuxnet's Creators Tried to Achieve." Arlington: Langner Group, 2013.
Lazar, Seth. "Necessity in Self-Defense and War." *Philosophy and Public Affairs* 40 (2012): 3–44.
Leverick, Fiona. *Killing in Self-Defence.* Oxford: Oxford University Press, 2006.
Lin, Patrick, George Bekey, and Keith Abney. *Autonomous Military Robotics: Risk, Ethics, and Design.* Washington, DC: US Department of the Navy, Office of Naval Research, 2008.
Lucas, George R. Jr. "Postmodern War." *Journal of Military Ethics* 9 (2010): 289–98.
Lucas, George R. Jr. "Industrial Challenges of Military Robotics." *Journal of Military Ethics* 10 (2011): 274–95.
Lucas, George R. Jr. "Permissible Preventive Cyberwar: Restricting Cyber Conflict to Justified Military Targets." Presentation at Society of Philosophy and Technology Conference, University of North Texas (2011).
Lucas, George R. Jr. "Ethics and Cyber Conflict: A Response to JME 12:1 (2013)." *Journal of Military Ethics* 13 (2014): 20–31.
McMahan, Jeff. *Killing in War.* Oxford: Oxford University Press, 2009.
NetSecurity.org. "Consumers' Bad Data Security Habits Should Worry Employers." February 26, 2014. Accessed July 11, 2014: http://www.net-security.org/secworld.php?id=16437.
Norcross, Alastair. "Harming in Context." *Philosophical Studies* 123 (2005): 149–73.
Orend, Brian. *The Morality of War.* 2nd ed. Peterborough: Broadview Press, 2013.
Plaw, Avery. "Sudden Justice." Paper presented at Seventh Annual Global Conference on War and Peace, Prague (2010).
Purves, Duncan. "Accounting for the Harm of Death." *Pacific Philosophical Quarterly* 95 (2014). doi: 10.1111/papq.12031.
Rawls, John. *A Theory of Justice.* Rev. ed. Cambridge: Harvard University Press, 2009.
Rid, Thomas. *Cyber War Will Not Take Place.* Oxford: Oxford University Press, 2013.
Rodin, David. *War and Self-Defense.* Oxford: Oxford University Press, 2005.
Ross, W. D. *The Right and the Good.* Oxford: Clarendon Press, 1930.
Rowe, Neil C. "The Ethics of Cyberweapons in Warfare." *International Journal of Technoethics* 1 (2010): 20–31.
Rowe, Neil C. "Towards Reversible Cyberattacks." In *Leading Issues in Information Warfare and Security Research*, ed. J. Ryan. Vol. 1, 145–58. Near Reading: Academic Publishing International, 2011.
Rowe, Neil C., Simson L. Garfinkel, Robert Beverly, and Panayotis Yannakogeorgos. "Challenges in Monitoring Cyberarms Compliance." *International Journal of Cyber Warfare and Terrorism* 1 (2011): 1–14.
Schmitt, Michael N., ed. *Tallinn Manual on the International Law Applicable to Cyber Warfare.* New York: Cambridge University Press, 2013.
Shiffrin, Seanna. "Wrongful Life, Procreative Responsibility, and the Significance of Harm." *Legal Theory* 5 (1999): 117–48.
Sparrow, Robert. "Killer Robots." *Journal of Applied Philosophy* 24 (2007): 62–77.
Sparrow, Robert. "Predators or Plowshares? Arms Control of Robotic Weapons." *IEEE Technology and Society Magazine* 28 (2009): 25–29.
SplashData. "'Password' Unseated by '123456' on SplashData's Annual 'Worst Passwords' list." Accessed July 11, 2014: http://splashdata.com/press/worstpasswords2013.htm.
Steinhoff, Uwe. "What Is War—And Can a Lone Individual Wage One?" *International Journal of Applied Philosophy* 23 (2009): 133–50.
Strawser, Bradley Jay. "Moral Predators: The Duty to Employ Uninhabited Aerial Vehicles." *Journal of Military Ethics* 9 (2010): 342–68.
Symantec, "Flamer: Urgent Suicide." (2012): http://www.symantec.com/connect/blogs/flamer-urgent-suicide.
Thomson, Judith. "More on the Metaphysics of Harm." *Philosophy and Phenomenological Research* 82 (2011): 436–58.
United Nations General Assembly. Definition of Aggression, Resolution 3314 (1974).

United States. International Strategy for Cyberspace: Prosperity, Security, and Openness in a Networked World, available at: http://www.whitehouse.gov/sites/default/files/rss_viewer/international_strategy_for_cyberspace.pdf; 2011.

Walzer, Michael. *Just and Unjust Wars*. 4th ed. New York: Basic Books, 2006.

Walzer, Michael. "Regime Change and Just War." *Dissent* 53 (2006): 103–8.

Wierenga, Edward. "Intrinsic Maxima and Omnibenevolence." *International Journal for Philosophy of Religion* 10, no. 1 (1979): 41–50.

Yannakogeorgos, Panayotis. "Internet Governance and National Security." *Strategic Studies* 103 (2012): 102–25.

Zenko, Micah. "Reforming U.S. Drone Strike Policies." 2013. Authored by, *Council on Foreign Relations Press*, Council Special Report No. 65, available at: http://www.cfr.org/wars-and-warfare/reforming-us-drone-strike-policies/p29736.

6

Postcyber

Dealing with the Aftermath of Cyberattacks

BRIAN OREND

There are cyberattacks, and—perhaps—there are cyberwars. We can at least imagine the latter, and the case of Russia's 2007 sharp, sustained, and crippling cybercampaign against Estonia may well count as one.[1] In either event, we can discern three phases to any cyberattack or cyberwar (just as with any other "merely" physical/kinetic attack or war): a beginning, a middle, and an end. In the ethics, law, and strategy of warfare, disproportionate attention is paid to the first two phases: an outbreak of violence; and the means of deploying armed force. I have long argued in favor of the notion that the termination phase of war deserves equal time and is of equal importance.[2] Since it seems we are on the precipice of a new era of war fighting—one that at least integrates cyberattacks into its means and methods, and in which perhaps entire wars might be executed through cybermeans and in cyberspace[3]—I want to try to ensure that, at an early period in this era, we pay sustained and thoughtful attention to the *aftermath* of cyberattacks and cyberwars.

We truly live in a "postcyber" moment, one in which the era of cyberwarfare is upon us, and there is no going back. Thus, we should not ignore the other sense of "postcyber": what political communities ought to do in the wake of a severe cyberstrike or, even more so, in the wake of a substantial cyberwar. This essay is devoted to advancing our understanding of precisely this subject. It shall try to do so in two ways: first by discussing the *jus post bellum* ("justice after war") project in general terms; and then by applying its insights specifically to instances of cyberwarfare, above all seeking out general moral and even legal principles to guide actors who find themselves within such situations.

---

[1] Singer and Friedman, *Cybersecurity and Cyberwar*.
[2] Orend, "*Jus Post Bellum*," 117–37; Orend, *Morality of War*.
[3] Clarke, *Cyber-War*.

The need for, and importance of, such a postcyber project is underlined by considering the two most detailed and up-to-date attempts at constructing authoritative legal principles to regulate cyberwarfare. These would be Heather Harrison Dinniss's *Cyber Warfare and the Laws of War* (2012);[4] and perhaps the single most authoritative source thus far, *The Tallinn Manual on the International Law Applicable to Cyber Warfare* (2013), edited by Michael N. Schmitt (drawing upon the expertise of dozens of authorities, including Dinniss).[5] In the first work, there is no mention whatsoever of any postbellum norms in the context of a cyberstrike or cyberwar: everything is focused upon the outbreak of (cyber)hostilities and how cybertechnology might be employed legitimately as a weapon or tactic within armed conflict. The second work, by contrast, *does contain precisely five rules* (numbers 85 and 87–90) that *might* be relevant to postconflict situations. This is encouraging, and progresses beyond Dinniss's work. But, to provide context: this is only five rules, embedded in a sea of ninety-five rules devoted to cyberwarfare overall (i.e., less than 6 percent are devoted to postconflict; the overwhelming majority—over 94 percent—are still devoted to the twin pillars of outbreak and conduct). And, again, some of these five are only unclearly and indirectly related to poststrike moments. They can seem, as we shall see, to have as much to do with proper fighting as with proper war termination. Still, we shall make what profitable use we can of these (nevertheless helpful and authoritative) rules, and strive indeed to go beyond them, improving on our mutual understanding of *what justice and law ought to require of good-faith actors in the aftermath of cyberwar*.

Now, of course, not every belligerent *is* a good-faith actor and, to the extent to which there are unprincipled belligerents here, they act unjustly in the postconflict moment and ought to face some kind of sanction for it—just as with those who violate the rules regarding the outbreak, and conduct, of armed conflict. The point here remains *the construction of the ideal principles*, and then preferably to see such principles someday codified into a binding legal document.[6]

6.1 *Jus Post Bellum*

How should wars end? It might seem surprising, but only very recently has the issue of justice after war—or *jus post bellum*—come into the prominence it deserves. Historically, it was assumed that, as the old saying goes, "to the victor go the spoils of war." As a result of this widespread belief, there is actually no clear international law regulating the termination phase of war.[7] This vacuum of ethical and legal principles

---

[4] Dinniss, *Cyber Warfare*.
[5] Schmitt, *Tallinn Manual*.
[6] Walzer, *Just and Unjust Wars*; Orend, "Justice after War," 175–96.
[7] In *War and International Justice*, 218–23, I refer to some pieces of international law in the Hague Convention that *do* mention war termination, but such laws are: (1) very dated, almost 100 years

regarding the end of war is a decrepit state of affairs that should be remedied for both conceptual and concrete historical reasons. Consider these:

- *Completion.* There are many international laws regulating both the start of war (*jus ad bellum*, the justice of war) and the middle of war (*jus in bello*, justice in war). Moreover, many of these laws make sound strategic, and good moral, sense.[8] Thus, to complete our analysis of war's many impacts on international life, we need to consider the ending phase of war. Bottom line: if it is important to guide both *the start* and *the middle*, it is just as important to guide *the end*.
- *Focus.* The practical task of drafting, and then ratifying, a binding legal document on this issue—or even just achieving consensus on the ethical principles—would focus international attention on doing something constructive and improving on war in general, thus taking *jus post bellum* out of abstract theory and into the concrete reality of global politics.
- *Guidance.* The function of any kind of law is to guide behavior, hopefully in a way useful, advantageous, and improving for all. The laws of *jus ad bellum* and *jus in bello* are designed to guide the behavior of all belligerents. The rules of *jus post bellum* could likewise guide both the winner and the loser in the aftermath of armed conflict. (This is assuming there even *is* a clear-cut winner and loser, which sometimes isn't the case, such as with the Iran-Iraq War of 1979–89, when the belligerents just stopped fighting after—eventually—realizing that neither of them could win.)[9] *Both winners and losers would gain by there being clear postwar rules.* The losers, of course, could be assured that they would not be subjected to cruel, vindictive treatment at the hands of a gloating, arrogant winner. And the winners could get a clear understanding of their rights and obligations during the aftermath of war. In particular, winners would appreciate being able to point to such rules and say, "Look, we've done what we're duty-bound to do, and now we are out of here." Rules provide assurances and expectations for everyone, plus clear ways of proceeding, and all parties benefit from such clarity and can put greater confidence in the process moving forward.
- *Ending the Fighting.* Failure to regulate war termination probably prolongs fighting on the ground. Since they have few assurances, or firm expectations, regarding the nature of the settlement, belligerents will be sorely tempted to keep using force to jockey for position. Since international law imposes very few constraints upon the winners of war, losers can conclude it is reasonable for them to refuse to surrender and, instead, to continue to fight. Perhaps, they think, "We might get

---

old now; and (2) radically thin and lacking in substance. See the section below on Occupation Law, plus n. 43.

[8] See especially the Hague Conventions (1899–1907), the UN Charter (1945), and the Geneva Conventions (1949 and 1977). See also Orend, *Morality*, 299–300 and Appendix A; Roberts and Guelff, *Documents on the Laws of War*; Solis, *Law of Armed Conflict*.

[9] Murray and Woods, *Iran-Iraq War*.

lucky and the military tide will turn. Better that than just throw ourselves at the mercy of our enemy." Many observers felt this reality plagued the Bosnian civil war (1992–95), which had many failed negotiations and a three-year "slow burn" of continuous violence as the very negotiations took place.[10]

- **Restraining the Winner.** Failure to construct principles of *jus post bellum* is to allow unconstrained war termination. And to allow unconstrained war termination is, indeed, to allow the winner to enjoy the spoils of war. This is dangerously permissive, since winners have been known to exact peace terms that are draconian and vengeful. The Treaty of Versailles, terminating World War I in 1918–19, is often mentioned in this connection. It is commonly suggested that the sizable territorial concessions and steep compensations payments forced upon Germany created hatred and economic distress, opening a space for Hitler to capitalize on, saying in effect, "Let's vent our rage by recapturing our lost lands, and let's rebuild our economy by refusing to pay compensation, and by ramping up war-related manufacturing."[11]

- **Preventing Future Wars.** When wars are wrapped up badly, they sow the seeds for future bloodshed. Some people, for example, think that America's failure to remove Saddam Hussein from power after they first beat him in 1991 prolonged a serious struggle and eventually necessitated the second war, of regime change, in 2003. *Would the second war have happened at all had the first been ended differently*—that is, more properly and thoroughly, with a longer-range vision in mind? Many historians ask the exact same question of the two World Wars and the recent, related Serb wars, first in Bosnia and then over Kosovo (1999).[12]

Peace treaties should still, of course, remain tightly tailored to the historical realities of the particular conflict in question. There is much nitty-gritty detail that is integral to each peace treaty. But admitting this is *not* to concede that the search for general guidelines, or universal standards, is futile or naïve. There is no inconsistency, or mystery, in holding particular actors, in complex local conflicts, up to more general, even universal standards of conduct. Judges and juries do that on a daily basis, evaluating the factual complexities of a given case in light of general moral and legal principles. We should do the same regarding war termination.[13]

---

[10] Reiff, *Slaughterhouse*.

[11] Boemeke, *Treaty of Versailles*. Though we shouldn't make the mistake of believing that the Treaty exonerates the German people from allowing Hitler to come to power. The Treaty may have been *a* factor in laying the groundwork for World War II but it can hardly be considered the only one. The failure of rival German elites to challenge Hitler, the ruthless thuggery of the Nazis, and the electoral appeal of simple solutions in a time of complex crisis were all important domestic factors in Germany. For an excellent study, see Margaret MacMillan, *Paris 1919*.

[12] Martel, *Victory in War*.

[13] Cimbala, *Strategic War Termination*; Pillar, *Negotiating Peace*; Albert and Luck, *On the Endings of Wars*; Hampson, *Nurturing Peace*.

6.2 Important Starting Assumptions behind *Jus Post Bellum*

Now, some have argued that the "termination phase" of war often does not have a clear starting point, and/or that the notion of victory itself can be quite vague.[14] This is important, as: when then do the norms of *jus post bellum* kick in? And out, as it were? Locating the exact "end" of war may well, at times, be quite difficult; and it may turn out that victory is only one possible outcome (alongside defeat, indeterminacy, military victory but political loss, vice versa, etc.). Thus, *we should think of the war termination phase, crucially, as a kind of process*—along a continuum—wherein there aren't razor-sharp distinctions between the *jus in bello* phase and the *jus post bellum* phase; and the end result may be a number of things, with victory for one side being only one—albeit important—option.

The reason why most conceptions of postwar justice focus on (or even assume) victory is not simple-mindedness, or a failure to realize that wars can end in truly messy and muddled ways.[15] Rather, the focus is methodological: it's just easier to start with more ideal cases, and then move to messier ones. Thus, when we talk about *jus post bellum*, it's easier to take the ideal case where the just side (in terms of *jus ad bellum*, e.g., the victim of an aggressive invasion) has achieved a clear victory (e.g., over the prior Aggressor whose actions triggered the war), and then ask about rights and duties in those situations. This turns out to be surprisingly hard in itself, and laden with controversy, as we'll see below. But it needs to be done before moving on to the more complex cases. Think of it as directly analogous to the *jus ad bellum*: in the ideal case there, you have the vital, foundational concepts of aggression and defense from aggression. These two binary concepts are absolutely central in both the morality and law of the *jus ad bellum*.[16] We think through the rights and duties in such an ideal case, and then use them to wade into such trickier, more advanced concepts as: preventive war and anticipatory attack; civil war; humanitarian intervention; asymmetrical war against nonstate actors; and so on. Again, we are here firmly in the realm of the ideal, trying to make sense of the best concepts and principles and then apply them to more muddled and difficult cases. Hence, there are some simplifying starting assumptions (*already embedded in existing international law*)[17] about Aggressor and Victim; loss and victory; and—as we will see—retribution versus rehabilitation.[18]

---

[14] I am indebted to Michael Walzer for pushing me on this.
[15] I thank Cian O'Driscoll for pushing me on this.
[16] Walzer, *Wars*, 1–126; Orend, *Michael Walzer on War and Justice*.
[17] Reisman and Antoniou, *Laws of War*.
[18] As a helpful initial claim, I'd say that victory has been achieved when the goal for which the war was started has been achieved: e.g., if the goal was to repulse an Aggressor and kick it out of the Victim country which it had invaded, then the *post bellum* phase begins when that has been achieved. There may still be some actual fighting going on at that time, but clearly the

120 BINARY BULLETS

6.3 Which Principles Regulate Postwar Situations?

So, there is a legal vacuum in the postwar moment, and it should be filled (for reasons stated above). But with what principles (as guided by the starting assumptions)? Historically, there is a forceful clash between opposing postwar theories of retribution and rehabilitation, and I have already written much about them, and explained their principles and historical cases in considerable detail.[19] There is no need to repeat all that here, so that we can get to the special case of cyberwar. Let us then take a summary-style bird's-eye view of the essence of these rather familiar theories, and note—as I've come to appreciate more fully recently[20]—that there is actually something of an overlapping consensus between the Two Big Binaries.

The Three Models of Postwar Settlement:

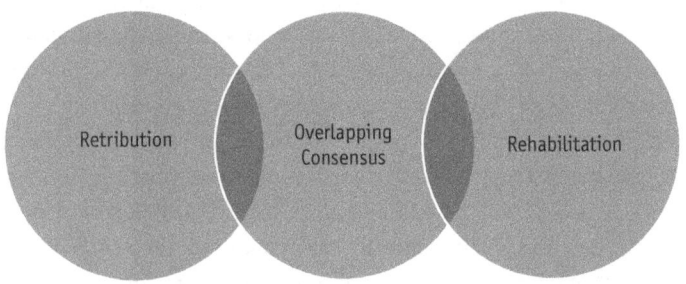

(1) *Overlapping Consensus: The Thin Theory*
- **GOAL**: vindicating the rights whose violation triggered the war, forcing the defeated Aggressor to accept a proportionate policy on surrender that includes:
- Public terms of settlement

victorious country will then move to shut down the violence after that point. This would be a grey area between the *jus in bello* and the *jus post bellum*. Once the violence has subsided (usually accompanied by a cease-fire or surrender)—and emphatically if the regime in Aggressor has collapsed—then it becomes progressively clearer that we've entered an importantly new and different phase of war termination and possibly postwar occupation and regime transformation. Not to pretend that such is a complete definition of victory: merely a useful, prima facie one for the purposes of moving our discussion onward.

[19] Orend, *Morality*, 185–208.
[20] Contrast, for instance, the second edition of my *Morality* with the first (seven years earlier), and with this growth in discernment of, and appreciation for, a middle ground an overlapping consensus emerges. And "overlapping consensus" is used in the sense that, whether the theorist be more of

- Mutual exchange of prisoners-of-war (POWs)
- Aggressor to apologize
- Aggressor to relinquish any unjust gains won during the war
- Aggressor to (at least partially) demilitarize
- War crimes trials (*jus ad bellum* trials for Aggressor; *jus in bello* trials for all sides)[21]

(2) *Thick Theory #1: RETRIBUTION*
- **GOAL**: to make the defeated Aggressor *worse off* than prior to the war (as backward-looking punishment).
- **MEANS**: all of the thin theory above, plus:
- Compensation payments from Aggressor to Victim, and possibly to the international community more broadly
- Sanctions put on Aggressor, to hamper its future economic growth
- No aid or assistance with postwar reconstruction; this is left up to the locals, with no forcible regime change imposed on Aggressor[22]

(3) *Thick Theory #2: REHABILITATION*
- **GOAL**: to make the defeated Aggressor *better off* than prior to the war (as forward-looking reconstruction).
- **MEANS**: all of the thin theory above, plus:
- No compensation payments
- No sanctions
- Aid and assistance with postwar reconstruction, including forcible regime change imposed on defeated Aggressor
- Follow ten-step "Rehabilitation Recipe" (below), with best efforts over ten to fifteen years postwar, to realize in the defeated former Aggressor a new, and minimally just, society[23]

The goal of justified postwar regime change is the timely construction of a minimally just political community. Such a community makes every reasonable effort to: (1) avoid violating the rights of other minimally just communities; (2) gain recognition as being legitimate in the eyes of the international community and its own people; and (3) realize the human rights of all its individual members.[24]

---

a retributivist, or more of a rehabilitator, they all tend to agree that—at minimum—postwar justice needs to involve these basic elements first.

[21] For more detail on each of these principles, and some relevant facts and cases, and the complexities of such, see Orend, *Morality*, 185–201.

[22] For more detail on each of these principles, and some relevant facts and cases, and the complexities of such, see ibid., 201–14.

[23] For more detail on each of these principles, and some relevant facts and cases, and the complexities of such, see ibid., 215–48.

[24] Ibid., 37–43, 215–48; Orend, *Human Rights*. See also Shue, *Basic Rights*; Rawls, *Law of Peoples*.

We can see the means for achieving that goal through what could be called the "Rehabilitation Recipe":

> The successful force will adhere diligently to the laws of war during the regime takedown and occupation. In line with this, they will purge much of the old regime, and prosecute its war criminals. Next, they will seek to disarm and demilitarize the society.

Following this they must then provide effective military and police security for the whole country. A new, rights-respecting constitution that features checks and balances will be developed and implemented in a way that actively involves working with a cross-section of locals.

Concomitantly, this partnership between the winning force and local representatives will allow other, nonstate associations, or "civil society," to flourish. In line with this, if necessary, they must revamp educational curricula to purge past propaganda and cement new values. The occupying force must forgo compensation and sanctions in favor of investing in and rebuilding the economy. In terms of justice and equality, they must ensure that the benefits of the new order will be both concrete and widely, not narrowly, distributed. Finally, they must follow an orderly, not-too-hasty exit strategy when the new regime can stand on its own two feet.[25]

6.4 What to Do with These Three Models?

Prior *post bellum* research has extensively considered the nature and justification of these three models, alongside relevant historical cases (e.g., World War I and the Persian Gulf War of 1991 as examples of retribution; and World War II, and Afghanistan and Iraq 2003, as examples of rehabilitation).[26] I have long argued in favor of two things. The first is the superiority of the model of rehabilitation over that of retribution. And the second is that, therefore, the principles of rehabilitation should be codified into a new treaty dealing exclusively with postwar justice.

I still believe in the superiority of rehabilitation over retribution, as an ideal *general* model of postwar settlement. But the *particular* problems with implementing rehabilitation recently, in Afghanistan and Iraq over the past dozen years, seem to show two important things. First, in many complex contemporary cases, we may in postwar situations have to settle for a version of Aristotle's "second-best solution,"[27]

---

[25] Orend, *Morality*, 215–48. See also: Dobbins et al., *America's Role in Nation-Building*; Dobbins et al., *United Nations' Role in Nation-Building*; Dobbins et al., *Europe's Role in Nation-Building*; Dobbins et al., *Beginner's Guide to Nation-Building*.

[26] Orend, *Morality*, 185–248.

[27] Aristotle, *Politics*.

and realize that *the limit of what we can realistically do is to make all best efforts at realizing the rehabilitation model over a ten- to fifteen-year period,* and then rest content with the fact that the nonideal result at that point may only be a mixed bag of success and failure (in contrast to the shining success of the best historical practices seen post–World War II in West Germany and Japan). Second, there will be little political appetite, in the near future, for including substantial requirements and burdens of rehabilitation in any hoped-for new international law treaty devoted exclusively to postwar justice. As a result, any such legal reform should instead first focus on codifying, and achieving, the thin, overlapping consensus on postwar settlement that does exist between retribution and rehabilitation.

Before leaving this summary of *jus post bellum* in general, it is worthwhile to list quickly the major pros and cons of the models of retribution and rehabilitation, for two reasons. First, this helps us to see the grounds for the claimed preference of rehabilitation. But, second, some of these reasons will play importantly in our conception of poststrike, or postwar, cyberjustice. And so:

RETRIBUTION

Pros: It is simple and straightforward; it is the historical norm (and I guess there's some value in long-standing habitual practice); there's no messing around with complex, costly, and controversial regime change and postwar reconstruction; and, if one believes that justice requires retribution as a matter of principle, then one is going to be deeply moved by retribution's claims.

Cons: It can seem like angry revenge and not balanced justice; the reparations and sanctions very often wind up hurting innocent civilians, and thus violate the established principle of discrimination; and, in historical cases where retribution has been employed, there often has been imperfect—or even quite shoddy—results (i.e., the revenge creates bad feelings and new generations of enemies, and the aggressive regime is left in place, combining together over time to produce—rather regularly—a second war). This negates any supposed "savings" of not trying to rehabilitate the decrepit regime.[28]

REHABILITATION

Cons: It takes a lot of time, effort, and resources; and there's controversy attaching to the imposition of values and new institutions. The recent cases in Afghanistan and Iraq show that the best results may not always be achieved.[29]

---

[28] For more, see Orend, *Morality,* 185–248.

[29] Tondini, *Statebuilding and Justice Reform;* US Government, *Afghanistan Reconstruction;* US Special Inspector General, *Hard Lessons;* Bridoux, *American Foreign Policy and Post-War Reconstruction.*

Pros: First, in historical cases where it has worked well, the rehabilitation model has been an amazing success, well beyond what the retribution model could ever have dreamed of achieving. (Indeed, in some historical cases, such as Germany, the rehabilitation model has been employed *to clean up the messes left behind* by the retribution model.) Second, the rehabilitation model does not leave behind rights-violating regimes that often serve as the causal agents of a second, and worse, war. The rehabilitation model does not leave the aggressive regime in place, nor does it trust that measures of punishment and/or containment will suffice to handle the issue of the bad regime. It views the bad regime as itself the main problem, and thus as needing tackling, through rehabilitative measures. Third, rehabilitation cannot be accused of creating a new generation of enemies, nor can it plausibly be seen as sowing the seeds of a second war. Clearly, this model is trying to help the people in the defeated Aggressor and, though complex emotions might be created by such actions, the desire for revenge is not typically one of them. And, fourth, against the accusation that rehabilitation involves the infliction of narrow or parochial values, contention has already been made that rehabilitation ought to be limited to the construction only of a minimally just state, one characterized by values that are not parochial, narrow, or harmful: the values underlying the structure of a minimally just society.[30]

6.5 Application of *Jus Post Bellum* to Cyberstrikes and Cyberwars, Step One: The *Tallinn Manual* Principles and Their Implications

Since cyberwar, or "information warfare,"[31] is so new and developing, it may be premature to expect that we can, as yet, construct in a satisfying way finished, contrasting models of retribution and rehabilitation in its connection, complete with well-analyzed case studies. But what seems readily achievable is that we can draw (or "cherry-pick") from these models, and meaningfully advance the state of the art, offering some compelling principles that take us further down the road. First, though, we should describe what that state of the art is: namely, I refer to those five postcyber rules from the *Tallinn Manual*.

The *Tallinn Manual* is so named because Tallinn is the capital of Estonia. In 2007, Russia launched a systematic and severe cyberattack on Estonia, a tiny neighboring country. There was a dispute between them regarding the movement of a war statue of great meaning to the Russians. (The Estonians viewed the statue as a symbol of

---

[30] For more, see Orend, *Morality*, 185–248.

[31] Floridi and Taddeo, *Ethics of Information Warfare*; Floridi, *Information*; Floridi, *Cambridge Handbook of Information and Computer Ethics*.

Soviet-era oppression; the Russians as a potent World War II–era symbol of their enormous sacrifice fighting off the Nazis.) When the Estonians moved the statue, Russia responded with a crippling cyberattack on the websites of the Estonian government, its media, and its richest banks. For nearly a week, these institutions could not conduct any business online, nor could their citizens/customers contact them electronically or access anything through their computer and electronic networks. The attack came to an end only when Russia decided to release its grip. Ever since then, "Tallinn" has become synonymous with the need for cyberdefense and cyberjustice.[32]

The *Tallinn Manual* is an attempt, by world-renowned experts in international law, to try to codify rules for regulating the prosecution of cyberwar—treating it as any other distinct and important kind of damage-causing political conflict. Presently, there is no international law whatsoever regarding informational warfare. In 2011 America, China, and Russia got together for a high-level meeting of officials, one branded as being "talks about talks" regarding a possible negotiated treaty between them on the acceptable methods and means of cyberwar—analogous to the many such treaties on kinetic warfare, like The Hague and Geneva Conventions. But the talks fell apart, amid bitter mutual accusations.[33] Thus the *Tallinn Manual* is an important kind of "as if" document: what *might* such an international treaty on cyberwarfare look like, based on existing principles of law—and assuming that we *could* actually get the great powers to negotiate them seriously? It does have some authority since, even though it is not itself a binding treaty, the *Tallinn Manual* is nevertheless a very serious and expert document, referring both widely and deeply to existing principles of law to which the community of nations has already consented, and trying to apply them as clearly and uncontroversially as possible to the new circumstances of the new technology.[34]

I should add quickly that the principles of the *Tallinn Manual*, like other existing pieces of the laws of armed conflict (LOAC), make simplifying assumptions of the kind mentioned above in section 6.3: that there has been some kind of unjustified cyber-strike, or otherwise some act of aggression, and the perpetrator of such has been identified, stopped in the moment, and even more broadly defeated. We admitted, back in section 6.3, that this is not always the case in the real world. But the law deals with ideals (or, at least, norms of behavior that usually set the bar at least slightly above average) and, to the extent to which actors fall short of these ideals, it simply means that justice fails to be done in the real world case. In any event, it is worth noting how all the major countries have simply (and resoundingly) declared that—as

---

[32] Rosenzweig, *Cyber-Warfare*; Karatzogianni, *Cyber-Conflict and Global Politics*.
[33] Betz and Stevens, *Cyberspace and the State*, 12–23.
[34] Schmitt, *Tallinn*, passim.

a matter of their foreign policy—they will consider any "severe" cyberattack against them as a casus belli.[35]

The five cyberrules enumerated by the *Tallinn Manual* that may, or do, have application to poststrike or postwar moments, include the following (numbers theirs):

1. "Rule 85: Collective punishment by cyber means is prohibited."[36] In light of the above, one can see how I myself would be delighted in the expert endorsement of this principle, as it seems to reject the retributivist model in general in favor of the rehabilitative model of postwar justice. Or, I should say, more literally and precisely: it does reject the retributivist model—and stops there, agnostic as yet regarding rehabilitation.
2. Furthermore, note how this principle only bars measures of *collective* punishment by cybermeans: it doesn't ban all retribution. Discriminating punishment—punishment of those who deserve it, and who have been proven to deserve it—is *not* prohibited by this principle.
3. Thus, I would argue that two important things flow from this principle. The first is an endorsement of **Discrimination**: that is, any postwar punishment, including cyberpunishment (e.g., banning of emails, electronic freezing of assets), can only be focussed on those demonstrably culpable for aggression (or cyberaggression), and that any more indiscriminate postwar punishment that impacts the entire civilian population is strictly prohibited. And then second, as war crimes trials are called for *après la guerre* (by the thin "overlapping consensus" outlined above), it would seem that cybercriminals need to be held accountable, and investigated for charges, following a cyberstrike or cyberwar.
4. Such **Trials for Cybercriminals** would serve to underline and enforce the seriousness of their actions as well as the attitude of the international community toward things like theft of intellectual property, espionage, and especially harm-causing acts of sabotage. These things are, at the least, not excluded by Rule 85.

But then note how, if the above is true, it presumes something else: that there be an **Investigation and Attribution** of any given serious cyberstrike, or certainly of a cyberwar. The investigation must focus on who is to blame for the cyberattack. And, of course, there are three kinds of such harm: *espionage* (i.e., using the Internet, and related advanced computer technologies, to gather information that a country has taken steps to protect as a matter of national security, such as secret, confidential, or classified information); *the spread of disinformation*, via the same means, in

---

[35] By this I mean a cause for war or a reason to resort to war—either of the new cyber kind, or else of the traditional physical kind, depending on what they are facing. See also: Carr, *Inside Cyber-Warfare*; Lynn, "Defending a New Domain," 97–108.

[36] Schmitt, *Tallinn Manual*, 195.

a manner that harms the security of the target country; and/or *sabotage* (i.e., using these means to bring about the nonfunctioning, or destruction, of various systems that are integral to the basic interests of a political community). With regards to sabotage, the systems most often mentioned include: electricity and power; water and fuel distribution; computerized parts of manufacturing facilities; transportation systems, such as air or rail; banking and the stock market; and even the Internet itself, or at least the most used websites (like Google or Facebook), Internet service providers, and/or the most basic operating systems.)[37]

Cyberattacks would then refer to any *specific* use of any of these as tools within an overall cyberwarfare strategy. The countries most frequently mentioned today, with reference to cyberwar technology, include Britain, China, France, India, Israel, Pakistan, Russia, and the United States. But, of course, the list is expanding daily.[38] And, of the great powers, it is of interest to note that China seems to have shown a preference for espionage thus far, whereas the United States and Russia have shown a penchant for sabotage. We've given the example of Russia versus Estonia already. What about Chinese and American tactics?

The Chinese seem to prefer espionage of both commercial and political varieties. Many of the top US high-tech firms, such as Google, Microsoft, Apple, and various weapons companies, have complained of sustained cyberattacks from China[39] that have accessed huge amounts of their highest-security information, including especially product design and patent information (as well as, intriguingly, human resources data, such as personal information about top executives). The companies have pressed the US government to respond, but thus far all that have been issued are verbal warnings.[40]

As for US cybersabotage, consider two major cases. In 1982, during the height of the Cold War, a Canadian oil and gas company thought they had a Soviet (Russian) spy in their midst. They contacted the United States; military. The Canadians and Americans launched a joint scheme: they would let the spy steal what he was after: a computer-control system for regulating the flow of oil and gas. (The Russians wanted this to modernize their pipeline system in Siberia.) But the Americans programmed the computer system with a "logic bomb," designed to make the pipelines malfunction and eventually explode after it was implemented. And that is exactly what happened, with some loss of life and a substantial setback for a key sector of the Soviet economy.[41]

---

[37] US Congressional House, *Computer Security*; Brenner, *Cyber Threats*.

[38] Libicki, *Cyberdeterrence and Cyberwar*.

[39] Perhaps the most well-known claims about Chinese espionage are found in "Exposing One of China's Cyber Espionage Units," frequently referred to as the Mandiant Report. See Mandiant, "Exposing One of China's Cyber Espionage Units."

[40] Gross, "Enter the Cyber-Dragon," 220–34; Wu, *Chinese Cyber-Nationalism*.

[41] "Special Report on Cyberwar: War in t; Gross, "Fog of Cyber-War," 155–98. Note, however, that some like Thomas Rid reject the claim that this was in fact a cyberattack. Rid, *Cyber War Will Not Take Place*, 5–6.

And in 2010, Iran was attacked by a computer virus or "worm" commonly believed to have been the joint creation of both the United States and Israel (nicknamed "Stuxnet"). A piece of malware, this very sophisticated computer virus was planted in a German-made component of one of Iran's nuclear reactors. When it was activated, the virus eventually disabled the reactor, forcing it to shut down—lest it melt down and cause enormous damage—for an unspecified time (thought to be at least for months, and perhaps even over a year). The goal, reputedly, was to set back Iran's progress toward developing nuclear weapons.[42]

Back to the issue of attribution, or determining who is responsible for such attacks—whether espionage, disinformation, or sabotage. Experts in the field repeatedly talk of the "attribution problem," noting how cyberattackers—especially those suspected to be linked, in some way, with China—go out of their way to hide their tracks and conceal the ultimate source of the strike.[43] This is of great concern, as it would no doubt color our judgment of whom it is permissible to strike back at. Yet, while being ignorant of the sophisticated details of how these things are determined, I note that eventually—and rather quickly, actually—the cybercommunity seems to have been able, thus far, to come up with pretty reliable attributions in connection with such famous cyberstrikes as the Estonian case (Russia as source) or the Stuxnet case (United States/Israel as joint source). It is said, after all, that even though the perpetrators try to "hide their tracks," such cybertracks are, in fact, always there—and can be unearthed with sufficient time and resources.[44] Is cyber-strike attribution really so different from, and so much more difficult than, say, the investigations that went into determining who was responsible for the kinetic 9/11 attacks (i.e., al Qaeda), and how the then-government of Afghanistan (the Taliban) was complicit in them as well, by offering safe harbor and state resources to al Qaeda?[45] Or, more recently, determining who shot down the Malaysian airliner over contested Ukrainian airspace (i.e., probably the pro-Russian Ukrainian separatists, likely with some kind of Russian state support)?

The second relevant *Tallinn Manual* principle is: "Rule 87: Protected persons in occupied territory must be respected and protected from the harmful effects of cyber-operations."[46] This is the first rule to be mentioned within chapter 6 of the *Tallinn Manual*, which deals with "Occupation." This is manifestly a postwar situation—an important kind wherein the victor has crushed the losing power's regime, and then taken over effective control and administration of that country. There is a body of so-called occupation law, one of the very few and finite pieces of international law,

---

[42] Gross, "Stuxnet Cyber-Weapon," 152–98; Lucas, "Permissible Preventive Cyberwar," 73–84.

[43] See, e.g., Randall Dipert, "Ethics of Cyberwarfare," 384–410; Cook, "'Cyberation' and Just War Doctrine," 417–22.

[44] Shakarian et al., *Introduction to Cyberwarfare*.

[45] Orend, Morality, 71–78; *Congressional Commission Report on the Attacks of 9/11*.

[46] Schmitt, *Tallinn*, 197.

which deals with postwar moments. As Simon Chesterman explains, occupation law demands that: any occupation be temporary; the occupation forces be able to defend themselves; and the occupier provide for civilian well-being, especially in connection with vital needs and basic infrastructure.[47] I would argue that two important things follow from Rule 87 and the occupation law in which it is explicitly nested: a clear reaffirmation of the aforementioned principle of discrimination and noncombatant immunity when it comes to measures of postwar punishment; and, out of Chesterman's third inferred principle, we see some kind of rehabilitative norm of **Cleanup and Aid with Restoration** of the most vital cyberservices following a severe cyberstrike, or more generally within a postwar situation in which one finds oneself in control of a vulnerable population. In modern economies, as mentioned above in connection with sabotage, we have seen that cyberoperations can be essentially, and completely, integrated into some of the most vital physical and social infrastructure, including electricity generation and distribution; water and sewage treatment systems; flows of oil and gas; banking and access to money; and even communication access to officials who occupy positions of public authority and power. Indeed, at Rule 87, subsection (7), the *Tallinn Manual* summarizes: "The Occupying Power shall . . . ensure the continuance of computer operations that are essential to the survival of the civilian population of the occupied territory."[48]

The final three *Tallinn Manual* principles offer further explanation of, and support for, this idea of what the civilian population needs in a postwar situation, and what any occupying power can legitimately do, by way of trying to meet those needs:

> Rule 88: The Occupying Power shall take all measures in its power to restore and ensure, as far as possible, public order and safety.[49]
>
> Rule 89: The Occupying Power may take measures necessary to ensure its own general security, including the integrity and reliability of its own cyber-systems.[50]
>
> Rule 90: To the extent the law of occupation permits the confiscation or requisition of property, taking control of cyber-infrastructure or systems is likewise permitted.[51]

---

[47] "Occupation law" is much in dispute, even regarding whether it is still in force. In addition to the three principles mentioned, it stridently stipulates that the occupier *not* change any of the existing laws or structures within the occupied society. So, Chesterman observes, occupation law "provides little support for regime change" of the sort endorsed by the rehabilitation theory. See Chesterman, "Occupation as Liberation," 226–27, regarding the complex relationship between occupation law and the rehabilitative theory of postwar justice.

[48] Schmitt, *Tallinn Manual*, 198.

[49] Ibid.

[50] Ibid., 199.

[51] Ibid., 200–201.

This is to say that poststrike/postwar cleanup and aid with restoration of cyberoperations needs to be done both as a matter of vital human need and in terms of the maintenance (Rule 88) of public order and safety (e.g., traffic lights, and sewage treatment, are often heavily regulated by computer technologies—and thus damage to such can manifestly jeopardize public health and safety). And in order to be efficacious with regard to any poststrike or postwar cyberoperations cleanup and restoration, it follows (Rule 89) that the war winner needs to maintain the proper functioning of its own cybersystems and prevent them from being attacked, hacked, or corrupted and rendered inert. This might, for example, imply, in certain circumstances, the securing of electricity first for its own cyber-operations, so that such can then be used to restore those of the general population.

The final principle (Rule 90) attracted the least consensus among the Tallinn experts. Still, it exists, and it was felt that occupation law does provide some basis for an occupying war-winner to take over the state-owned computer property of the defeated nation, in order to render efficacious the other duties above. Some experts felt this was obvious and a matter of causal necessity. Others urged caution. Both groups stressed that the property referred to was *state-owned*, and probably assumed that the state or government in question had fallen as a result of the war activities—and so, who else to operate it except the occupying power? In any event, it was stressed how the taking of control of *private* property was *not* permitted under this principle (unless such was being used for demonstrably criminal activities, such as fraud, theft, drug- or human-trafficking, or incitement of revolution or any other clear postwar threat to public health, order, and safety).

6.6 Application of *Jus Post Bellum* to Cyberstrikes and Cyberwars, Step Two: Further Principles, from the Three Theories of Postwar Justice

Perhaps it shouldn't surprise us that the principles present in, or implied by, the *Tallinn Manual* combine aspects of both retribution and rehabilitation. On the retribution side, discriminating punishment is permitted, and some public cyberproperty might be forfeited. And, on the rehabilitation side, a norm of cleanup and restoration is present and endorsed, at least to the extent to which such cyberoperations restoration is causally necessary for the fulfilment of the vital human needs of the civilian population.

Consistent with this blend of the Two Big Binary theories, we might ask whether anything's missing from the *Tallinn* principles that might still plausibly be recommended—either in light of one of the two big theories, or else the thin overlapping consensus? Two principles leap out from the overlapping consensus. The first is a norm of publicity, or public accountability; and the second is consideration of potential "decyberization" of the defeated state (compare, "demilitarization"). We have encountered shades of the publicity rule already, in connection with the attribution problem above. Let us strive to summarize, and condense publicity and attribution into one:

Public Accountability: After a cyberstrike or cyberwar, there needs to be public accountability for such. If one committed the cyberstrike, but for justifiable reasons (as, e.g., the United States might say in connection with Stuxnet), one needs to both admit this and give a public accounting as to why such a strike was justified, according to commonly accepted ethical and legal principles reading armed conflict. If, on the other hand, one is the victim of such a strike, or coming to the aid of such, one must—in light of the attribution problem above—do the due diligence required to determine who, in fact, carried out the attack. And then consideration of an appropriate response would kick in.[52]

The second added principle would be **Decyberization:** In light of the logic of demilitarization, and perhaps as permitted by *the Tallinn Manual*'s Rule 90 on cyberproperty forfeiture, a war-winning occupier may justly seize and partially dismantle those public cyberassets that were used by the now-defeated state to launch a severe and unjustified cyberstrike or information war. The limit of such seizing and decommissioning would be set at that point where the target country could no longer commit cyberaggression of a similar magnitude, and for a reasonable amount of time afterward. But such seizing and/or decommissioning could never be of such an extent that it might jeopardize the public health, safety, and order of the ordinary civilian population within the target country.

Another principle that might be called for, by the thin overlapping consensus, relates to the norm of the Aggressor having to give up any unjust gains that it made during the war. If some of the gains included valuable information—let's say, intellectual property information (like weapons design, hardware- or software-product design, etc.)—and, if it makes sense to say that such can be retrieved (or at least destroyed, so no one but the rightful owner can make use of it), then such ought to be done in the postwar moment as well. Now, this might not be possible—as some knowledge, once known, cannot be "taken back"—but, again, if it is possible, it seems called for by this principle. We could phrase such as something like: **Reversal of Cybertheft**.[53]

I think the only two things that remain to be talked about, at least for now, concern those bolder and more controversial aspects of retribution, as applied to the cyberworld, that is, **Compensation and Sanctions.** In general, argument was given above as to how, very often, the application of such has backfired historically, and

---

[52] And on that further question, consult Orend, "Fog in the Fifth Dimension," 3–23.
[53] Baase, *Gift of Fire*.

wound up hurting civilians in a way violative of the principle of discrimination. Those concerns should be repeated here. Yet, perhaps, if a way could be found to enforce a discriminating use of this principle, then it would have to be considered. For example, consider the Russia/Estonia case, where Estonians may have suffered real (mainly financial) hardship during their week of being blocked out from their banks, not having access to government services, and so on. Some kind of actual monetary restitution might be in order (from the government of Russia). Enforcing that, though, may in the end prove very difficult.

Another option might be to slap so-called targeted sanctions on the Aggressor, which are measures of economic and other noncooperation that are aimed at hurting only the culpable elites in a given society, and avoiding damage done to ordinary civilians. This might be permissible, though again there's the issue of enforcement, plus the raw fact that, usually, targeted sanctions only work when the inflictor has very substantial power over, and thus leverage on, the country suffering the targeted sanctions. In such cases as the United States versus the leaders of the undemocratic coup d'état in Haiti in 1994, targeted sanctions worked; in such other cases as weak western sanctions on targeted Russian individuals in light of Russia's recent expansion into the Crimea and Eastern Ukraine, success is not exactly to be expected.

To summarize: if such things can actually be done effectively, and avoid civilian damage, then they are probably permissible. But we need to be sanguine as to how difficult and unlikely such a combination can be: strong enough to be effectively enforced and leverage-inducing, yet not so strong as to hurt ordinary civilians.[54] Now, the new cyberera may actually hold out some further, potentially more effective, promise with regards to these sorts of measures. For instance, it may now be possible to employ cybertools to directly access the bank accounts of those elites found culpable of unjustified cyberattacks. Or it might be possible to set up barriers to trade through cybermeans (again, perhaps focusing on using cybertools to block transfers of money, or payments, for the cross-border shipping of goods). Thus, hoping, or waiting, for leverage to kick in, in the old ways, may not be required, or might be a problem for which we can find new cyberways to circumvent.[55]

6.7 The Postcyber Seven

Whenever there is a cyberstrike, or a cyberwar, there will be an aftermath to such. The aftermath of armed conflict is woefully underregulated, and we should avoid that situation when it comes to cyberconflict. Using existing theories of postwar justice, plus the principles of occupation law and the *Tallinn Manual*, it was suggested

---

[54] Orend, *Introduction to International Studies* 110–20.
[55] I thank the editors for raising this point with me.

that a well-grounded conclusion to a serious cyberstrike or cyberwar ought to include adherence to the following rules or principles. Let us, in an effort to make them stick in the mind, dub them the *Postcyber Seven*:

- *Public accountability* (whether transparent disclosure of one's own actions, or else the discovery via attribution of who probably did what).
- *Discrimination and noncombatant immunity.* Any postwar measures, especially of punishment or retribution, must not impact negatively on the civilian population.
- *Reversal, if possible, of any cybertheft.*
- *Punishment for cybercriminals.* Those culpable of unjustified cyberattacks or aggression should be put on trial for such, as any other war criminal.
- *Decyberization.* A cyberaggressor can reasonably be stripped of any cyberassets that would allow it to repeat its aggression in the foreseeable future (compare to demilitarization).
- Possibly terms of *reasonable compensation and targeted sanctions*, but *only where such do not violate* the prior principle of discrimination and noncombatant immunity.
- *Rehabilitative aid with cleanup and restoration of cyberoperations*, including, emphatically, all those cybertools required to meet the vital human needs of the civilian population.

The construction of these norms, drawing upon existing and authoritative principles and sources, is designed at the least to advance our thinking—and urge further thought—about all aspects of a cyberattack or cyberwar. These rules are plausible; and in the past the international community has found it both advantageous and principled to agree upon certain values and measures in connection with damage-causing hostilities. This is to say that in connection with both the outbreak of war, and conduct during war, countries have found it to be both prudentially advantageous and morally principled to come up with a set of general guidelines as to how these sorts of fraught situations ought to be constrained and prevented from becoming exercises in mutual ruin. The same holds true not merely for the aftermath of war in general, but also for the complex consequences of this brave new world of cyberaggression and cyberdefense.

Bibliography

Albert, Stuart, and Edward Luck, eds. *On the Endings of Wars*. London: Kennikat, 1980.
Aristotle. *The Politics*. Trans. Carnes Lord. Chicago: University of Chicago Press, 1985.
Baase, Sara. *A Gift of Fire: Social, Legal and Ethical Issues for Computing Technology*. 4th ed. Saddle River, NJ: Prentice Hall, 2012.
Betz, David, and Tim Stevens. *Cyberspace and the State*. New York: Routledge, 2012.
Boemeke, Manfred, et al., eds. *The Treaty of Versailles*. Cambridge: Cambridge University Press 1998.

Brenner, Susan. *Cyber Threats: The Emerging Fault Lines of the Nation-State*. Oxford: Oxford University Press, 2009.
Bridoux, Jeff. *American Foreign Policy and Post-War Reconstruction: Comparing Japan and Iraq*. New York: Routledge, 2012.
Carr, Jeff. *Inside Cyber-Warfare*. 2nd ed. London: O'Reilly, 2011.
Chesterman, Simon. "Occupation as Liberation: International Humanitarian Law and Regime Change." *Ethics and International Affairs* (2004): 51–64.
Cimbala, Stephen. *Strategic War Termination*. New York: Praeger, 1986.
Clarke, Richard. *Cyber-War: The Next Threat to National Security and What to Do About It*. New York: Ecco, 2012.
The Congressional Commission Report on the Attacks of 9/11. Washington, DC: US Government Printing Office, 2004.
Cook, Martin. "'Cyberation' and Just War Doctrine." *Journal of Military Ethics* (2010): 417–22.
Dinniss, Heather Harrison. *Cyber Warfare and the Laws of War*. New York: Cambridge University Press, 2012.
Dipert, Randall. "The Ethics of Cyberwarfare." *Journal of Military Ethics* (2010): 384–410.
Dobbins, James, et al. *America's Role in Nation-Building: From Germany to Iraq*. Santa Monica, CA; RAND, 2003.
Dobbins, James, et al. *Europe's Role in Nation-Building: From the Balkans to Congo*. Santa Monica, CA: RAND, 2008.
Dobbins, James, et al. *The Beginner's Guide to Nation-Building*. Santa Monica, CA; RAND, 2009.
Dobbins, James, et al. *The United Nations' Role in Nation-Building: From Congo to Iraq*. Santa Monica, CA: RAND, 2005.
Floridi, Luciano. *Information*. Oxford: Oxford University Press, 2010.
Floridi, Luciano, ed. *The Cambridge Handbook of Information and Computer Ethics*. Cambridge: Cambridge University Press, 2010.
Floridi, Luciano, and Mariarosaria Taddeo, eds. *The Ethics of Information Warfare*. Cham, Switzerland: Springer, 2014.
Gross, Michael. "Enter the Cyber-Dragon." *Vanity Fair*, September 2011, 220–34.
Gross, Michael. "The Fog of Cyber-War." *Vanity Fair*, April 2011, 155–98.
Gross, Michael. "The Stuxnet Cyber-Weapon." *Vanity Fair*, April 2011, 152–98.
Hampson, Fen Osler. *Nurturing Peace: Why Peace Settlements Succeed or Fail*. Washington, DC: US Institute of Peace, 1996.
Karatzogianni, Athina, ed. *Cyber-Conflict and Global Politics*. London: Routledge, 2008.
Libicki, Martin. *Cyberdeterrence and Cyberwar*. Santa Monica, CA: RAND, 2009.
Lucas, George. "Permissible Preventive Cyberwar." In *The Ethics of Information Warfare*, ed. L. Floridi and M. Taddeo. Cham, Switzerland: Springer, 2014, 73–84.
Lynn, W. J. "Defending a New Domain: The Pentagon's Cyberstrategy." *Foreign Affairs* (September/October 2010): 97–108.
MacMillan, Margaret. *Paris 1919*. New York: Random House, 2003.
Martel, William. *Victory in War*. Cambridge: Cambridge University Press, 2011.
Murray, Williamson, and Kevin Woods. *The Iran-Iraq War*. Cambridge: Cambridge University Press, 2014.
Orend, Brian. "Fog in The Fifth Dimension: The Ethics of Cyberwar." In *The Ethics of Informational Warfare*, ed. L. Floridi and M. Taddeo. Cham, Switzerland: Springer, 2014, 3–23.
Orend, Brian. *Human Rights: Concept and Context*. Peterborough, Canada: Broadview Press, 2002.
Orend, Brian. *Introduction to International Studies*. Oxford: Oxford University Press, 2013.
Orend, Brian. "Jus Post Bellum." *Journal of Social Philosophy* (spring 2000): 117–37.
Orend, Brian. "Justice after War: Towards a New Geneva Convention." In *Ethics Beyond War's End*, ed. E. Patterson. Washington, DC: Georgetown University Press, 2012, 175–96.
Orend, Brian. *Michael Walzer on War and Justice*. Cardiff: University of Wales Press, 2000.
Orend, Brian. *The Morality of War*. 2nd ed. Peterborough, Canada: Broadview Press, 2013.

Orend, Brian. *War and International Justice: A Kantian Perspective*. Waterloo, ON: Wilfrid Laurier University Press, 2000.
Pillar, Paul. *Negotiating Peace: War Termination as a Bargaining Process*. Princeton: Princeton University Press, 1983.
Rawls, J. *The Law of Peoples*. Cambridge, MA: Harvard University Press, 1999.
Reiff, David. *Slaughterhouse: Bosnia and the Failure of the West*. New York: Simon & Schuster, 1995.
Reisman, W. M., and Chris Antoniou, eds. *The Laws of War: A Comprehensive Collection of Primary Documents Governing Armed Conflict*. New York: Vintage, 1994.
Rid, Thomas. *Cyber War Will Not Take Place*. London: Hurst, 2013.
Roberts, Adam, and Richard Guelff, eds. *Documents on the Laws of War*. 3rd ed. Oxford: Oxford University Press, 2000.
Rosenzweig, Paul. *Cyber-Warfare*. New York: Praeger, 2013.
Schmitt, Michael N., ed. *The Tallinn Manual on the International Law Applicable to Cyber Warfare*. New York: Cambridge University Press, 2013.
Shakarian, Paulo, et al. *Introduction to Cyberwarfare*. New York: Syngress, 2013.
Shue, Henry. *Basic Rights*. 2nd ed. Princeton: Princeton University Press, 1996.
Singer, Peter, and Allan Friedman. *Cybersecurity and Cyberwar*. Oxford: Oxford University Press, 2014.
Solis, Gary. *The Law of Armed Conflict*. Cambridge: Cambridge University Press, 2010.
"Special Report on Cyberwar: War in the Fifth Domain." *The Economist* (July 1, 2010), 18–26.
Tondini, Matteo. *Statebuilding and Justice Reform: Post-Conflict Reconstruction in Afghanistan*. New York: Routledge, 2010.
U.S. Congressional House. *Computer Security: Cyber-Attacks and War without Borders*. Washington, DC: Books LLC, 2011.
U.S. Government. *Afghanistan Reconstruction: Despite Some Progress*. Washington, DC: Books LLC, 2011.
U.S. Special Inspector General. *Hard Lessons: The Iraq Reconstruction Experience*. Washington, DC: US Independent Agencies and Commissions, 2009.
Walzer, Michael. *Just and Unjust Wars*. New York: Basic Books, 1977.
Wu, Xu. *Chinese Cyber-Nationalism*. New York: Lexington Books, 2007.

Part III

ETHOS OF CYBERWARFARE

7

Beyond *Tallinn*

The Code of the Cyberwarrior?

MATTHEW BEARD

I will begin with a broad claim: thus far, discussions of cyberwar have failed to engage in serious conversations about the morality of cyberwarriors. For those of us familiar with military ethics this is curious, as there is widespread debate within the literature surrounding the morality of soldiers themselves. It is important to discuss the ethics of soldiering, of course, because soldiers are asked to kill or harm other people, which ordinary moral thinking holds to be a prima facie wrong. Thus, we require a nuanced account to explain how soldiers' killings can be justified. If cyberwar is in any way analogous to conventional war, one would expect similar discussions regarding the profession of the cyberwarrior—who is the soldiers' moral equivalent in the cybersphere—and a nuanced account justifying cyberwarriors' actions.

When ordinary soldiers—here understood as those engaged in combat—go to war, they are bound by a series of ethical principles that proscribe particular actions and forbid others. The laws of armed conflict (LOAC) and Codes of Conduct[1] have been developed by various military institutions to teach and enforce LOAC to their soldiers. However, soldiers are also bound by a more informal, less explicit set of beliefs about what it is honorable or shameful for warriors to do in times of war: the warrior code.[2] Both LOAC and the warrior code cooperate together in motivating soldiers to perform their professional roles with moral distinction. Until now,

---

[1] Throughout this chapter I will use capital letters to indicate formally written documents that list the moral and legal duties of professionals, and lowercase letters to discuss "codes"—such as codes of honor—more generally.

[2] It is beyond the scope of this paper to consider exactly who the warrior code does and does not apply to within the military. Not all military personnel are actually engaged in fighting, and therefore they may not be governed by the warrior code. The warrior code is, as I will show, a code for combatants. This chapter aims to consider whether an analogous code might be developed for cybercombatants. It is another question as to exactly which cyberoperatives should be considered cyberwarriors.

discussions of cyberwar have been limited to the discussion of how LOAC should be applied to cyberwar, or what new laws should be developed to govern war in the digital space. As worthwhile as this project is, it should not be done in isolation; equally important is the development of a code of honor that applies directly to cyberwarriors. Without a complementary "code of the cyberwarrior," LOAC are at risk of lacking the motivating force they require, and military ethics also risks playing catch-up in a rapidly advancing field.

As well as being governed by principles of *jus in bello*, laws of war, and statements of professional responsibility, military personnel tend to see themselves as being bound by what Shannon E. French calls a "warrior code," or code of honor— "an amalgam of specific regulations, general concepts... history and tradition that adds up to a coherent sense of *what it is to be a Marine*."[3] This code, which amounts to a normative tradition handed down over generations, tends to be more effective in eliciting moral compliance than externally prescribed statements of moral responsibility. Thus, in this chapter I will argue that those involved in fighting cyberwars—"cyberwarriors"—require a similar traditional and informal code of honor to help cultivate morally good conduct and to develop a robust and unique normative identity.

How, though, should we conceive of the code of the cyberwarrior? We might begin by looking to the *Tallinn Manual on the International Law Applicable to Cyber Warfare*, published in 2013. That publication, and the academic discussion that preceded it, recognized the need to identify "the scope and manner of international law's applicability to cyber operations."[4] Laudably, the *Tallinn Manual* identified that "the *jus ad bellum* and *jus in bello* apply to cyber operations,"[5] meaning that existing moral and legal governances over military operations can also be intelligibly applied to the cyberrealm. This position (which I have argued for elsewhere)[6] recognizes that cyberwar does not represent an *essential* change in the nature of warfare even if it does present unique and novel moral challenges. However, as a *legal* discussion, the *Tallinn Manual* does not provide explicit insights into what sort of traits those engaged in cyberwar should characterize.

Further, in this chapter I will take it for granted that because conventional war and cyberwar differ dramatically with regard to the day-to-day operations of the military personnel fighting those wars, it is unlikely that either formal Codes of Conduct or informal codes of honor will be symmetrical between conventional war and cyberwar. Each will reflect the different activities, duties, and skills implicit in each profession. In this chapter I will argue that developing a "cyberwarrior code" should go beyond determining the rights and duties of

---

[3] French, *Code of the Warrior*, 15. Emphasis in original.
[4] Schmitt, *Tallinn Manual*, 3.
[5] Ibid., 5.
[6] Beard, "Cyberwar and Just War Theory," 1–12.

cyberespionage, cybersabotage, and the direct and intentional killing of human beings with cyberweapons—what I will call "cyberassassination." In addition, this code must explore questions of the character, identity, and virtue of those engaged in these practices directly. That is, it should ask what kind of person the cyberwarrior should be in order to perform his or her roles without violating any of his or her moral duties. I will argue that any proper cyberwarrior code will address two questions: first, what are the moral principles that determine the minimum standard of ethical behavior for cyberwarriors? And second, what virtues, dispositions, and values should define the morally excellent cyberwarrior? Most will accept from the outset that cyberwarriors are rightly distinguished from conventional warriors, but how are these differences reflected in the moral nature of their profession?

This chapter will work from the premise that it is of great importance that cyberwarriors develop a code of honor, and that it is necessary that this code be distinct from that of traditional warriors in order to reflect the differences in practice between them. Military institutions should begin to foster a culture that honors and valorizes virtues such as creativity and discretion to enable cyberwarriors to develop a normative identity that will motivate adherence to the moral requirements of their craft.

7.1 Warriors and the Code of Honor

Before moving to discuss cyberwarriors directly, I will establish why *any* breed of warrior will require more than formally described codes of ethics to elicit morally upstanding conduct. Although formal codes of ethics are useful in describing objectively the deontologically prescribed normative limits of a profession, they fail to recognize the intimate connection between what a person does and the type of person he or she is; or, put more simply, what it is that connects a rule with the type of behavior that obeys it. In order for a principle to be obeyed, it must appear to be real and binding to the people it applies to. One way in which rules can be made *real* in this sense is through a code of honor.

In her book, *The Code of the Warrior*, Shannon E. French argues that warrior cultures scattered throughout history and geography have developed codes of honor: a commonly held standard of what the ideal warrior does and does not do that bears normatively on each warrior within the culture.[7] Forming close associations with their peers, warrior communities develop understandings about what personality traits and behaviors should be awarded praise, and which should be awarded shame. Going well beyond LOAC or any formal set of rules, the warrior code of honor sees morally good conduct as being inextricably connected to the normative identity

[7] French, *Code of the Warrior*, 3.

of the warrior. If a person sees him- or herself as a warrior, he or she will want to be a particular type of person. The type of person he or she wants to become will be determined by what the rest of the warrior community honors. In making her case, French borrows an incident from Mark Osiel's book, *Obeying Orders: Atrocity, Military Discipline, and the Law of War*, in which he describes an incident in which a Marine refrained from killing a noncombatant after receiving the simple rebuke, "Marines don't do that."[8]

"Marines don't do that" is not merely shorthand for "Marines don't shoot unarmed civilians; Marines don't rape women; Marines don't leave Marines behind; Marines don't despoil corpses," even though those firm injunctions and many others are part of what we might call the Marines' Code. What Marines internalize when they are indoctrinated into the culture of the Corps is an amalgam of specific regulations, general concepts (e.g., honor, courage, commitment, discipline, loyalty, teamwork), history, and tradition that adds up to a coherent sense of *what it is to be a Marine*.[9]

In this particular situation, one might assume that the Marine in question had not forgotten nor was he unaware of his professional duty to refrain from killing civilians. Rather, the emotions, cognitive and psychological demands, physical exhaustion, and other mitigating factors that accompany war meant that, at that particular point in time, LOAC did not appear *real* to this Marine: in that particular moment, it just didn't matter what the law said he could or couldn't do. What did matter, we discover, is how his peers would perceive him, and how he would perceive himself.

Importantly, French argues that "[t]he [warrior] code is not imposed from the outside. The warriors themselves police strict adherence to these standards, with violators being shamed, ostracized, or even killed by their peers."[10] The reason for warrior codes (there are many different codes, as French's work shows) being developed by the warriors themselves is because an internally developed code assented to by peers has a more powerful binding force than externally imposed rules. As Lt. Gabriel Bradley argues:

> Law is the judgment of the community at large, but the impetus for ethical conduct among warriors must come from other warriors. The real challenge for commanders is not just to teach their troops about the law of armed conflict but to inculcate in their troops the ethos of the professional warrior —to instill [sic] an abiding sense of honor. It is not enough for soldiers to know the rules, or even to follow them. Without deep reserves of character and psychological strength, troops in high-stress

---

[8] Ibid., 15.
[9] Ibid. French's story comes from Osiel, *Obeying Orders*, 25.
[10] French, *Code of the Warrior*, 3.

battlefield situations may fall prey to undisciplined impulses. Honor, not law, is the key to battlefield discipline.[11]

Warrior codes of honor utilize the powerful sense of self-identity that a profession like soldiering provides as a means to ensuring that soldiers adhere to the moral and legal restrictions on their conduct. They consist of two types of elements: first, aretaic elements—the development of psychologically influential beliefs regarding the type of person the morally excellent warrior ought to be; and second, teleological elements, which concern the ultimate purpose of the warrior profession. These two sets of elements pertain to two separate questions regarding the normative identity of the warrior: How ought a warrior to stand? And what ought a warrior to stand *for*?

7.2 A Professional Code of Conduct for Cyberwarriors?

I will begin answering these two questions, perhaps ironically, by turning to formal Codes of Conduct for cyberwarriors and cyberoperatives.[12] These codes, although of limited use at this point in time, will help to indicate similarities and differences in conventional warriors and cyberwarriors and provide important insights into the ways in which cyberwarriors' and cyberoperatives' normative identities may differ from their conventional equivalents.

To my knowledge there is, at the time of writing, little by way of *national* policies on the subject. Perhaps the closest thing, the United States Army's Field Manual on "Cyber Electromagnetic Activities," *FM 3-38*, addresses "activities leveraged to seize, retain, and exploit an advantage over adversaries and enemies in both cyberspace and the electromagnetic spectrum, while simultaneously denying and degrading adversary and enemy use of the same and protecting the mission command system," including cyberspace operations and electronic warfare.[13] However, this document is dedicated primarily toward operational considerations, and there is little in it to indicate that there are particular ethical or legal norms of conduct that apply to cyberoperatives.

---

[11] Bradley, "Honor, Not Law."

[12] Here I distinguish between these two groups (although the distinction I draw is broad and, in some ways, problematic): cyberwarriors are *military* personnel who use cyberweapons to conduct war; cyberoperatives, by contrast, are those nonmilitary personnel who use cybertechnology in the interests of national defense, or those military personnel who use cyberweapons for nonmilitary purposes. I acknowledge that a more detailed discussion of this distinction is important, but it is beyond the scope of this chapter to address.

[13] United States Department of Army, *FM 3-38 Cyber Electromagnetic Activities*.

Given the absence of national documents regarding cyberconduct, we must turn to international policy. In this arena, there has been talk of developing a formal code of conduct, but it has not been finalized (nor even, to my knowledge, begun) at the time of writing. At this stage, the *Tallinn Manual* remains the most authoritative source for international policy on cyberwar, including the regulation of the conduct of cyberwarriors. Notably, the *Tallinn Manual* argues that cyberoperations are ultimately governed by LOAC as they are represented in documents such as The Hague and Geneva Conventions.[14] Thus, we can expect the principles underpinning the cyberwarrior's code of conduct to be analogous to the principles governing the conduct of traditional soldiers.

It is worth noting, however, that the *Tallinn Manual* is not intended to be a code of conduct. Although aspects of it appear similar in nature to a typical code of conduct, it is not a comprehensive summary of the normative duties of a particular person or profession. There are, however, principles inherent in the document that will bear on all cyberoperatives over whom the *Tallinn Manual* has jurisdiction. I found eight principles that one would expect to see on a formal code of conduct for cyberwarriors and which are present in the Manual:

1. **Jurisdiction:** The laws of armed conflict vis-à-vis cyberwar apply to all those who "belong to," or serve, a party to a particular conflict.[15]
2. **Existing LOAC and cyberwar:** "The law of armed conflict applies to the targeting of any person or object during armed conflict irrespective of the means or methods of warfare employed. Consequently, basic principles ... will apply to cyber operations just as they do to other means and methods of warfare."[16]
3. **Presumption of civilian status:** If there is doubt as to whether a person is a civilian or not, that person will be presumed to be a civilian,[17] if there is doubt as to whether an *object* is serving a military or civilian purpose, it may only be attacked following careful assessment and monitoring.[18]
4. **Minimized suffering:** Cyberwarriors must minimize suffering and injury that occurs as a consequence of their actions to the absolute minimum. There can be no permissible "superfluous injury or unnecessary suffering."[19]
5. **Discrimination:** Cyberwarriors must not utilize methods that are indiscriminate, meaning methods that are unable to (i) be directed toward a specific military target, or (ii) be limited and controllable in their effects.[20] Furthermore, as

---

[14] Schmitt, *Tallinn Manual*, 75.
[15] Ibid., rule 26.7, 97.
[16] Ibid., 105.
[17] Ibid., rule 33, 114.
[18] Ibid., rule 40, 137.
[19] Ibid., rule 42, 143.
[20] Ibid., rule 43, 145.

well as avoid the use of weapons that are inherently indiscriminate, cyberwarriors must avoid using legitimate weapons in an indiscriminate *manner*; that is, they must avoid directly targeting nonmilitary personnel or objects in the performance of their duties.[21]

6. **Responsible weapons development:** Given that (i) cyberwarriors must avoid both the use of per se indiscriminate weapons and the use of any weapon in an indiscriminate manner, and (ii) states are required to ensure any cyberweapons developed within their borders comply with LOAC, it follows that cyberwarriors (who, because of their expertise, are likely to have a hand in both the development and deployment of cyberweapons) must refrain from development of inherently indiscriminate (or otherwise illegal) weapons.

7. **Avoid excessiveness:** Rule 51 of the *Tallinn Manual* prohibits cyberattacks that can be expected to cause incidental (collateral) damage to civilian personnel or infrastructure in levels that exceed the military benefits of the attack.[22] Thus, cyberwarriors are required not only to avoid the intentional targeting of civilians, but to avoid excessive levels of collateral fallout while attacking military targets. This is frequently referred to as the principle of proportionality.

8. **Constant caution and care in verification:** From the requirement to avoid excessiveness, it follows that cyberwarriors must take constant consideration for the well-being and security of civilian personnel throughout the entire decision-making process of cyberwarfare. This includes ensuring that decision-makers are equipped with technological expertise and understanding of cyberweapons and their potential effects,[23] and that all potential targets are submitted to a thorough vetting process prior to attack.[24]

These principles are supplemented by two principles guiding the conduct of cyberespionage, which is distinct from the actual waging of war by cybermeans. These are:

1. **Restrained perfidy:** One of the benefits of cyberespionage is the ease with which one can remain anonymous or use the identity of another person for one's own ends. This type of conduct *can* be legitimately employed by cyberoperatives in order to obtain, destroy, damage, or control enemy information, systems, or infrastructure.[25] However, it may *not* be used to kill or injure enemy personnel.[26] It is perfectly legitimate to employ creative

---

[21] Ibid., rule 49, 156.
[22] Ibid., rule 51, 159.
[23] Ibid., rule 52, 166.
[24] Ibid., rule 53, 167.
[25] Ibid., rule 60, 182–83.
[26] Ibid., rule 60, 180.

means to treacherously gain the confidence of a target for ulterior ends, but cyberoperatives must limit their use of perfidious means toward nonhuman targets.
2. **Respect for the integrity of indicators:** Rules 62–65 prohibit the use of various indicators for espionage by cyberoperatives, including the United Nations emblem,[27] the Red Cross indicators, the flag of truce,[28] impersonating enemy indicators,[29] or identifying oneself as a neutral state.[30] All of these are concerned with maintaining the ongoing safety of impartial personnel during conflict; if, for instance, the Red Cross indicators were frequently abused, the safety of Red Cross personnel would be jeopardized because of the increased suspicion with which they would be treated. Further, the impersonation of neutral states risks escalation and further damage to both civilian and military persons and infrastructure.

My reason for collecting these principles together is not to demonstrate the existence of a fully fledged, comprehensive, or coherent code of conduct. Rather, I aimed to reveal how the *Tallinn Manual* demonstrates some principles of conduct that will bear on cyberwarriors and cyberoperatives in different ways and over different contexts. It demonstrates how, despite the complexity of the cyberdomain as a theater of war, espionage, and international relations, there can emerge basic principles to restrain those persons responsible for representing their nations in the cyberdomain from doing whatever is necessary, expedient, or advantageous in the pursuit of the national interest.

This list of principles implies certain characteristics that the ethically upstanding cyberwarrior will be required to possess or demonstrate. For instance, the principle of restrained weapons development suggests not only the need for self-control, empathy for noncombatant populations, and temperance against the temptation to do what is expedient rather than what is right, but also the need for cyberwarriors, or whomever is responsible for the development of cyberweapons, to be *creative*. Richard de George, for instance, argues that a fundamental principle of the development of "smart weapons" (of which cyberweapons are one) is "a morally obligatory smart arms race," that is, "a race to develop weapons that will do as little damage as possible to innocent non-combatants."[31] In order to develop such weapons, developers will need to creatively apply their knowledge of both the moral limits of war and the practical limits of technology to the question of how to develop militarily efficacious weaponry.

---

[27] Ibid., rule 63, 187–88.
[28] Ibid., rule 62, 185–86.
[29] Ibid., rule 64, 188–91.
[30] Ibid., rule 65, 191–92.
[31] De George, "Post-September 11," 183–90, 186–87.

Similarly, the principle of constant care for civilian risks in the conduct of cyberoperations requires that cyberwarriors and their commanders possess a sophisticated level of understanding of the technologies they are using in order to ensure that they can accurately predict the consequences of a cyberattack. As the authors of the *Tallinn Manual* note:

> Given the complexity of cyber operations, the high probability of affective civilian systems, and the sometimes limited understanding of their nature and effects on the part of those charged with approving cyber operations, mission planners should, where feasible, have technical experts available to assist them in determining whether appropriate precautionary measures have taken place.[32]

A similar process could be undertaken with many, if not all, of the principles listed above, indicating some of the aretaic elements of the morally excellent cyberwarrior.

Those familiar with traditional principles of *jus in bello* will note the similarities between the principles described by the *Tallinn Manual* and those that just war theory states are applicable to conventional warriors. This is most clearly evidenced by the shared importance of proportionate responses to threat and attack in both just war theory and the *Tallinn Manual*, as well as the shared insistence on discrimination between combatants and noncombatants.[33] It is noteworthy that the *Tallinn Manual*, so far the most comprehensive, fully fledged code of conduct for cyberwarriors, presents a strong similarity between the professional duties of cyberwarriors and conventional warriors. Given this, it is worth examining whether an independent code of honor for cyberwarriors is necessary, or whether they are governed by the same codes as conventional warriors.

7.3 Why Not the Traditional Warrior Code?

There are several reasons why we cannot simply apply a conventional warrior code of honor to cyberwarriors. However, before explaining these I will respond to the apparent consistency between the codes indicated by the *Tallinn Manual*. It is not the case that the manual is mistaken in seeing a close similarity between conventional and cyberprinciples of conduct; indeed, as Charles J. Dunlap notes, "existing law has ready applicability to cyber operations."[34] Rather, the *Tallinn Manual* is focused only on *macro* principles that can and do apply to *all*

---

[32] Schmitt, *Tallinn Manual*, 166.
[33] For a concise introduction to the just war tradition, see Quinlan, "Justifying War," 7–15.
[34] Dunlap, "Intersection of Law and Ethics in Cyberwar," 2.

modes of warfare, including both conventional war and cyberwar. The project of the *Tallinn Manual* is to show how existing principles of warfare, such as the principles of discrimination and proportion, apply practically in the cyberdomain. Because these moral principles are derived from the basic *telos* of war—namely, the achievement of a just and lasting peace, and the security of the rights of innocent persons—and are shared by conventional war and cyberwar, one should not be surprised to see similarities in the normative requirements of military personnel in both instances.

The reason for the apparent commonality between the professional duties of conventional warriors and cyberwarriors, then, is that they do have a shared purpose and a common justification for their professions. Ultimately, all just wars are in part justified by the moral goodness of the just peace at which they aim.[35] Presumably then, at least on occasions when cyberwar is fought as an addendum to, or substitute for, conventional war, it will—or at least ought to—aim to influence the international domain in the same way, by substituting an unjust, rights-violating set of relationships with just and nonviolent ones. Thus, conventional war and cyberwar can, and often will, share a common teleology; and similarly, conventional warriors and cyberwarriors may at times be committed to achieving the same goals: a just and lasting peace in which the rights of all persons—in particular the innocent—are safe from further violation.[36]

While the teleological explanation reveals some areas of commonality between cyberwarriors and conventional warriors, it also reveals the first reason why the warrior code of honor is not readily translatable to the cyberdomain: *not all cyberoperatives fight wars*. Even the *Tallinn Manual* allows for the legitimate use of cyberweapons, cyberstrategies, and cybertechnologies for nonmilitary ends such as espionage. Thus, an exclusively military-oriented code of honor will fail to capture the breadth of cyberoperations. Of course, not all military personnel fight wars either: some are engineers, medics, chaplains, and so on. In these cases too, I believe a warrior code is ill-suited for the professions because the warrior code is a code concerned primarily with combat.

Further complicating any attempt to closely align the normative identity of conventional warriors with those of cyberwarriors is the fact that many cyberoperatives are not members of the military or intelligence communities at all, but private

---

[35] This claim was originally made by medieval just war theorist Thomas Aquinas, who believed war could only be morally justified where it aimed to restore a state of peace that had been broken. Compare Aquinas, *Summa Theologica*, II-II, Q. 40, Art. 1; the same was argued by Francisco di Vitoria. Compare di Vitoria, *On Civil Power*, Q.3, Art. 4. In the modern age, this idea is present in Walzer, *Just and Unjust Wars*, 121; Quinlan, "Justfying War," 9; and Orend, who describes the aims of *jus post bellum* as a "just peace" in *Morality of War*, 181.

[36] For a defense of this approach to *jus post bellum*, see Orend, *Morality of War*, 161–89; May, *After War Ends*.

contractors—essentially civilians whose professional roles will vary sharply from their uniformed colleagues.[37]

Most significantly though, consider that the warrior code has a very specific intention; namely, to preserve the humanity of warriors and distinguish them from mere murderers. To be so distinguished, French contends that warriors "must learn to take only certain lives in certain ways, at certain times, and for certain reasons. Otherwise, they become murderers and will find themselves condemned by the very societies they were created to serve."[38] Indeed, as I noted at the beginning of this chapter, much of the concern for *jus in bello* stems from the belief that the killing of another human being is a prima facie wrong. This is to say that the majority of ethical consideration of conventional warriors is concerned with the fact that the profession of soldiering is likely to require one to directly kill or harm other people.

French argues that part of the purpose of a warrior code is to prevent warriors from becoming "mere murderers."[39] Both murderers and warriors kill other people, but one is justified by (i) the cause for which he or she kills; (ii) the *manner* in which he or she kills; and (iii) who it is that he or she aims to kill. The same can be said with regard to harming: warriors harm only certain people in certain ways for a certain cause. Cyberwarriors, by contrast, are unlikely to harm to anywhere near the same extent as conventional warriors. Furthermore, much of the harm they commit will be incidental to acts of sabotage; that is, the target of their attack will be a *thing*, not a person.

For these reasons, the warrior code is of limited use for cyberwarriors. Although it is true that some cyberwarriors can—and probably will—use cyberweapons to kill other people, it is not clear that killing is central to the profession in the same way as it is for conventional soldiers. Indeed, the most widely popularized cyberattacks we have seen employed to date (including the Stuxnet virus, the blockade against Estonia, and the Russian offensive against Georgia) have been used to attack systems and infrastructure rather than people.[40] It appears conceivable that one might be able to conduct a cyberwar without ever having to take another human life[41]—something that cannot be said for conventional war. As such, the close alignment of killing and the warrior code makes it a bad fit for cyberwarriors.

Thus far I have claimed that soldiering—perhaps the archetypal military profession—is not only morally governed by formally written laws and codes of

---

[37] Some question whether civilian personnel ought to be conducting state cyberoperations. The *Tallinn Manual* allows for civilian contractors and mercenaries, although it denies them combatant status (rules 28 and 29). On this basis, Dunlap argues that civilian involvement in cyberwar may be imprudent. See Dunlap, "Intersection of Law and Ethics in Cyberwar," 5–7.

[38] French, *Code of the Warrior*, 3.

[39] Ibid., 1–2.

[40] Rid, *Cyber War Will Not Take Place*.

[41] This point is discussed in more detail by Ryan Jenkins in this volume.

conduct but also by informal beliefs about what constitutes a morally praiseworthy, honorable warrior. This warrior code of honor helps warriors to internalize the professional duties that bear on them by making moral conduct an intrinsic element of the warrior identity. While the development of legal governance of cyberwar is relatively new, and ethicists are only now beginning to unpack the moral responsibilities of cyberwarriors, there are nevertheless basic principles of conduct for cyberwarriors, as evidenced by those I discovered in the *Tallinn Manual*, above. However, despite similarities between the professional duties of warriors and cyberwarriors vis-à-vis the teleological elements of the professions, the warrior code of honor is, in many ways, inadequate for cyberwarriors. In what remains of this chapter I will argue that cyberwarriors can indeed possess a unique code of honor that reflects both their disembodied status and the particular roles they are required to perform.

7.4 Cyberwarrior: Spy, Saboteur, and Assassin

In order to develop the beginnings of a code of honor for cyberwarriors, I will distinguish between what I believe are the three central roles of the cyberwarrior. Before I do, however, I emphasize that the following discussion is the beginning of what I hope to be a long and engaging conversation between scholars, practitioners, and lawyers regarding the professional responsibilities and moral excellences of cyberwarriors. If the professional responsibilities of cyberwarriors change, so too will their code of honor, as the code is derived directly from the nature of their work. With this said, I believe we can categorize the roles of today's cyberwarriors into three different categories: (i) collection, manipulation, and protection of data (espionage); (ii) invasion, manipulation, interruption, and destruction of critical systems and infrastructure (sabotage); (iii) use of cyberweapons to bring about the deaths of targeted persons (assassination).[42]

This threefold division emerges from Herbert Lin's division of offensive operations in cyberspace into two areas: cyberattack, "the use of deliberate activities to alter, disrupt, deceive, degrade, or destroy computer systems or networks used by an adversary or the information and/or programs resident in or transiting through these systems or networks"; and cyberexploitation, "activities designed to penetrate computer systems or networks used by an adversary, for the purposes of obtaining information resident on or transiting through these systems or networks."[43] However, the growing potential for cyberweapons to be able to harm people as well

---

[42] I should also note that it is possible—even likely—that cyberwarriors will work in only one of these three roles, in which case only certain duties and virtues will be relevant to them. However, as this chapter aims only to make some opening remarks on the matter of a cyberwarrior code, I will assume that there is a possibility that the same person may be required to perform each of these three roles.

[43] Lin, "Cyber Conflict and International Human Law," 518–19.

as systems warrants dividing cyberattacks into two areas: attacks on systems and attacks on people. Thus, the profession of the cyberwarrior is divided into the three areas I outlined above.

The division is also similar to that of Thomas Rid, who argues that acts of cyberwar will typically be of three different kinds: sabotage, espionage, and subversion.[44] I have not included subversion on my list because most of Rid's examples of subversion—"the deliberate attempt to undermine the authority, the integrity, and the constitution of an established authority or order"[45]—are in fact instances of politically motivated sabotage ("hacktivism"), such as the Anonymous attacks on the Church of Scientology and HBGary Federal and DDoS attacks in Estonia, Georgia, and Israel.

Subversion in the form of hacktivism is not likely to be a chief role of cyberwarriors, and is more likely to be undertaken by private citizens. Whilst there are a number of militarily significant functions that individuals can play in the cyberdomain, hacktivism, for instance, which Dorothy Denning's essay "Cyberwarriors: Activists and Terrorists Turn to Cyberspace" focuses exclusively on,[46] these are not definitive of cyber*warriors*. Hacktivists are no more cyberwarriors than the participants in a crowd riot are warriors. For the sake of this discussion, cyberwarriors will be understood as being those acting from within the national defense organization of a legally recognized state, or who have been contracted to perform in an equivalent role. What defines a cyber*warrior*, I will argue, is his or her involvement in cyberespionage, cybersabotage, and cyberassassination; and what defines an *honorable* cyberwarrior is participation in those activities with particular virtues and in accordance with LOAC, the teleological elements of the profession, and adherence to the *in bello* norms of warfare.

Let us begin with the espionage component: the *Tallinn Manual* stipulates that cyberespionage and other forms of digital information gathering are not violations of LOAC,[47] and although that alone does not mean it will always be morally justifiable, it does mean we should expect states to engage frequently and readily in espionage in the cyberdomain. One suspects that this is nothing new to international relations, and that states frequently engage in espionage against not only hostile nations but neutral and perhaps allied nations as well through the use of analysts and—more popularly—intelligence operatives, or "spies." Intelligence operatives deal not only in obtaining information, but also in discriminating between information that is relevant and useful, and information that is useless or false. Thus, the good spy is discerning and prudent. He or she is able to identify relevant information quickly and use it to the benefit of his nation.

---

[44] Rid, "Cyber War Will Not Take Place," 5–32.
[45] Ibid., 22.
[46] Denning, "Cyberwarriors: Activists and Terrorists Turn to Cyberspace," 70–75.
[47] Schmitt, *Tallinn Manual*, 192–93.

However, operatives are usually sent to extract information from sources that have already been identified (by other operatives, as well as through other forms of intelligence) as having a high chance of possessing valuable information. Tony Pfaff and Jefferey R. Tiel argue that "the feature that will make persons subject to espionage operations is that they possess or are likely to possess secrets that threaten national security."[48]

This is often not the case in cyberespionage, where security operatives have access to huge amounts of information that is collected indiscriminately. Edward Snowden's leaks regarding the United States' NSA Management Directive #424 have recently demonstrated this aspect of cyberespionage. However, rather than undermining the need for virtues of discernment, this fact makes them all the more pressing. The good cyberwarrior will be able to determine what, from a huge array of information, is worth pursuing in depth and what is worth setting aside. As George R. Lucas notes, reflecting on the Snowden leaks:

> It would be an abuse ... for example, for a human "cyber warrior" to listen in to Skype conversations between private adults engaged in no crime, or to read their emails for amusement.[49]

Rather than undermining the need for discernment, the cyberaspect of espionage serves to enhance it. Privacy, the good most at stake in espionage (so important that Anita L. Allen argues that "[e]ven wrongdoers' expectations of privacy matter *prima facie* when selecting the means of information gathering"),[50] must be a critical consideration of the discerning cyberwarrior.

Discernment introduces another virtue of the good cyberwarrior: discretion. Within their work, cyberwarriors are likely to come across a host of sensitive information of little relevance to their purposes. Cyberwarriors must consider it a matter of honor never to disclose any information that they do not judge explicitly relevant to national security. This includes, I would argue, at least some forms of personal information that can be used to blackmail people into becoming "false-flag" agents (agents working against their own nation).[51] One reason why blackmail and coercion should not be justified under a cyberwarrior code is because it, like the warrior code, is subject to universally applicable moral laws and LOAC. If coercion and exploitation are unjustifiable under broader moral principles, they should not be lauded by a code of honor.

Another, more compelling, reason why blackmail and reckless violations of privacy are violations of the cyberwarrior code of honor is because they contradict the

---

[48] Pfaff and Tiel, "Ethics of Espionage," 6.
[49] Lucas, "NSA Management Directive #424," 37.
[50] Allen, "Virtuous Spy."
[51] Perry, "'Repugnant Philosophy': Ethics, Espionage, and Covert Action."

teleological justification of the profession of the cyberwarrior: the pursuit of a just and lasting peace, and the defense and restoration of the rights of innocent civilians. Needless or reckless violations of civilian privacy cannot reasonably be described as being in the interests of those same civilians, even in the interests of peace. Such a peace would, as Larry May describes, "[provide] security from external threats, but . . . fail to protect human rights within the society."[52] These are rights, in this case, to privacy that, if regularly violated, can hardly be said to exist at all. Thus, the aretaic element of cyberespionage (and which ought to be represented in any formal code of conduct), that is, the virtues of discretion and discernment, are directly linked to the telos of cyberespionage that morally justifies the profession.

The second central role of cyberwarriors is their use in undermining cybersystems of their enemies. These can be used to cripple an opponent's own ability to launch cyberattacks, disrupt enemy communications before one's own (physical) military strike, and so on. Here, the function of the cyberwarrior is to sabotage the infrastructure of an enemy nation through "deliberate attempt[s] to weaken or destroy an economic or military system."[53] Like physical saboteurs, the primary moral concerns here are ones that warriors share: that civilian harms from such attacks are minimized, and that the damage is proportionate to the offense committed by the target. As Rid notes, "[i]f violence is used, things are the prime targets, not humans."[54]

The honorable cyberwarrior will go further than merely to ensure each cyberattack he or she orchestrates is proportionate. Rather, because, as Ryan Jenkins notes, part of the appeal of cyberwar is its potential to sharply reduce harm,[55] the cyberwarrior honor code should aim to exploit this moral benefit as much as possible, aiming to reduce collateral damage from "proportional" to "none." One way of describing this particular virtue of the cybersaboteur is as *cleanliness:* the objective being that that each cyberattack be as precise as cyberweapons promise to be and thus enhance the moral promise of cyberwar to reduce risk to civilian lives. Doing this in a way that is effective will almost certainly make weapons development more difficult and time consuming, and although this fact does not undermine the moral importance of clean operations, it may be that time-sensitive operations have some licence to be "messier" than less pressing ones.

Although the appeal of certain acts of cybersabotage are appealing because of their cleanliness, the sheer mass of interconnected and dual-use systems in existence today make the possibility of unforeseen consequences of an attack a constant risk. For this reason, the technical proficiency of cyberwarriors must be a matter of honor. As Dunlap argues, "one of the key ethical responsibilities of cyberwarriors

---

[52] May, *After War Ends*, 50
[53] Rid, "Cyber War Will Not Take Place," 16.
[54] Ibid.
[55] Jenkins, "Is Stuxnet Physical? Does it Matter?" 74. See also Jenkins's chapter in this volume.

is the virtue of *competence*."[56] The technical understanding of cyberweapons, cybertechnology, and the means of using it for benefit and exploitation is as important to cyberwarriors as the physical strength, marksmanship, or tactical nous of a conventional warrior. Although this is in part the stuff of competence, cyberwarriors will also—more so than their conventional equivalents—esteem *creativity and innovation*. Unlike conventional warriors, the best cyberwarriors develop tools, systems, and programs that can discover exploitable bugs in computer code faster than they could do so themselves. In a confidential 2013 interview on *Infoworld*, an anonymous cyberwarrior explained that he spent years developing better "fuzzers"—programs that check code for exploitable bugs.[57] This cyberwarrior's ability to develop better tools than his peers made him "one of the elite, even in a group of elites."[58] Although large and complex cyberweapons like Stuxnet are developed by large teams,[59] the smaller, day-to-day hacking undertaken by cyberwarriors will be largely autonomous, and more creative cyberwarriors are likely to enjoy greater success.

Because of their technical expertise (and the fact that high-level military commanders are likely to be far less expert than they are), cyberwarriors are likely to be required to identify vulnerabilities and develop attack strategies themselves. Ethical and strategic oversight from senior command figures will, at least in these early stages, not be equipped with technical knowledge or prowess remotely equivalent to the cyberwarrior him- or herself. Thus, the cyberwarrior code will afford much more honor to autonomy and independent thinking than the conventional warrior code.

Cybersabotage will not always be perfectly clean: at times, attacks on military targets, factories, warships, or a host of other possible targets may cause death or harm to military or civilian personnel. However, so long as the target of attack was a system or object, such fallout can be morally justified as a side effect of a morally just action.[60] However, there may be times when cyberweapons are deployed with the express intention of killing a person or group of persons. In such cases, the cyberwarrior responsible is involved in cyberassassination rather than cybersabotage. Although I believe this is a possible function of the cyberwarrior, I do not believe it can be morally justified, because, as was noted earlier, the *Tallinn Manual*

---

[56] Dunlap, "Intersection of Law and Ethics in Cyberwar," 12. Emphasis added.

[57] Grimes, "In His Own Words: Confessions of a Cyber Warrior."

[58] Ibid.

[59] For a detailed explanation of the process of developing Stuxnet, see Rid, "Cyber War Will Not Take Place," 17–20.

[60] This position, usually described as the doctrine of double-effect, has long been the position of Catholic moral theologians and has more recently found a secular basis in military ethics. For a clear introduction to the justification and limitations of collateral damage, see Ford, "Morality of Obliteration Bombing," 261–309. Note that others disagree with the very notion of the doctrine of double-effect. See, e.g., Steinhoff, *Ethics of War and Terrorism*.

currently forbids the use of perfidy as a means to killing or injuring other people. In addition to this, I believe that international law ought to extend the *Tallinn Manual*'s prohibition of perfidious means to forbid all targeted killing by cyberweapons.[61]

There are several reasons for this. The first is because, as Perry notes, the inability of victim nations to attribute responsibility in cases of assassination risks "misguided retaliation against a third party by allies of the assassination victim."[62] Given that one of the major challenges of cyberwar is the difficulty in attributing attacks,[63] a cyberassassination is doubly anonymous: a factor that is not only problematic from the perspective of international law but also makes legitimizing cyberassassination a potential force multiplier. A second objection is that, as Christian Enemark notes, "war necessarily involves some kind of contest."[64] Most modern accounts of the *in bello* legitimacy of killing in war are predicated on either the legitimacy of self-defense, the mutual threat that combatants pose to one another, or the moral justice of one's cause. However, the first two of these are unavailable to cyberwarriors, and the third is not something that cyberwarriors could be certain of all the time.[65] Thus, it seems to me that cyberwarriors should, as a matter of both law and honor, refuse to participate in targeted killings by way of cyberweapons. Instead, they should define their profession by way of the potential of cyberwar to be a near-bloodless mode of warfare.

Bibliography

Allen, Anita L. "The Virtuous Spy: Privacy as an Ethical Limit." *The Monist* (January 2008): https://www.law.upenn.edu/institutes/cerl/conferences/cyberwar/papers/reading/Allen.pdf.

Aquinas, Thomas. *Summa Theologica*. Trans. Fathers of the English Dominican Province. 1920. http://newadvent.org/summa/index.html

Beard, Matthew. "Cyberwar and Just War Theory." In *Applied Ethics, Risk, Justice and Liberty*, ed. Center for Applied Ethics and Philosophy. Hokkaido: Centre for Applied Ethics and Philosophy, 2013.

Bradley, Lt. Gabriel. "Honor, Not Law." *Armed Forces Journal* (March 2012). http://www.armedforcesjournal.com/honor-not-law/.

de George, Richard. "Post-September 11: Computers, Ethics and War." *Ethics and Information Technology* 5 (2003): 183–90.

Denning, Dorothy. "Cyberwarriors: Activists and Terrorists Turn to Cyberspace." *Harvard International Review* 23 (2001): 70–75.

di Vitoria, Francisco. *On Civil Power*. In *Vitoria: Political Writings*, ed. Anthony Padgen and Jeremy Lawrence. Cambridge: Cambridge University Press, 1991.

Dipert, Randall. "The Ethics of Cyberwarfare." *Journal of Military Ethics*, 9, no. 4 (2010): 384–410.

---

[61] This point is discussed in more detail by Heather Roff in this volume.

[62] Perry, *Partly Cloudy*, 187.

[63] Compare Dipert, "Ethics of Cyberwarfare," 385.

[64] Enemark, *Armed Drones and the Ethics of War*, 77.

[65] In this regard, they are akin to conventional warriors: neither warriors nor cyberwarriors can be confident of the justice of his or her cause. Though this is a point of commonality, it is still a point worth acknowledging in discussions of intentional killing by cyberwarriors.

Dunlap, Maj. Gen. Charles J. "The Intersection of Law and Ethics in Cyberwar: Some Reflections." *Air & Space Journal* 24 (2012): 1–17.
Enemark, Christian. *Armed Drones and the Ethics of War: Military Virtue in a Post-Heroic Age.* London: Routledge, 2014.
Ford, John C. "The Morality of Obliteration Bombing." *Theological Studies* 5 (1944): 261–309.
French, Shannon E. *The Code of the Warrior.* Lanham, MD: Rowman & Littlefield, 2003.
Jenkins, Ryan. "Is Stuxnet Physical? Does it Matter?" *Journal of Military Ethics* 12 (2013): 68–79.
Lucas, George R. "NSA Management Directive #424: Secrecy and Privacy in the Aftermath of Edward Snowden." *Ethics and International Affairs* 28 (2014): 29–38.
May, Larry. *After War Ends: A Philosophical Perspective.* Cambridge: Cambridge University Press, 2012.
Orend, Brian. *The Morality of War.* Toronto: Broadview, 2006.
Osiel, Mark. *Obeying Orders: Atrocity, Military Discipline, and the Law of War.* Piscataway, NJ: Transaction, 1999.
Perry, David L. "'Repugnant Philosophy': Ethics, Espionage, and Covert Action." *Journal of Conflict Studies* 15 (1995).
http://www.scu.edu/ethics/publications/submitted/Perry/repugnant.html.
Perry, David L. *Partly Cloudy: Ethics in War, Espionage, Covert Action, and Interrogation.* Plymouth, UK: Scarecrow Press, 2009.
Pfaff, Tony, and Jeffrey R. Tiel. "The Ethics of Espionage." *Journal of Military Ethics* 3 (2004): 1–15.
Quinlan, Michael. "Justifying War." *Australian Journal of International Affairs* 58, no. 1 (2004): 7–15.
Rid, Thomas. "Cyber War Will Not Take Place." *Journal of Strategic Studies* 35, no. 1 (2011): 5–32.
Schmitt, Michael N., ed. *Tallinn Manual on the International Law Applicable to Cyber Warfare.* Cambridge: Cambridge University Press, 2013.
United States Department of Army. *FM 3-38 Cyber Electromagnetic Activities.* (2014). http://www.fas.org/irp/doddir/army/fm3-38.pdf.
Walzer, Michael. *Just and Unjust Wars: A Moral Argument with Historical Illustrations.* 4th ed. New York: Basic Books, 2006.

8

Immune from Cyberfire?

The Psychological and Physiological Effects of Cyberwarfare

DAPHNA CANETTI, MICHAEL L. GROSS,

AND ISRAEL WAISMEL-MANOR

When noncombatants suffer bodily injury or loss of life during war, they experience harm in the most obvious way. While protected from direct or intentional harm, noncombatants may suffer proportionate collateral harm in the course of effective and necessary military operations. This is the principle of noncombatant immunity. To inflict *direct* harm upon noncombatants is to egregiously violate this principle and commit a crime of war against the innocent.

What then of cyberwar? What kind of harm does cyberwar inflict upon noncombatants? Do victims of cyberattacks suffer significant physiological harm or only some measure of mental suffering, distress, and anxiety? And, if the latter, does such suffering violate noncombatant immunity? Compared to death and injury, psychological harm appears far less grave. While one can certainly paint scenarios of cyberattacks that cause acute mental trauma, much of the suffering that cyberwarfare seems to bring lacks the pain and persistence of many physical injuries.

Following an overview that describes the challenge that cyberoperations pose for the principle of noncombatant immunity, the following sections map out and analyze the harms of cyberwarfare. Consider, first, physiological harm. Although no person has lost his or her life or suffered any kind of physical injury from a cyberattack at time of writing, the literature is replete with scenarios of death and devastation. These come in the course of cyberattacks on vital infrastructures that disrupt air and rail transportation or poison water supplies. In many ways, these are similar to the consequences of conventional war. For the most part, however, modern cyberwarfare causes no physical injury. As a result, one may reasonably ask whether noncombatants enjoy protection from cyberattacks that disrupt telecommunications; disable social media; or destroy, disclose, or steal financial data and personal information. The answer hinges upon the psychological harm that victims suffer, particularly if belligerents target civilians and civilian infrastructures directly. Extrapolating from studies of cyberbullying, identity theft, and ordinary burglary, and building upon the effects of simulated cyberterrorism in the laboratory, we explore the psychological harms of cyberwarfare. Cyberwarfare is not benign but causes stress, anxiety, and fear. Such mental suffering threatens to disrupt routine life, impair educational and workplace performance, impact significantly on the poor and elderly, and increase public pressure on the government to act. Although most forms of psychological suffering are not as intense, prolonged, or irreversible as bodily injury or loss of life, our analysis suggests that the psychological harm of cyberwar can affect well-being nonetheless

8.1 Noncombatant Immunity, Cyberwar, and Cyberterrorism

In conventional war noncombatants, that is, those who take no direct part in the hostilities, are protected from both direct and collateral injuries. Posing no threat, noncombatants may not be intentionally killed or injured. Everyone has long recognized, however, that noncombatants will die in war as belligerents disable military targets. When such deaths are necessary, unavoidable, and unintended, just war theory and international law make room for collateral or incidental civilian casualties as long as they are neither excessive nor disproportionate relative to the military gains a belligerent seeks. When too many civilians die in the course of military operations, states face condemnation for causing disproportionate harm. When parties to a conflict intentionally harm civilians, they face charges of murder and terrorism. To assess proportionality or terrorism, one must understand the harm noncombatants suffer. Observers usually measure harm in terms of civilian deaths. Injuries and property destruction may also weigh in, but psychological malaise enjoys little attention and rarely figures in calculations of proportionality. This leaves cyberwarfare—which targets facilities rather than individuals and has, to date, caused no immediate injuries—beyond considerations of proportionality and wide open to unconstrained use.

Cyberoperations, whether directed at military targets (cyberwarfare) or civilian targets (cyberterrorism), attack a wide range of infrastructures whose destruction may kill or injure noncombatants and whose penetration may lead to the theft, eradication, or disclosure of privileged data. Other operations may go after civilians directly by attacking personal cell phones or computers to steal identities, pilfer bank accounts, or threaten civilians with personal harm. Only a few of these attacks present a threat to life or limb (Table 8.1).

Table 8.1 **Types and Outcomes of Cyberattacks**

Target of Cyberattack	Outcomes	Possibility of Physical Harm to Life and Limb
Critical Infrastructures		
Public Security: Fire, Police	• Disclosure of confidential information • Disruption/cessation of service	No Yes
Water/Dams	• Disruption of water supply • Pollution • Flooding	No Yes Yes
Transportation Networks	• Disrupted schedules • Equipment failure (train/airplane crashes)	No Yes
Other Infrastructures		
Medical Infrastructures	• Disclosure of personal information • Alteration of medical records and prescriptions • Disruption of vital medical services: operating rooms, ventilators	No Yes Yes
Financial Networks	• No access to bank accounts • Stolen funds • Collapse of stock exchange	No No No
Public Records	• Disclosure of criminal records or classified court hearings (sexual abuse, national security cases, adoption) • Alteration of public records • Disclosure of biometric information	No No No
Personal Attacks		
Personal Computers/ Cell Phones	• Destruction/theft of data • Disclosure of personal information • Identity theft • Invasion of privacy	No No No No
Individual Users	• Cyberbullying: threaten individuals with harm	Yes

Table 8.1 suggests that most cyberoperations cannot cause physical harm. Those attacks that do bring death or injury have yet to materialize. Absent the prospect of injuries or loss of life in the course of cyberwar, the principle of noncombatant immunity faces two challenges. First, do cyberattacks harm noncombatants? If not, then noncombatants are appropriate targets of cyberoperations and cyberterrorism is permissible. If, on the other hand, cyberattacks bring harm to noncombatants, then cyberterrorism is impermissible but proportionate collateral harm is not. This raises the second challenge: Do cyberattacks against military targets cause disproportionate harm to noncombatants?

Injury and loss of life provide one metric to answer these two questions. Kinetic attacks that kill or injure cause sufficient harm to prohibit both direct and disproportionate attacks on noncombatants. But if cyberattacks do not kill or injure anyone, how might one evaluate direct or collateral harm? How do we know when cyberattacks violate the principle of proportionality? Psychological harm and mental suffering provide one criterion to address these questions. Kinetic terrorism offers the clearest example. When a suicide bomber, for example, kills twenty people and injures one hundred, terrorism violates noncombatant immunity in the most extreme way, first by intentionally killing small numbers of individuals (the primary victims of terror) and then by traumatizing large number of individuals (the secondary victims of terror). Terrorism causes intense anxiety and dread among those who fear that they or their loved ones will be next to die.

Cyberterrorism, however, works differently. While in some cases cyberattacks may cause death and injury, none at time of writing have done so. Instead, cyberterrorism disrupts civilian infrastructures and often targets noncombatant assets *directly*. There is no doubt that these attacks are intentional and direct. The question is: "Do they cause any harm?" *The Tallinn Manual on the International Law Applicable to Cyber Warfare*, for example, defines a cyberattack as "a cyber-operation, whether offensive or defensive, that is reasonably expected to cause injury or death to persons or damage or destruction to objects."[1] Understanding that not all cyberattacks can cause injury or death, the framers of the *Tallinn Manual* consider that cyberoperations cause sufficiently severe mental suffering to warrant condemnation:

> While the notion of attack extends to injuries and death caused to individuals, it is, in light of the law of armed conflict's underlying humanitarian purposes, reasonable to extend the definition to serious illness and *severe mental suffering* that are tantamount to injury. In particular, note that Article 51(2) of the Additional Protocol I prohibits "acts or threats

---

[1] Schmitt, *Tallinn Manual*, 106.

of violence the primary purpose of which is to spread terror among the civilian population." Since terror is a psychological condition resulting in mental suffering, inclusion of such suffering in this Rule is supportable through analogy.[2]

The *Tallinn Manual*'s conclusion is short an empirical anchor. We will review the psychological sequelae of terrorism in sections 8.2 and 8.3, but it is not at all clear that cyberterrorism of the most extreme, hypothetical kind will cause severe mental suffering. Instead, and in the worst case, the effects of cyberwarfare closely resemble acts of conventional warfare and economic sanctions that bring long-term damage to industrial, agricultural, utility, and water infrastructures but do not necessarily cause widespread death or injury. The same might be expected of many cyberattacks. Some cyberoperations may leave the economy decimated while others may leave people distressed and even terrified of losing data or money. They are not, however, necessarily "terrorized" if by that we mean fearful of losing life and limb.

In short, we need to know about psychological harm for two reasons. The first is to determine whether direct cyberattacks upon noncombatants constitute terrorism. The second is to assess the proportionality of collateral harm in the course of legitimate military strikes. Direct harm defines terrorism. Excessive harm constrains proportionality. If cyberattacks cause nothing but moderate inconvenience, they cannot be acts of terrorism or ever cause disproportionate harm. "Inconvenience," of course, is a very broad term and may include all kinds of hardship short of severe mental or physical suffering. If severe mental suffering turns on fear of death, then cyberoperations will usually fall short. The question remains whether cyberattacks substantially affect well-being in other ways.

8.2 Terrorism and Cyberterrorism: Confronting Bodily Injury and Loss of Life

The primary victims of kinetic terrorism die or suffer horrible injuries while secondary victims avoid physical harm but suffer psychologically. Terrorism gains purchase by posing a deadly, persistent, and unpredictable threat. The psychological effects of kinetic terrorism are well documented. Among the most severe is post-traumatic stress disorder (PTSD), a severe anxiety disorder that occurs following exposure to a traumatic event involving death or serious injury and to which individuals respond with "fear, helplessness, or horror."[3] Following terror attacks, PTSD victims reexperience their trauma through intrusive recollections, dreams, and hallucinations

---

[2] Ibid., 93 (emphasis added).
[3] Yehuda, "Post-Traumatic Stress Disorder," 108.

and suffer from insomnia, uncontrollable anger, and difficulty concentrating. PTSD can impair daily functioning and puts patients at increased risk for depression, drug and alcohol abuse, eating disorders, suicidal thoughts and actions, cardiovascular disease, chronic pain, and autoimmune diseases. Prior to 9/11, PTSD affected 5 to 6 percent of men and 10 to 14 percent of women in the United States.[4] Following the 9/11 attacks and other terror attacks around the world, studies demonstrate a significant increase in PTSD and other anxiety disorders.[5] In southern Israel, too, PTSD and related anxiety-disorder symptoms were common in the aftermath of rocket attacks that continued unabated from 2001 to 2008.[6]

Digging deeper, closer studies reveal two distinct groups of individuals suffering fear-related effects from terrorism. One group exhibits the common PTSD symptoms including psychological distress, insomnia, and exaggerated startle responses. The other group does not reexperience a *past* trauma but suffers instead from "anticipatory anxiety," that is, fear and dread associated with *future* attacks. These fears grow as the threat persists. While relatively few people suffer from PTSD following a terrorist attack, many more suffer from various degrees of debilitating fear.[7] These, too, are secondary victims of terrorism who experience no physical harm. Nor are the victims necessarily present at the site of a terror attack.

The differences between the incidence of PTSD and anticipatory anxiety are striking. While the incidence of PTSD dropped across a US sample from 17 percent two months after 9/11 to 5.8 percent six months after the attacks, 60 to 65 percent continued to fear future terrorist attacks and worry about harm befalling their family.[8] Widespread fear, then, rather than the specific incidence of PTSD is the more pervasive effect of terrorism and more accurately reflects the psychological malaise that accompanies war and terrorism. Terrorism fears correlate with anxiety, depression and insomnia, and feelings of incapacitating helplessness.[9] Random bombings and missile attacks lead to fear-induced changes in behavior. Victims of terrorism avoid public transportation, public forums, and confined venues such as restaurants, cafes, and theaters, while others often disparage those ethnic groups they identify with terrorists.[10] Others simply flee.[11] As a result of increased isolation, migration, and ethnocentrism, social intercourse diminishes. Terrorism brings

---

[4] Ibid.

[5] Sinclair, "Fears of Terrorism and Future Threat," 101–15; Yehuda et al., "Pathological Responses to Terrorism," 1793–1805.

[6] Gelkopf et al., "Protective Factors and Predictors of Vulnerability to Chronic Stress," 757–66.

[7] Sinclair and LoCicero, "Fearing Future Terrorism," 75–90.

[8] Silver-Cohen et al., "Nationwide Longitudinal Study of Psychological Responses to September 11," 1235–44.

[9] Sinclair and Antonius, *Psychology of Terrorism Fears*.

[10] Canetti-Nisim et al., "New Stress-Based Model of Political Extremism Personal Exposure to Terrorism," 363–89.

[11] Diamond et al., "Ongoing Traumatic Stress Response (OTSR) in Sderot, Israel," 19.

a constant sense of anxiety, fears about harm to family members, and heightened vigilance with regard to suspicious packages and people. Ruminations about recent attacks and fruitless efforts to predict future strikes in an atmosphere of acute fatalism become constant preoccupations. Workplace efficiency deteriorates, turnover and absenteeism increase, and performance, morale, and motivation suffer.[12] Civil society perseveres but community and economic life suffer in the wake of terrorism. Among many secondary victims of kinetic terrorism, personal well-being deteriorates in a most fundamental way.

Will cyberterrorism, even in its most extreme form, bring such consequences? This depends on the nature of the cyberattack. Consider the hypothetical scenarios that pervade the literature. Here, individuals may certainly die if trains derail or airplanes crash. In these cases cyberterrorism resembles suicide bombings. The primary victims will suffer death and injury while the secondary victims will endure psychological pain and suffering. Yet cyberattacks of even the most extreme type are far more focused than kinetic attacks. A terrorist can blow himself up anywhere, but cyberterrorism requires a computer network to attack. Public gathering places, a favorite venue for suicide bombers, would not make likely targets for cyberterrorists. As such, the random nature of terror, and with it the resulting anxiety, might be mitigated by avoiding vulnerable targets.

The psychological consequences of cyberterrorism diminish further when one considers less catastrophic assaults on other infrastructures. Flooding, pollution, and the destruction of utility networks are common scenarios. What physical and psychological harm do these attacks bring? While deaths may occur, far more prevalent is the economic devastation and deleterious long-term public health effects that come when farmlands flood, electrical grids collapse, or water treatment plants break down. While unknown in the world of cyber, such effects are common during and after armed conflict.

In the course of war, the destruction of property is not benign. Apart from immediate harm to persons are the longer-term effects that come when vital services collapse following armed attacks. This is particularly true when health, water, sanitation, manufacturing, and agricultural facilities are destroyed or damaged. During the 2006 Second Lebanon War, for example, observers documented extensive damage to airports, ports, water and sewage treatment plants, electrical plants, roads, fuel stations, bridges, overpasses, commercial properties, homes, cropland, and livestock in southern Lebanon. Unexploded ordnance further rendered large tracts of land untillable, while the destruction of fuel storage tanks caused a disastrous oil spill along the coast.[13] The indirect costs of armed conflict include capital

---

[12] Howie, "Terrorism Threat and Managing Workplaces," 70–78.
[13] Fattouh and Kolb, "Outlook for Economic Reconstruction in Lebanon after the 2006 War,"; 97–111; Darwish, Farajalla, and Masri, "2006 War and Its Inter-Temporal Economic Impact on Agriculture in Lebanon," 629–44.

flight, discouragement of investments, decreased tourism, emigration (of medical and other professionals in particular), inflation, and food insecurity.[14] Many of these outcomes mirror those predicted for cyberwarfare and cyberterrorism. Under these conditions, the civilian population suffers enormously. In the wake of the Second Lebanon War, financial hardship and "trauma exposure" significantly increased psychiatric morbidity among the southern Lebanese civilian population."[15]

If cyberoperations destroy infrastructures in ways similar to conventional war, one might expect similar psychological consequences. When some civilians lose their lives, others will fear massively for their own. When an economy is wrecked, many will suffer significant stress and anxiety. But civilians will also rebound. Following missile attacks in southern Israel, a large percentage (40 to 78 percent) of victims was symptom-free and "the emotional impact . . . fairly moderate," an outcome that did not change much after forty-four months of intermittent attack.[16] Researchers attribute resilience to "a habituation process and coping mechanisms," "self-efficacy," strong community networks, and social cohesion.[17] Following the 2006 Second Lebanon War, these same tight social networks prevented outbreaks of major disease or social unrest and mitigated the incidence of PTSD and depression. Lebanese communities without the requisite resources and social networks, on the other hand, experienced greater incidence of mental illness and dysfunction.[18]

One might expect similar effects from cyberoperations that target critical infrastructures whose destruction will lead to death, disease, and severe economic hardship. Most cyberoperations, however, lack this reach and aim instead to disable or disrupt facilities that support social networks, banking institutions, and public institutions. Other operations target civilians directly by stealing data and money or threatening personal harm. What psychological suffering do these acts bring? Are routine life and personal well-being affected as adversely when the threat to life and limb is absent?

8.3 Cyberterrorism: Confronting Psychological Harm and Severe Mental Suffering

In the worst cases, cyberoperations that disable or destroy critical infrastructures and cause physical injury and loss of life are nearly analogous to kinetic terror attacks.

---

[14] Lindgren, *Studies in Conflict Economics and Economic Growth*.

[15] Farhood, Dimassi, and Strauss, "Understanding Post-Conflict Mental Health."

[16] Zemishlany, "Resilience and Vulnerability in Coping with Stress and Terrorism," 307–9.

[17] Bleich, Gelkopf, and Solomon, "Exposure to Terrorism, Stress-Related Mental Health Symptoms, and Coping Behaviors among a Nationally Representative Sample in Israel," 612–20; Bleich et al., "Mental Health and Resiliency Following 44 Months of Terrorism," 21.

[18] Nuwayhid et al., "Summer 2006 War on Lebanon," 505–19; Farhood, Dimassi, and Strauss, "Understanding Post-Conflict Mental Health."

Although relatively few people may die, one might easily speculate that the secondary victims of cyberterrorism experience some measure of severe mental suffering. However, cyberattacks, unlike kinetic terrorist attacks, do not target individuals but infrastructures. The effects on individuals may be immediate (as when trains derail) or indirect and less lethal (as when water sources are poisoned or electrical grids rendered inoperative). As a result, the secondary target (i.e., the civilian population) may not suffer psychological harm akin to terrorization but something less acute as occurs with the collapse of many infrastructures during war more generally.

When cyberoperations target individuals, computerized networks, or facilities, there is no obvious reason to expect that such attacks will cause severe suffering at all. For this reason, perhaps, the *Tallinn Manual* dismisses many potentially harmful cyberattacks. Among these are cyberoperations that include "blocking email throughout the country," (§30.12); DDOS attacks, "mere economic coercion" (§11.2); "cyber psychological operations intended solely to undermine confidence in a government or economy" (§11.3); or, in one elaborate example, a tweet to cause panic "falsely indicating that a highly contagious and deadly disease is spreading rapidly throughout the population" (§36.3). Yet these do not rise to the requisite level of force to constitute an armed attack in the opinion of the manual's experts. Any such attack, therefore, will not constitute cyberterrorism if directed against noncombatants.

Addressing these serious lacunae in the evolving law of cyberwarfare requires a different conception of terrorism than that assumed by the *Tallinn Manual* as well as a better understanding of the psychological harm that cyberterrorism can cause. The framers of *Tallinn* believe, for example, that "the internet is not indispensable to the survival of the civilian population" (§81.5). Such a remark exhibits a complete lack of understanding of the growing role of cybernetworks in everyday life. Although not indispensable in the way food and water are, the Internet is the foundation of modern communications, banking, and other services and, for many, social connectivity. And while individuals will survive without the Internet (just as they can survive without many basics), they may suffer significant distress if the network is disrupted or destroyed. While cyberattacks may not cause serious harm, they nevertheless impair the faith citizens have in their governments. Until twenty or so years ago, citizens' sense of security was a derivative of safe streets and borders. In today's world, where individuals' lives largely take place online, a person may live in a safe nation and still feel high anxiety for his online safety.

This is precisely the supposition our research tests. There is some evidence to expect that cyberattacks and related assaults cause significant anxiety. Notwithstanding the fact that many Americans are victims of identity theft, there is little research on related psychological harm of cyberattacks. Studies by Sharp and his colleagues, for example, found that two weeks after learning of the identity theft, victims experienced irritability, anger, fear, anxiety and frustration, sleep deprivation, anxiety, nervousness, loss of appetite, weight changes, and headaches.[19] Twenty-six weeks later, emotional responses turned to severe distress and desperation, mistrust and paranoia, nervousness, gastrointestinal problems, and headaches. These are little different from the psychological trauma of ordinary burglary and nonviolent home invasion.[20] Qualitative research has suggested that fear of identity theft stokes fear of financial losses, damage to reputation, and loss of online privacy.[21]

Cyberbullying is an aggressive act that subjects targets to a barrage of degrading, threatening, and/or sexually explicit messages and images using websites, instant messaging, blogs, chat rooms, cell phones, email, and personal online profiles that is very difficult to supervise or detect.[22] Targets of cyberbullying experience intense anger, powerlessness, sadness, fear, loss of confidence, disassociation, a general sense of uneasiness, possible trauma, and aggressiveness.[23] These findings suggest that significant psychological suffering may be present even when the threat of physical harm is relatively minor, thereby reinforcing our perception of cyberterrorism as acts that do not necessarily entail death or injury but elicit fear by damaging personal property, creating civil disorder, or causing significant economic harm.[24] At the same time, the fear, anxiety, and mental suffering that cyberterrorism can bring belies any attempt to understand cyberterrorism as victimless. Quite the contrary, Hamas hacktivists, for example, recently used text messaging to deliver hostile, personal threats to intimidate Israeli civilians. When credible, such threats can raise fears of injury or death.

Extrapolating from these psychological data we expect two sorts of psychological suffering in the wake of cyberwarfare. First, individuals will experience the distress and anxiety that come with the disruption of everyday services when people cannot ensure their privacy, access their bank accounts, fill prescriptions in a timely way, travel as necessary, maintain communications, and run their computers. Realistic scenarios depicting the impact of cyberwarfare variously describe denial of service, the inability to enter websites, lost or stolen data, the unauthorized disclosure of confidential information, the destruction of computer infrastructures, and the collapse of social networks. While these disruptions are free of

---

[19] Sharp et al., "Exploring the Psychological and Somatic Impact of Identity Theft," 131–36.

[20] Beaton et al., "Psychological Impact of Burglary," 33–43; Brown and Harris, "Residential Burglary Victimization," 119–32; Maguire, "Impact of Burglary upon Victims," 261–75.

[21] Roberts, "Cyber Identity Theft," 542–57.

[22] Smith et al., "Cyberbullying: Its Nature and Impact in Secondary School Pupils," 376–85; Shariff and Gouin, "Cyber-hierarchies: A New Arsenal of Weapons for Gendered Violence in Schools," 33–41; Li, "Cyberbullying in Schools," 157–70; Milson and Chu, "Character Education for Cyberspace," 117–19.

[23] Sourander et al., "Psychosocial Risk Factors Associated with Cyberbullying among Adolescents," 720–28; Gini and Pozzoli, "Association between Bullying and Psychosomatic Problems," 1059–65; Hoff and Mitchell, "Cyberbullying: Causes, Effects, and Remedies," 652–65.

[24] Ariely "Knowledge Management, Terrorism, and Cyber Terrorism," 7–16.

the fears of injury or death that accompany kinetic terrorism, they would seriously impair people's ability to function effectively in a modern industrial society. These effects may be particularly severe among vulnerable groups such as the poor and elderly. But ordinary citizens may be no less affected. On January 21, 2014, South Koreans awoke to find that thieves had stolen the credit card numbers, names, and addresses of 40 percent of the population. The immediate result was widespread panic, system crashes, massive lawsuits, and a run on banks to cancel credit cards.[25] The culprit in Korea was an insider, but the effect upon the citizenry was no different than a cyberstrike. Attacks such as these feed a secondary but amorphous fear that comes with the constant assault by unknown, malevolent agents whose agenda is neither clear nor predictable. Cyberterrorism stokes anxieties about loss of control and unpredictability that might be as inescapable as those accompanying war and kinetic terrorism.

8.4 Assessing the Psychological Effects of Cyberterrorism in the Lab

To evaluate the psychological effects of cyberterrorism, we conducted a series of laboratory experiments to simulate cyberattacks on individuals. Our experimental attacks simulate those perpetrated by hacktivist and nonstate actors, whose goal is to disrupt the lives of individuals and, often, establish a platform for their grievances. In this way, cyberattacks share the aims of conventional, kinetic terror attacks. Both use short-lived but spectacular attacks to strengthen morale among compatriots, place their political cause squarely on the international agenda, and discomfit their enemies by underscoring their weaknesses.

The manipulations we chose simulated the way ordinary citizens may experience a cyberattack. These are individual, not mass casualty attacks. While mass casualty terrorism hopes to violently disrupt civil society and kill civilians, individual attacks will at best bring chaos, personal discomfort, or anxiety. The manipulation, therefore, had but one purpose—to generate among respondents the recognition that their private online identity was private no more. We did not cause or threaten specific harms such as loss of medical or financial information. Yet by using a video chat window, together with a threatening Anonymous logo and text message to their private phone, we sent the participant a clear message that she was not alone any more. "Anonymous" is a well-known, diffused global network of hackers responsible for hundreds of cyberattacks, from Tunisia and PayPal to the Scientology Church and Swiss financial institutions.[26] By making Anonymous the attacker, we steered away

---

[25] Lee, "South Koreans Seethe."
[26] Olson, *We Are Anonymous*.

from the particularity of a specific conflict, thereby making our findings generalizable. To simulate intrusiveness and breach of privacy, we conducted the experiment on a lab computer and the participants' private cell phones. The text message to the participants' cell phones cemented the feeling among participants that *they* were the target of the cyberattack, not the lab computer. Before and during the manipulation subjects provided a saliva sample to measure cortisol, a hormone associated with stress.

The study began with a battery of questions asking participants to describe their computer savviness and usage, probe political attitudes, and describe their overall psychological well-being.[27] After providing a saliva sample, respondents (n=100) continued the survey. At this stage respondents saw a pop-up screen with a message from Anonymous, which only the research assistant (RA) could unlock. If questioned about whether it was part of the experiment, the RA was instructed to reassure the student that she knew nothing and that subjects must ignore it and continue the study.

After a few additional questions, a Skype-like split video screen popped up where subjects could see themselves live and see and hear a suspicious-looking person typing. As before, only the RA could close the screen, and again, it was her role to reassure the respondent it must be a fluke. Finally, five subjects received an anonymous phone-text message, which stated that their personal data were hacked. If a respondent became uncomfortable, the RA again asked the participant to continue. Control group respondents completed the very same questions, but without the cyberthreat component.

At the completion of the survey, all respondents provided a second saliva sample, again reported their overall psychological well-being, and completed a battery of questions on cyberthreats and cyberpolicies. Upon completion, we debriefed all respondents. As expected, our exposure to cyberattacks has a psychological and physiological impact. Among the control group, the level of cortisol decreased by 7 percent, an outcome consistent with diurnal effects that cause a decrease of cortisol as the day progresses. The treatment group, however, experienced an average rise of 16 percent in cortisol, a clear indication that the cyberattack caused stress and anxiety. These physiological findings are further supported by data that show that individuals subjected to cyberterrorism are agitated and significantly more likely to fear imminent cyberattacks. When subjects were asked to what extent cyberattacks undermined (or harmed) their sense of personal security, those who had experienced simulated cyberterrorism reported a significantly greater sense of personal insecurity. These data confirm a positive relationship between *nonviolent* cyberterror (simulated attack on computers and cell phones), stress, dread, and personal insecurity.

---

[27] Canetti et al., "Streaming Terror: Cyber-Terrorism and its Global Threat," in preparation.

Together with the cortisol results, these findings demonstrate that cyber research, which is predominantly governed by security experts (national and computer), must not only take into account the number of casualties, computers, or mainframes affected but also the way in which individuals might be psychologically impaired following such an attack.[28] Cyber research, therefore, must emphasize *human* cybersecurity as well as *national* cybersecurity. As noted, the simulated attacks did not cause or threaten to cause permanent damage or harm to participants. Cyberattacks that steal identities, data, or money; disclose confidential information; or threaten individuals with random, personal harm are likely to cause significant fear, stress, and anxiety that can effectively impinge upon the rational decision-making that governments require from their citizens for good governance.

Acute stress disrupts decision making,[29] making people with higher levels of cortisol more sensitive to immediate rewards than those with lower levels.[30] The former are also more prone to making snap decisions, indicative of a loss of top-down control.[31] Beyond the stress response, there is evidence that the psychological and physiological reactions following exposure to threatening events such as political violence affect the immune system negatively and cause inflammations in the body in a way that can significantly radicalize political attitudes and behavior.[32] Investigating cortisol and inflammatory markers is of special interest to those concerned with protecting against politically related violence that comes with the militant and aggressive attitudes following cyberterrorism. While people may not necessarily be aware of the forces and conditions that underlie their reactions to cyberattacks, understanding the role of physiological reactivity markers fills a pressing need for objective data and empirically based generalizations about their effects on civilians.

Figure 8.1 describes the general outcome of our investigation into the psychological effects of cyberwarfare and cyberterrorism. The first row depicts a model that informs the framers of the *Tallinn Manual*, terrorists themselves, and nearly everyone else. Kinetic terrorism kills or injures small numbers of individuals (the

---

[28] The effects of these attacks are expected to be significantly larger when they take place outside a lab setting and when the person is the actual owner of the attacked computer.

[29] Keinan, Friedland, and Ben-Porath, "Decision Making under Stress," 219–28; Preston et al., "Effects of Anticipatory Stress on Decision Making in a Gambling Task," 257–63; Porcelli and Delgado, "Acute Stress Modulates Risk Taking in Financial Decision Making," 278–83.

[30] Piazza et al., "Corticosterone in the Range of Stress-Induced Levels," 11738–42; Adam and Epel, "Stress, Eating and the Reward System," 449–58; Newman, O'Connor, and Conner, "Daily Hassles and Eating Behaviour," 125–32.

[31] See Keinan, "Decision Making"; Porcelli, "Acute Stress."

[32] Graham. "Hostility and Pain Are Related to Inflammation in Older Adults," 389–340; Pace et al., "Innate Immune, Neuroendocrine and Behavioral Responses," 310–15; Canetti et al., "Inflamed by the Flames? The Impact of Terrorism and War on Immunity," 1–8.

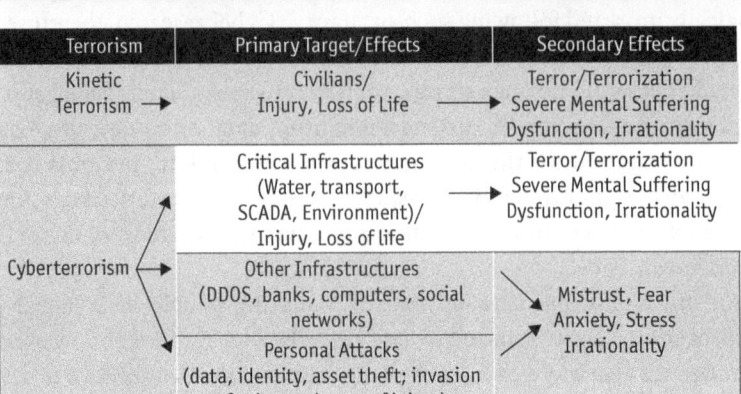

Figure 8.1 The Effects of Kinetic and Cyberterrorism

primary target) and terrorizes large numbers of individuals by inflicting severe mental suffering that disrupts daily life, skews rational decision-making, and hopes to bring the civilian population to pressure their government to take immediate steps to end the conflict and meet terrorists' demands.

Cyberterrorism and cyberwar are more complex. Cyberterrorism attacks critical and other infrastructures and individuals indiscriminately, while cyberwarfare harms the same civilians collaterally. The destruction of some critical infrastructures may bring loss of life that can have the same effects upon the civilian population as kinetic terrorism. This remains a matter of conjecture as no such attacks have yet occurred. Instead, cyberoperations will most likely disable other infrastructures or target individuals directly. In these cases, the civilian population will most likely suffer fear, anxiety, despair, loss of control, and mistrust. Some will lose medical or legal records, confidential information, email communications, or social networks. Others will find their identity or assets stolen or face physical threats from unknown assailants. Lives and businesses might be radically disrupted.

Psychological distress also shapes attitudes and political decision-making. Exposure to kinetic terrorism leads to "psychological insecurity that induces militant attitudes, and violent and non-conciliatory political responses."[33] Helping to explain this outcome, the Shattered Assumptions Approach argues that traumatic events

---

[33] Canetti et al., "An Exposure Effect? Evidence from a Rigorous Study on the Psychopolitical Outcomes of Terrorism," 193–212; Hobfoll et al., "Exposure to Terrorism, Stress-Related Mental Health Symptoms, and Defensive Coping among Jews and Arabs in Israel," 207–18.

undermine a person's basic assumptions about the world, triggering enhanced perceptions of "the world as threatening" and a correspondingly strong desire to reduce this threat through increased militancy.[34] Perceived threat, fear, and anxiety are the single best predictors of militarism.[35] Chronic exposure to war and terrorism not only harm personal well-being but also contribute to an ongoing cycle of violence as affected citizens harden their political viewpoints in an attempt to cope with stress.[36] By impinging on the public's well-being, cyberwar and cyberterrorism may affect political attitudes and public policy in a similar way, particularly as democratic leaders tend to follow public opinion when faced with a major public opinion shift.[37] In the wake of concerted cyberattacks, leaders will face a barrage of demands. Some demands might be reasonable (protective software products), others may be expensive (a strategic reserve of bandwidth and cybercapability), others intrusive (state monitoring of networks and systems, regulation and/or wiretapping), and others belligerent (kinetic attacks against cyberattackers). In the worst case, policy makers may have no choice but to retaliate and escalate the conflict rather than capitulate.

8.5 Cyberwarfare: Implications for Ethics and Law

Despite the far-reaching psychological effects of cyberwar and cyberterrorism, one cannot escape the thought that they are preferable to armed conflict and analogous to economic warfare, sanctions, and blockades. The psychological effects of economic warfare, like many of the worst forms of cyberwarfare, are long-term, diffuse, and of variable duration and intensity. Such indeterminate and mixed outcomes make it very difficult for commanders in the field or policy makers to weigh mental suffering as they wrestle with the principle of proportionality. In fact, it seems that the psychological sequelae of many forms of armed conflict merit no place at all when considering the ills that befall the civilian population.

It is no wonder then that international law is confused. While the Geneva Conventions, particularly Additional Protocol I, take a strong stand against terrorism-induced mental suffering, they take virtually no stand when the same suffering follows economic warfare. Unlike the collateral harm that follows when infrastructures are destroyed in the normal course of war, sanctions and blockades target civilians directly. Yet economic warfare remains beyond the purview of the

---

[34] Carnelley and Janoff-Bulman, "Optimism about Love Relationships," 5–20.
[35] Bonanno and Jost, "Conservative Shift among High- Exposure Survivors," 311–23.
[36] Bitterman at al., "Characterization of the Best Anatomical Sites in Screening for Methicillin-Resistant Staphylococcus Aureus Colonization," 391–97. Bonanno, "Conservative Shift."
[37] Burstein, "Impact of Public Opinion on Public Policy," 29–40; Page and Shapiro, "Effects of Public Opinion on Policy," 175–90.

law of war as long as blockades or sanctions do not create a "humanitarian crisis" that takes the lives of large numbers of innocent civilians and while reducing the rest to penury. Despite its legal cover, the sanctions imposed on Iraq by the international community after the First Gulf War brought precisely this sort of crisis. There, notes Gottstein, "50,000 children under the age of five died each year, a quarter of all emergency patients in the hospitals could not be saved due to missing medicines and about 40% of the Iraqi people went hungry having received a food ration that provided only 25% of their vital needs."[38]

Note the conspicuous absence of any reference to mental suffering in these descriptions of humanitarian crises. In contrast to death and debilitating injuries, mental suffering is the proper goal of sanctions and blockades and, therefore, carelessly ignored. By inflicting pain and hardship, one nation hopes to squeeze the civilian population of another so it pressures its government to desist from aggression. Economic warfare stops short of armed conflict, and so it is lauded as the penultimate resort that gives war legitimacy as the last resort. States, in other words, are often encouraged to wage economic warfare before resorting to armed force when they face aggression. It is tempting, therefore, to consider cyberwarfare and even cyberterrorism as nothing but another form of economic warfare. As such, any resulting harm, whether direct or collateral, is of little consequence unless it brings a humanitarian crisis. Few cyber scenarios hold such potential.

From the perspective of economic warfare, then, it is easy to conclude that most cyberoperations neither violate the principle of noncombatant immunity nor constitute terrorism. On the contrary, cyberoperations may save a nation from the ravages of war. Before rushing to judgment, however, consider that there are several ways to view terrorism. One, that the *Tallinn* experts and many others adopt, turns on manifest terrorization accompanied by the ever-present threat of death and, as Hannah Arendt describes it, "the bestial, desperate terror which, when confronted by real, present horror, inexorably paralyzes everything that is not mere reaction."[39] Noting that Arendt's view represents the most extreme outcome of terrorism, Jeremy Waldron suggests that terrorism also turns on less violent and coercive means. "The idea that I am pursuing," writes Waldron, "is that a government might be coerced by the loss of something it values very highly—indeed, something indispensable for its status as government—namely, the ability to command and mobilize a large civilian population. By rendering or threatening to render the population mindless with terror, the intimidator deprives the target regime of something it needs, a population capable of rational choice."[40] However, there is no need that a population be "mindless with terror" to undermine its rational decision-making capability. And in fact,

---

[38] Garfield, "Impact of Economic Sanctions"; Gottstein, "Peace through Sanctions?"
[39] Arendt, *Origins of Totalitarianism*.
[40] Waldron, "Terrorism and the Uses of Terror," 21.

Waldron looks to something short of "bestial desperate panic" to include "a state or condition that governments cannot afford to let their populations fall into or languish in for long." Examples include the "collapse of economic morale," feelings of insecurity, apprehension, and disruption of social intercourse and daily life.[41] These are precisely the effects we can expect of most cyberoperations.

If by "terrorism" we mean abject terrorization, then cyberoperations are not acts of terror or acts that violate the principle of noncombatant immunity. But if we think a little out of the box, we can easily imagine how cyberoperations can cause terrorism of a more pervasive and no less dangerous kind by undermining well-being, morale, public trust, and governability. To accomplish this end, one need not commit horrific acts of murder. In a modern society it is enough attack the foundations of everyday life. Among these, cybernetworks stand out. As critically, one cannot forget that many cyberoperations, however nonlethal they can be, place civilians in the crosshairs. By targeting civilians and civilian infrastructures, cyberoperations knowingly seek out noncombatants to demoralize the civilian population and bring pressure upon a government to meet their demands. Noncombatants, however, are not the proper objects of attacks that significantly impair their physiological or psychological well-being. Not only do noncombatants pose no threat, but singling them out for any intentional or disproportionate harm whatsoever constitutes a grave affront to human dignity to which noncombatants are entitled. Noncombatants are not instruments of war and, for this reason, economic warfare, although often lawfully permissible, has earned the justifiable wrath of many moral philosophers. For this reason, too, noncombatants deserve every protection from cyberwarfare and the harms it brings.

Bibliography

Adam, Tanja C., and Elissa S. Epel. "Stress, Eating and the Reward System." *Physiology & Behavior* 91 (2007): 449–58.
Arendt, Hannah. *The Origins of Totalitarianism*. New ed. New York: Harcourt Brace Jovanovich, 1973.
Ariely, Gil. "Knowledge Management, Terrorism, and Cyber-terrorism." In *Cyberwarfare and Cyber-terrorism*, ed. L. Janczewski and A. Colarik, 7–16. Hershey, PA: IGI Global, 2008.
Beaton, Alan, Mark Cook, Mark Kavanagh, and Carla Herrington. "The Psychological Impact of Burglary." *Psychology, Crime and Law* 6, no. 1 (2000): 33–43.
Bitterman, Haim, Arie Laor, Sarah Itzhaki, and Gabriel Weber. "Characterization of the Best Anatomical Sites in Screening for Methicillin-Resistant Staphylococcus Aureus Colonization." *European Journal of Clinical Microbiology and Infectious Diseases* 29, no. 4 (2010): 391–97.
Bleich, Avi, Marc Gelkopf, Yuval Melamed, and Zahava Solomon. "Mental Health And Resiliency Following 44 Months of Terrorism: A Survey of an Israeli National Representative Sample." *BMC Medicine* 4, no. 1 (2006): 21. http://www.biomedcentral.com/1741-7015/4/21
Bleich, Avraham, Marc Gelkopf, and Zahava Solomon. "Exposure to Terrorism, Stress-Related Mental Health Symptoms, and Coping Behaviors among a Nationally Representative Sample in Israel." *JAMA* 290, no. 5 (2003): 612–20.

---

[41] Ibid., 21–23.

Bonanno, George A., and John T. Jost. "Conservative Shift among High-Exposure Survivors of the September 11th Terrorist Attacks." *Basic and Applied Social Psychology* 28, no. 4 (2006): 311–23.

Brown, Barbara B., and Paul B. Harris. "Residential Burglary Victimization: Reactions to the Invasion of a Primary Territory." *Journal of Environmental Psychology* 9, no. 2 (1989): 119–32.

Burstein, Paul. "The Impact of Public Opinion on Public Policy: A Review and an Agenda." *Political Research Quarterly* 56, no. 1 (2003): 29–40.

Canetti, Daphna, Carmit Rapaport, Carly Wayne, Brian Hall, and Stevan E. Hobfoll. "An Exposure Effect? Evidence from a Rigorous Study on the Psychopolitical Outcomes of Terrorism." In *The Political Psychology of Terrorism Fears*, ed. Samuel Justin Sinclair and Daniel Antonius, 193–212. Oxford: Oxford University Press, 2013.

Canetti, Daphna, Eric Russ, Judith Luborsky, and Stevan E. Hobfoll. "Inflamed by the Flames? The Impact of Terrorism and War on Immunity." *Journal of Traumatic Stress* 27 (2014): 1–8.

Canetti-Nisim, Daphna, Eran Halperin, Keren Sharvit, and Stevan E. Hobfoll. "A New Stress-Based Model of Political Extremism: Personal Exposure to Terrorism, Psychological Distress, and Exclusionist Political Attitudes." *Journal of Conflict Resolution* 53, no. 3 (2009): 363–89.

Carnelley, Katherine B., and Ronnie Janoff-Bulman. "Optimism about Love Relationships: General vs Specific Lessons from One's Personal Experiences." *Journal of Social and Personal Relationships* 9, no. 1 (1992): 5–20.

Darwish, Ragy, Nadim Farajalla, and Rania Masri. "The 2006 War and Its Inter-Temporal Economic Impact on Agriculture in Lebanon." *Disasters* 33, no. 4 (2009): 629–44.

Diamond, Gary M., Joshua D. Lipsitz, Zvi Fajerman, and Omit Rozenblat. "Ongoing Traumatic Stress Response (OTSR) in Sderot, Israel." *Professional Psychology: Research and Practice* 41, no. 1 (2010): 19.

Farhood, Laila, Hani Dimassi, and Nicole L. Strauss. "Understanding Post-Conflict Mental Health: Assessment of PTSD, Depression, General Health and Life Events in Civilian Population One Year after the 2006 War in South Lebanon." *Journal of Trauma and Stress Disorders Treatment* 2, no. 2 (2013). https://scholarworks.aub.edu.lb/bitstream/handle/10938/9704/understanding_post_conflict_mental_health.pdf?sequence=1.

Fattouh, Bassam, and Joachim Kolb. "The Outlook for Economic Reconstruction in Lebanon after the 2006 War." *MIT Electronic Journal of Middle East Studies* 6 (2006): 97–111. http://web.mit.edu/cis/www/mitejmes.

Galea, Sandro, David Vlahov, Heidi Resnick, Jennifer Ahern, Ezra Susser, Joel Gold, Michael Bucuvalas, and Dean Kilpatrick. "Trends of Probable Post-Traumatic Stress Disorder in New York City after the September 11 Terrorist Attacks." *American Journal of Epidemiology* 158, no. 6 (2003): 514–24.

Galea, Sandro, Jennifer Ahern, Heidi Resnick, Dean Kilpatrick, Michael Bucuvalas, Joel Gold, and David Vlahov. "Psychological Sequelae of the September 11 Terrorist Attacks in New York City." *New England Journal of Medicine* 346, no. 13 (2002): 982–87.

Garfield, R. "The Impact of Economic Sanctions on Health and Well-Being." *Relief and Rehabilitation Network*, London. http://www.essex.ac.uk/armedcon/story_id/The%20Impact%20of%20Econmoic%20Sanctins%20on%20Health%20abd%20Well-Being.pdf. 1999.

Gelkopf, Marc, Rony Berger, Avraham Bleich, and Roxane Cohen Silver. "Protective Factors and Predictors of Vulnerability to Chronic Stress: A Comparative Study of 4 Communities after 7 Years of Continuous Rocket Fire." *Social Science & Medicine* 74, no. 5 (2012): 757–66.

Gini, Gianluca, and Tiziana Pozzoli. "Association between Bullying and Psychosomatic Problems: A Meta-Analysis." *Pediatrics* 123, no. 3 (2009): 1059–65.

Gottstein, Ulrich. "Peace through Sanctions? Lessons from Cuba, Former Yugoslavia and Iraq." *Medicine, Conflict and Survival* 15, no. 3 (1999): 271–85.

Graham, Jennifer E., Theodore F. Robles, Janice K. Kiecolt-Glaser, William B. Malarkey, Michael J. Bissell, and Ronald Glaser. "Hostility and Pain Are Related to Inflammation in Older Adults." *Brain, Behavior and Immunity* 20 (2006): 389–400.

Hennessy, John W., and Seymour Levine. "Stress, Arousal, and the Pituitary-Adrenal System: A Psychoendocrine Hypothesis." *Progress in Psychobiology and Physiological Psychology* 8 (1979): 133–78.
Hobfoll, Stevan E., Daphna Canetti-Nisim, and Robert J. Johnson. "Exposure to Terrorism, Stress-Related Mental Health Symptoms, and Defensive Coping among Jews and Arabs in Israel." *Journal of Consulting and Clinical Psychology* 74, no. 2 (2006): 207.
Hoff, Dianne L., and Sidney N. Mitchell. "Cyberbullying: Causes, Effects, and Remedies." *Journal of Educational Administration* 47, no. 5 (2009): 652–65.
Howie, Luke. "The Terrorism Threat and Managing Workplaces." *Disaster Prevention and Management* 16, no. 1 (2007): 70–78.
Lachow, Irving, and Courtney Richardson. *Terrorist Use of the Internet: The Real Story*. National Defense University, Washington, DC, 2007: http://oai.dtic.mil/oai/oai?verb=getRecord&metadataPrefix=html&identifier=ADA518156.
Huddy, Leone, Stanley Feldman, Charles Taber, and Gallya Lahav. "Threat, Anxiety, and Support of Antiterrorism Policies." *American Journal of Political Science* 49, no. 3 (2005): 593–608.
Huddy, Leone, Stanley Feldman, Theresa Capelos, and Colin Provost. "The Consequences of Terrorism: Disentangling the Effects of Personal and National Threat." *Political Psychology* 23, no. 3 (2002): 485–509.
Keinan, Giora, Nehemia Friedland, and Yossef Ben-Porath. "Decision Making under Stress: Scanning of Alternatives under Physical Threat." *Acta Psychologica* 64, no. 3 (1987): 219–28.
Lee, J. "South Koreans Seethe, Sue as Credit Card Details Swiped." *Reuters*, January 21, 2014. http://www.reuters.com/article/2014/01/21/us-korea-cards-idUSBREA0K05120140121
Li, Qing. "Cyberbullying in Schools a Research of Gender Differences." *School Psychology International* 27, no. 2 (2006): 157–70.
Lindgren, G. 2006. *Studies in Conflict Economics and Economic Growth*. Report No. 72. Department of Peace and Conflict Research. Uppsala: Uppsala University
Maguire, Mike. "Impact of Burglary upon Victims." *British Journal of Criminology* 20 (1980): 261–75.
Milson, Andrew J., and Beong-Wan Chu. "Character Education for Cyberspace: Developing Good Netizens." *Social Studies* 93, no. 3 (2002): 117–19.
Newman, Emily, Daryl. B. O'Connor, and Mark Conner. "Daily Hassles and Eating Behaviour: The Role of Cortisol Reactivity Status." *Psychoneuroendocrinology* 32, no. 2 (2007): 125–32.
Nuwayhid, Iman, Huda Zurayk, Rouham Yamout, and Chadi S. Cortas. "Summer 2006 War on Lebanon: A Lesson in Community Resilience." *Global Public Health* 6, no. 5 (2011): 505–19.
Olson, Parmy. *We Are Anonymous: Inside the Hacker World of LulzSec, Anonymous, and the Global Cyber Insurgency*. New York: Hachette Digital, 2012.
Pace, Thaddeus W., Lobsang Tenzin Negi, Teresa I. Sivilli, Michael J. Issac, Steven P. Coled, Daniel D. Adamee and Charles L. Raison. "Innate Immune, Neuroendocrine and Behavioral Responses to Psychosocial Stress Do Not Predict Subsequent Compassion Meditation Practice Time." *Psychoneuroendocrinology*, 35, no. 2 (2010): 310–15.
Page, Benjamin I., and Robert Y. Shapiro. "Effects of Public Opinion on Policy." *American Political Science Review* 77, no. 1 (1983): 175–90.
Piazza, Pier V., Veronique Deroche, Jean Marie Deminiere, Stefania Maccari, Michael Le Moal, and Herve Simon. "Corticosterone in the Range of Stress-Induced Levels Possesses Reinforcing Properties: Implications for Sensation-Seeking Behaviors." *Proceedings of the National Academy of Sciences of the United States of America* 90, no. 24 (1993): 11738–42.
Porcelli, Anthony J., and Mauricio R. Delgado. "Acute Stress Modulates Risk Taking in Financial Decision Making." *Psychological Science* 20, no. 3 (2009): 278–83.
Preston, Stephanie D., Tony W. Buchanan, Robert B. Stansfield, and Antoine Bechara. "Effects of Anticipatory Stress on Decision Making in a Gambling Task." *Behavioral Neuroscience* 121 (2007): 257–63.
Roberts, Lynne D. "Cyber Identity Theft." In *Handbook of Research on Technoethics*, ed. R. Luppicini and R. Adell, 542–57. Hershey, PA: Information Science Reference, 2008.

Schlenger, William E., Juesta M. Caddell, Lori Ebert, B. Kathleen Jordan, Kathryn M. Rourke, David Wilson, Lisa Thalji, J. Michael Dennis, John A. Fairbank, and Richard A. Kulka. "Psychological Reactions to Terrorist Attacks: Findings from the National Study of Americans' Reactions to September 11." *JAMA* 288, no. 5 (2002): 581–88.

Schmitt, Michael N. *Tallinn Manual on the International Law Applicable to Cyber Warfare.* New York: Cambridge University Press, 2013.

Shariff, S., and R. Gouin. "Cyber-hierarchies: A New Arsenal of Weapons for Gendered Violence in Schools." In *Combating Gender Violence in and around Schools*, ed. C. Mitchell and F. Leech, 33–41. London: Trentham Books, 2006.

Sharp, Tracy, Andrea Shreve-Neiger, William Fremouw, John Kane, and Shawn Hutton. "Exploring the Psychological and Somatic Impact of Identity Theft." *Journal of forensic sciences* 49, no. 1 (2004): 131–36.

Silver, Roxane Cohen, E. Alison Holman, Daniel N. McIntosh, Michael Poulin, and Virginia Gil-Rivas. "Nationwide Longitudinal Study of Psychological Responses to September 11." *JAMA* 288, no. 10 (2002): 1235–44.

Sinclair, J. S., and D. Antonius. Personal communication, 2012.

Sinclair, Justin, and Daniel Antonius. *The Psychology of Terrorism Fears.* Oxford: Oxford University Press, 2012.

Sinclair, Justin, and Daniel Antonius, eds. *The Political Psychology of Terrorism Fears.* Oxford: Oxford University Press, 2013.

Sinclair, Justin. "Fears of Terrorism and Future Threat: Theoretical and Empirical Considerations." In *Interdisciplinary Analyses of Terrorism and Political Aggression*, ed. D. Antonius, 101–15. Newcastle upon Tyne: Cambridge Scholars, 2010.

Sinclair, Samuel J., and Alice LoCicero. "Fearing Future Terrorism: Development, Validation, and Psychometric Testing of the Terrorism Catastrophizing Scale (TCS)." *Traumatology* 13, no. 4 (2007): 75–90

Smith, Peter K., Jess Mahdavi, Manuel Carvalho, Sonja Fisher, Shanette Russell, and Neil Tippett. "Cyberbullying: Its Nature and Impact in Secondary School Pupils." *Journal of Child Psychology and Psychiatry* 49, no. 4 (2008): 376–85.

Sourander, Andre, Anat Brunstein Klomek, Maria Ikonen, Jarna Lindroos, Terhi Luntamo, Merja Koskelainen, Terja Ristkari, and Hans Helenius. "Psychosocial Risk Factors Associated with Cyberbullying among Adolescents: A Population-based Study." *Archives of General Psychiatry* 67, no. 7 (2010): 720–28.

Spielberger, Charles D., Richard L. Gorsuch, Robert Lushene, Peter R. Vagg, and Gerald A. Jacobs. *Manual for the State-Trait Anxiety Inventory.* Palo Alto, CA: Consulting Psychologists Press, 1983.

Yehuda, Rachel. "Post-Traumatic Stress Disorder." *New England Journal of Medicine* 346, no. 2 (2002): 108–14.

Zemishlany, Zvi. "Resilience and Vulnerability in Coping with Stress and Terrorism." *Israeli Medial Association* 14, no. 5 (2012): 307–9.

9

Beyond Machines

Humans in Cyberoperations, Espionage, and Conflict

DAVID DANKS AND JOSEPH H. DANKS

9.1. The Importance of the Human

It is the height of banality to observe that people, not bullets, fight kinetic wars.[1] The machinery of kinetic warfare is obviously relevant to the conduct of each particular act of warfare, but the reasons for, and meanings of, those acts depend critically on the fact that they are done by humans. Any attempt to understand warfare—its causes, strategies, legitimacy, dynamics, and resolutions—must incorporate humans as an intrinsic part, both descriptively and normatively. Humans from general staff to "boots on the ground" play key roles in all aspects of kinetic warfare, and the literature about it reflects this focus (e.g., the emphasis on understanding the adversary's goals and constraints when developing battle plans).[2] In contrast, many discussions of cyberwarfare and cyberconflict focus principally on the technical aspects of machines, systems, and data,[3] and human agents are included only as collateral effects (e.g., in discussions about the impact of disabling an adversary's electrical

---

[1] Thanks to the editors, as well as Susannah Paletz and Alan Mishler (both at CASL), for their valuable comments on an earlier version of this chapter. DD was partially supported by a James S. McDonnell Foundation Scholar Award. JHD's work was supported, in whole or in part, with funding from the US government. Any opinions, findings and conclusions, or recommendations expressed in this material are those of the authors and do not necessarily reflect the view of the University of Maryland, College Park, and/or any agency or entity of the US government.

[2] Examples range from the exhortations to know both oneself and one's enemies in Sun Tzu's *The Art of War*, to the emphasis on emotions and other motivations in Clausewitz's *On War*, to quite contemporary work, such as the detailed cognitive analyses of military decision-makers in many chapters of Zsambok and Klein, *Naturalistic Decision Making*.

[3] Examples can be found in many of the papers cited below, as well as Paul Cornish et al., *On Cyber Warfare*,; Rosenzweig, *Cyber Warfare*; Andress and Winterfeld, *Cyber Warfare*. It should be noted that much of the present volume is a welcome exception to this trend.

grid), or as loci of moral responsibility (e.g., providing the ground for the moral justification of a cyberattack).

In some respects, this technical focus is unsurprising: many cybercapabilities are completely novel in the history of warfare, and it is only natural to focus on the new and original. But this focus has come at a significant cost, as these debates have largely ignored the fact that cyberactions are typically designed, initiated, and responded to by fallible, cognitively bounded human agents. In particular, the humans that actually engage in cyberwarfare and cyberconflict are not what many analyses assume—perfectly rational, fully self-aware agents who can automatically bear true moral responsibility, whether praise or blame, for their actions. Instead, cyberwarfare and cyberoperations are conducted by human agents who suffer from cognitive constraints and biases; who are often unaware of their own beliefs and desires (present or future); and who frequently exhibit failings that undermine their ability to be full moral agents. As we show below, our understanding of the ethics, conduct, and performance of cyberwarfare and cyberoperations changes when we bring the humans back "in the loop."

There are at least four different, morally salient roles for which we should attend to the cognitive limits and features of human agents engaged in cyberwarfare and cyberoperations. Brief examples may help to see the importance of considering cognitively realistic human agents in our analyses. First, humans are the developers of cyberactions, whether attacks or exploits, and so are arguably responsible for foreseeable outcomes of those actions. As we discuss in section 9.3, however, the complexity of cybersystems will frequently exceed our cognitive abilities.[4] In those cases, we will frequently not know the likely outcomes of our actions, even though we have—or rather *should* have, given the large psychological literature on the topic (see citations in section 9.3)—metaknowledge that we do not know. As a result, we are arguably culpable—at least somewhat—for our ignorance, which thereby reduces or eliminates the mitigating moral power of our ignorance. Our cognitive limitations, coupled with our knowledge of those very limits, potentially imply that many cyberactions have uncertain moral legitimacy.

Second, human agents are the targets of cyberactions. The essential humanity of the targets is relevant, for example, in thinking about whether a particular cyberaction is a *cyberattack*. The definition of a "cyberattack" proposed in Rule 30 of the *Tallinn Manual*[5] refers in part to "damage ... to objects," which is only a helpful definition if all relevant parties have a shared understanding of what counts as "damage." Unfortunately, it is not clear that there is such an understanding; a categorization of some event as "damage" depends partly on the

---

[4] See also Danks and Danks, "Moral Permissibility of Automated Responses During Cyberwarfare," 18–33.

[5] Schmitt, *Tallinn Manual*.

expectations and culturally shaped perceptions of the putative "target." As just one instance of this dependence—discussed further in section 9.4—information extraction could be either an illegitimate, damaging intrusion or appropriate dissemination and sharing, depending on the (partly culturally determined) "ownership" of that information. The very same cyberaction could be an attack or just publicity, depending on the target's culturally shaped understanding of the relevant information.

Third, human agents defend against cyberattacks, cyberintrusions, and cyberespionage, and the moral legitimacy of a defensive action will depend partly on cognitive factors, such as the agents' ability to predict the complex sequence of interactions that could result, or the agents' interpretation of the damage due to the attack. For example, a standard ethical principle about the conduct of warfare is that responses should be proportionate to the triggering event, so application of that principle requires an assessment of the intensity (or degree, or amount) of both the attack and the defense. If human agents were all perfectly rational and fully self-aware, then there could arguably be public, shared assessments of outcome damage. People are not such agents, however; perceptions of the costs and valuations of different actions are sensitive to various cognitive and emotional biases. Our evaluations of cyberdefenders' actions must recognize the complex cognitive origins of the beliefs and perceptions that drive their decisions and actions, particularly as those origins can explain potentially unexpected or seemingly irrational choices by defenders.

Fourth, human agents act as third parties who adjudicate disputes, shape (and sometimes form) public opinion, and generally provide many of the constraints and backgrounds against which cyberconflicts occur. The people who form these third parties are rarely fully rational, fully self-aware, or fully transparent about their values and goals. Any incorporation of third parties into an analysis of cyberwarfare or cyberoperations must therefore understand them not as ideal agents, but rather as human actors with all of their cognitive, conceptual, and cultural biases, tendencies, and shortcomings.

We have already signaled some of the issues that we will address, such as the morally questionable nature of automatic cyberactions in light of our cognitive limitations and biases (discussed in section 9.3), and the importance of incorporating cultural and cognitive features in any analysis of the fuzzy cyberwarfare versus cyberespionage "boundary" (in 9.4). We begin in section 9.2, however, with a simpler case: the importance of including the human in any solution to the attribution problem. Discussions about how to attribute cyberactions have focused on technical challenges, but actual cyberattribution inevitably relies heavily on the imputed motives, constraints, goals, and capacities of various potential adversaries. We thus find our first example of the theme of this chapter, as attribution that fails to be sensitive to human cognitive biases and limits can easily be attribution that goes wrong.

We will refer throughout to three case studies—the 2007 distributed denial-of-service (DDoS) attacks in Estonia, the Stuxnet operation against the Iranian nuclear program, and the apparent Chinese cyberespionage documented in the Mandiant Report—but it is important not to become fixated on the details of any particular case. Our more general moral is simply that our understanding of cyberwarfare and cyberoperations must include human agents with their cognitive failings and foibles, not only as interchangeable, identity-less "units" or fully rational, fully self-aware idealizations. Moreover, this is *not* simply a manifestation of the general idea that cognitive details can matter. Rather, analyses in the cyberdomain face novel, distinctive issues and challenges if they presuppose a purely rationalist, idealized understanding of the humans making the cyberdecisions. The complexities of real emotion and cognition matter when thinking about not only the physical bullets of kinetic warfare but also the binary bullets of cyberwarfare, though differently in the two domains.

9.2 Motives Matter

Perhaps the most-discussed challenge in cyberoperations, particularly in cyberattacks and cyberwarfare, is the attribution problem:[6] How do we attribute responsibility for some action or event? This question obviously arises in the kinetic domain as well, but is particularly challenging in cybercontexts. Most discussions of the attribution problem have focused on technical questions:[7] What are the technical conditions under which cyberattacks can be traced back to their source? Can we determine whether an action originated from a machine that had been compromised in various ways? How should we assign responsibility for distributed attacks? And there are many other questions. The overall theme of our chapter, however, is to instead ask about the human element. In particular, we contend that attribution is not a purely technical matter, but should also be based on information about motives, opportunities, and behavioral patterns. This additional "human information" not only provides a basis for deciding between equiprobable attributions but can also lead to attributions that are improbable given solely technical information. In this regard, there are many similarities with so-called signature strikes by drones, which should—to be legal or legitimate—depend on not just observed (patterns of) behavior, but also knowledge about potential targets' motives and

---

[6] For more detailed discussions see, e.g., Dipert, "Preventive War and the Epistemological Dimension of the Morality of War," 32–54; Dipert, "Ethics of Cyberwarfare," 384–410; Rowe, "Ethics of Cyberweapons in Warfare," 20–31.

[7] As just a limited sample, consider Clark and Landau, "Untangling Attribution"; Hunker, Hutchinson, and Margulies, *Role and Challenges for Sufficient Cyber-Attack Attribution*; Waxman, "Cyber-Attacks and the Use of Force," 421–59.

intentions.⁸ Similarly, a criminal law conviction (i.e., attribution of a crime to an individual) requires more than simply demonstration of means and opportunity; one must also incorporate "human information" (e.g., motives, intentions) about the suspect.

We can easily see why human motives and opportunities matter by thinking about attribution as an inference problem. In almost all cybersituations, our information set fails to determine the responsible party with certainty. Our attribution inferences are noisy and defeasible because of errors and indeterminism in both our observations and people's decisions and actions. Now suppose that we have only technical knowledge about a particular cyberaction C. In this case, we can infer only that C must be due to some actor or actors who have the technical capabilities to perform C, including (perhaps) the ability to access and control particular machines. In some cases, this inference might be sufficient, as it might uniquely identify the actor. More commonly, though, there will be a nontrivial set of actors who could be responsible for C. This underdetermination can be resolved by (i) prior information about which actors were a priori most likely to want to perform C, as well as (ii) further information about the particular motives or intentions of different actors. In both cases, this information is about the human actors—their motives, intentions, biases, tendencies, and predilections. Of course, we could be mistaken in our assessment of others' reasons, intentions, and desires; our prior beliefs and subsequent information-gathering are obviously not infallible. Moreover, we must focus on the agents' actual preferences and intentions, not their (possibly misleading) publicly stated ones. Nonetheless, the general point stands: correct attribution requires the integration of both technical assessments and judgments about the (potential) human actors for both of the key terms.⁹

This interaction can be seen in two different examples. Consider first the 2007 DDoS attack against Estonia that occurred during protests and riots by Russian nationalists, triggered by the Estonian government moving a Soviet-era statue/memorial. These cyberattacks targeted multiple Estonian government agencies and the Estonian financial sector, principally using attack techniques in which key servers were overwhelmed by queries and requests from multiple computers,

---

⁸ Heller, "'One Hell of a Killing Machine': Signature Strikes and International Law," 89–119.

⁹ More formally minded readers will undoubtedly notice the conceptual resemblance of this discussion to a Bayesian analysis. In fact, one could straightforwardly construct a full Bayesian model of the attribution problem to both formalize these intuitions and also explore the quantitative interactions between the different types of knowledge. We do not provide such a model here due to both space constraints and a desire to focus on conceptual matters. For those interested in the formal details, the key observations are that (i) $P(A_i) \neq P(A_i \mid PH)$ for actors A_i and prior human information PH (i.e., the priors change when we incorporate human information); and (ii) $P(C \mid A_i, T) \neq P(C \mid A_i, T, SH)$ for technical information T and subsequent human information SH (i.e., the likelihoods change as well).

many of which were presumably compromised themselves.[10] Attribution on purely technical grounds was particularly challenging in this case for two different reasons. First, the attack was highly distributed and employed a large number of compromised machines, so enormous expense would be required to trace a significant subset of the commands back to their original source(s). Second, the attack required relatively minimal technical skill and capabilities; many different actors—state, private, and even individual—could have designed and executed the attack. As a result, the technical features alone massively underdetermine the source of the attack. Instead, successful attribution requires the incorporation of information about the "humans in the loop," and in particular, knowledge of which actors would be most probable to launch an attack at this time. In this particular case, that knowledge about the relevant humans did not turn out (at least at first) to uniquely identify the attackers, largely because there were too many plausible candidates. It did, however, significantly narrow the pool of possibilities from the incredibly large list of those with sufficient technical capabilities.

A second example of the importance of human information in attribution arises in the Stuxnet operation. The Stuxnet worm was introduced into Iranian uranium-enrichment facilities from outside—perhaps via a USB drive—and then proceeded to map the facilities' internal structures, change the operations of the enrichment centrifuges to slowly damage them, and finally alter the centrifuge readouts so that everything appeared normal.[11] The Iranian nuclear program was set back as a result, though the impact may have been more limited than initially thought.[12] As with the previous example, attribution could not be resolved on purely technical grounds. Obviously, there were many fewer actors who had the technical capability to carry out the attacks; at the very least, significant state backing—whether official or unofficial—was presumably necessary. Nonetheless, multiple actors, all of whom appear to have engaged in many cyberactivities, arguably had the technical knowledge and capabilities to have performed the attack. Technical features of the attack are thus insufficient to answer the attribution challenge. Instead, one must employ information about the relevant humans in order to determine both (i) the likelihood that actors (with the relevant technical capabilities) would engage in a cyberattack against the Iranian nuclear program; and (ii) for each possible actor, the likelihood that each actor would choose this particular type of attack—given that they do attack. Arguably, both of these factors point toward a (very) small set of probable actors. Moreover, much of the speculation about who performed the Stuxnet attack seems to have employed exactly this type of reasoning and human information;

---

[10] See also: Herzog, "Revisiting the Estonian Cyber Attacks," 49–60.
[11] Kushner, "Real Story of Stuxnet,"; Langner, "Stuxnet's Secret Twin."
[12] Albright, Brannan, and Walrond, *Stuxnet Malware and Natanz*.

it seems to have been widely, though implicitly, understood that attribution requires thinking about human motives, interests, and reasoning. In particular, much of the public reporting about Stuxnet for the two years after its discovery[13] argued that the US government was likely involved, principally because of imputed or assumed motives.[14]

The importance of "human information" for attribution purposes in both of these examples is so obvious as to proceed almost unnoticed. In fact, although many high-level discussions of the attribution problem focus on purely technical challenges,[15] the practical reporting, speculation, and discussion of attribution for concrete, particular cyberattacks almost always invoke human factors.[16] Unfortunately, however, the human is often introduced into the discussion in ways that fail to account for the messiness of actual human cognition. In particular, many attribution discussions assume that the relevant actors (on both sides) are rational agents of the type often studied in decision theory, game theory, or (classical) microeconomics. Such agents have full knowledge of their current desires and interests; can accurately predict their future values and goals; and can design and implement a plan of action that maximizes their chances of success. This rationalist view of the agents and actors is often not explicitly articulated, but is implicitly assumed in order to attribute cyberactions. Inferences of the form "Agent A has reason to do action C. C occurred. Therefore, A is the (or a likely) cause of C" depend on exactly this type of (implicit) view of agents: A must be aware of, understand, and appropriately respond to her rational reasons for action for this inference to be justified.

The past forty years of research in cognitive psychology, behavioral economics, and cognitive neuroscience have given us many reasons to doubt all of these assumptions about actual humans, at least at the individual level: people often do not understand their own reasons for action;[17] they have great difficulty in predicting future desires and preferences;[18] and they will often act against their own (future) best interests, particularly in situations in which they face significant

---

[13] Until the publication of Sanger, *Confront and Conceal: Obama's Secret Wars and Surprising Use of American Power*.

[14] Broad and Sanger, "Worm Was Perfect for Sabotaging Centrifuges"; Sanger, "Iran Fights Malware Attacking Computers."

[15] See, e.g., Shackelford and Andres, "State Responsibility for Cyber Attacks: Competing Standards for a Growing Problem."

[16] E.g., Sanger, "Iran Fights Malware Attacking Computers"; Nakashima, "Iranian Hackers Are Targeting U.S. Officials through Social Networks, Report Says"; Herzog, "Revisiting the Estonian Cyber Attacks: Digital Threats and Multinational Responses."

[17] See, e.g., the voluminous literature following from: Nisbett and Wilson, "Telling More Than We Can Know: Verbal Reports on Mental Processes," 231–59.

[18] Loewenstein, O'Donoghue, and Rabin, "Projection Bias in Predicting Future Utility," 1209–48.

threats.[19] The (implicit) rationalist assumption in many attribution discussions is simply false, precisely because cyberactors are real humans with all of their cognitive biases and foibles. Of course, individual ignorance and irrationality is perfectly consistent with group-level or organizational awareness and rationality, and so perhaps the (implicit) rationalist assumption holds of the groups that, in practice, make many of the most important cyberdecisions. The limited empirical research on the rationality of group planning and decision-making suggests, however, that groups are not necessarily any more rational than the individuals that compose them:[20] some groups can (sometimes) act rationally and in their best interests, but much depends on internal group dynamics.

The importance of recognizing the complexity of human cognition, and the ways that attribution can go wrong when based on rationalist assumptions, can be seen in the Estonia case study. One preliminary attribution for the cyberattacks, including by the Estonian foreign minister, was the Russian government. This attribution seems to have been based principally on the belief that the Russian government had been engaged in systematic retaliation against the Estonian government, and a DDoS attack would have been a rational way to increase the pressure. There seems to have been little initial thought about the possibility that Russian nationalists inside Estonia might be responsible, seemingly because such private citizens were assumed to be uninterested in employing such tactics. It now seems likely that the cyberattacks actually did come from the second group, driven in large part by rage at perceived attacks that they believed were due to their minority status in Estonia.[21] Standard decision-theoretic or rationalist-attribution analyses that assume agents are sensible, self-interested actors would arguably miss such an emotion-based motivation that can lead to disproportionate responses. In contrast, an analysis that incorporates a cognitively sophisticated model of actors and decision-makers could be sensitive to the possibility that some agents will act in seemingly irrational (or at least, arational) ways. Successful attribution depends on understanding how the world appears to different agents, all of whom exhibit various biases and idiosyncrasies; a "one size fits all" approach, particularly a strongly rationalist one, will not work for attribution.

The moral of this section is not that attribution is impossible, but rather that many of our current discussions about it are incomplete. Proper attribution requires consideration of not only technical features of the cyberattack but also human information and factors about the motives, goals, interests, and constraints of the actors who might be responsible. Moreover, incorporation of this human information is a

---

[19] Many of the key studies in these areas are collected as references in Ariely, *Predictably Irrational*; Kahneman, *Thinking, Fast and Slow*.

[20] This literature spans forty years of research; see, e.g., Campbell, "Individual versus Group Problem Solving in an Industrial Sample," 205–10; Woolley et al., "Bringing in the Experts," 352–71.

[21] Herzog, "Revisiting the Estonian Cyber Attacks."

nontrivial task: one cannot simply assume that all actors know their own motives and act (rationally) on that basis. Rather, attribution must be based on the full messiness of the humans, groups, and organizations that could potentially have been responsible for some cyberattack. Without this additional complexity, one runs significant risk of misattribution, and so the possibility of a host of ethical, legal, and political problems.

9.3 Bypassing the (Essential) Human

One key difference between kinetic attacks and cyberattacks is their speed. Kinetic events often unfold over minutes, hours, days, or longer, while extended, significant sequences of cyberevents can occur in less than a second. This difference in timing leads to a difference in human involvement: human decision-makers are almost always part of kinetic decisions or actions that are morally, politically, or psychologically significant, but not necessarily for corresponding cyberdecisions and cyberactions. Essentially all ethical norms and principles about the conduct of warfare—open or covert—and espionage assume that humans are "in the loop" to make the morally salient decisions: both the *jus ad bellum* and *jus in bello* aspects of just war theory rely on rational actors having the capacity to consider and make decisions in real time. This assumption is simply false in many cybercontexts: many significant cyberactions occur without any human endorsing the specific actions in real time.[22] Instead, those actions are the products of decisions made substantially earlier to institute automated responses that are triggered whenever relevant preconditions occur.[23] That is, the morally relevant decisions often occur substantially prior to the cyberevent itself.[24]

In the 2007 attacks against Estonia, for example, some servers were receiving, by orders of magnitude, more requests per second than they previously had; that type of speed increase cannot plausibly be achieved by humans "in the loop." The decision to launch the attack was presumably made by a human, but the form of that attack could easily change in the short timespan of cyberactions into something quite different than what the decision-maker anticipated. Or consider instead much of the cyberespionage described in the Mandiant Report.[25] That report alleged that there have been systematic, targeted intrusions into a range of companies, governments, and other organizations by members of China's

---

[22] See also Sparrow, "Killer Robots," 62–77.

[23] See also Heather M. Roff in this volume.

[24] There are also kinetic armaments, most notably landmines, where a decision to deploy is made significantly before the arms are triggered. Cyberweapons are notably different, however; in particular, neither problem we discuss below arises for these kinetic armaments.

[25] *APT1: Exposing One of China's Cyber Espionage Units.*

People's Liberation Army (PLA). The alleged intrusions involved many different initial attacks, malware families, and other techniques in order to establish entry points, extract information, and leave behind backdoors.[26] Although humans were involved in many stages of the process, a certain level of autonomy was required on both sides in order to respond and adapt to threats and defenses with sufficient speed. For example, a standard cyberdefense involves blocking or ignoring all signals from machines that exhibit certain characteristics; the Mandiant Report and other cybersecurity bulletins provide "attack signatures" for exactly this reason. More precisely, many different programs provide cyberdefense by watching for behavior that matches an attack signature and then automatically adjusting the target machine in prespecified ways (e.g., ignoring further requests from particular IP addresses). On the other side, multiple well-known cyberattacks require a series of rapid actions by the attacking machine, each of which is dependent upon particular responses from the target. The details are not particularly important here; the key is simply that cyberattack, cyberdefense, cyberespionage, and cyberwarfare all require that humans be "*out* of the loop" at key moments in the sequences of events.[27]

Automaticity poses an ethical challenge precisely because we normally think that the machine cannot itself be a locus of moral responsibility. Software is not morally responsible for some outcome (good or bad); the human who designed, implemented, or used the software bears the moral praise or blame. We do not intend this to be a controversial claim, as we take it to be a natural constraint that responsibility for actions must ultimately inhere in a human decision-maker.[28] The natural question is thus: who bears the responsibility for some automatic action *A*? The natural response is: the individual who decided to implement the automatic response.[29] This answer would be quite sensible if people were rational and fully self-knowledgeable. If that background assumption were true, then the spatial/temporal/systems gap between the human decision-maker and automatic machine action would be irrelevant, as we could justifiably infer that the decision-maker still endorsed the action at that later spatiotemporal point. The problem, of course, is that humans are neither fully rational nor fully self-aware, and so we can easily have automatic actions—whether cyberattack, cyberdefense, or cyberespionage—for which no human is morally responsible, simply because no human endorses that action at that

---

[26] Ibid.

[27] Automaticity is arguably also a feature of certain types of kinetic events: so-called doomsday devices provide one infamous example; heat-seeking missiles are a more prosaic case. The observations in this section should apply equally well to automatic responses across domains, though the spatiotemporal separation might well be greater in the cyberdomain than in the kinetic one.

[28] Of course, there might be no one responsible for some *outcome*, as it could be due to luck or some unforeseeable external factor.

[29] Sparrow, "Killer Robots," examines a number of possible loci for moral responsibility.

moment.[30] The situation is somewhat analogous to—though different in key ways from—the kinetic case of a tripwire that triggers a response, but where no human actually endorses the particular response at that moment (e.g., because the wire was tripped by a noncombatant).

More specifically, there are at least two challenges from cognitive science to the necessary background assumptions of rationality and full self-awareness. The first is what we have previously called the "chain reaction challenge."[31] Cybersystems involving multiple automatic actions can readily exhibit complex dynamics that exceed human cognitive abilities for reasoning and inference. Even if we have full knowledge of the different cybersystems involved in some scenario, we will frequently not be able to adequately predict the likely outcomes due to the complex interrelationships among the components, though we might be able to predict the set of possibilities. And of course, we rarely have anything close to full knowledge, particularly in adversarial contexts in which parties are attempting to hide their systems from one another, or for prediction contexts in which our previous learning about the system was driven by more specific goals or needs.[32] We are thus unable to act as fully rational agents in making decisions about cybersystems, precisely because we are unable to predict or understand the likely impacts of our actions. Events can rapidly escalate or spiral out of control in ways that are predictable for a fully rational agent, but unpredictable for any actual human. As a result, a human decision-maker might endorse, at time t, the decision to implement an automatic response R, but only because she fails to understand the role that R can play in a later sequence of events. As a result, the spatiotemporal gap between decision and implementation can be meaningful, as the decision-maker could acquire new information about the system dynamics that lead her to no longer endorse the previous decision. This problem is actually ubiquitous in cybercontexts, including nonwarfare ones: we face a similar difficulty, for example, in assigning responsibility for financial shocks due to surprising and deleterious behaviors of automated trading systems in the complex financial cyberecosystem.[33]

The second cognitive science challenge is also a prediction problem, but this time about the decision-maker herself. Consider a decision made at time t_1 about a plan that will be put into action at some later time t_2. The decision-maker must predict at t_1 the preferences, goals, and values she will have at t_2. That is, the decision now should involve the predicted later desires. Of course, one's current preferences and goals could lead to a decision now that constrains one's future self to act somewhat against her future interests. Leaving aside these complexities, it is clear that one's future desires and goals should matter in making current decisions. The problem is that people are

---

[30] Danks and Danks, "Moral Permissibility of Automated Responses during Cyberwarfare."
[31] Ibid.
[32] Danks, *Unifying the Mind*.
[33] Popper, "Flood of Errant Trades Is a Black Eye for Wall Street."

not necessarily good at predicting their own future preferences, beliefs, desires, and attitudes, and so they can make decisions at t_1 based on incorrect beliefs about what they will want at t_2. For example, people tend to act as if they believe that transient features of their current context will persist indefinitely.[34] As a result, the decision at t_1 can lead to later automatic actions or responses that no one endorses at t_2 and, more importantly, would not have been endorsed at t_1 if the decision-maker had known her likely future desires. We have elsewhere called this the "future self-projection bias challenge."[35] This challenge often does not arise in the kinetic domain, as the human decision-maker is plausibly close enough in space and time to her own future self to make an accurate prediction of her future preferences and desires.

One might object that these so-called challenges are not actually problems for the ethics of automatic cyber-responses, but rather modern instances of the old observation that people can act in morally problematic ways because of ignorance. Moreover, the resulting problematic behavior is often excused on the grounds that the agent "could not have known better"; ignorance actually can be an excuse. An important caveat on these excusings, however, is that the agent not be culpable for the ignorance that led to the action,[36] particularly when the action could have significant consequences.[37] That is, ignorance is only an excuse if one also could not have known that one was ignorant. Both of our challenges—our less-than-full rationality impairing predictions of system dynamics, and our less-than-full self-knowledge and self-awareness leading to incorrect predictions about our future desires—are based on substantial and well-known empirical evidence from the cognitive sciences. Given the importance of many decisions about automatic cyber-responses, ignorance is no excuse, and so the challenges maintain their moral force. Instead, we must recognize our own limitations, even if that implies more conservative decision-making than we might otherwise expect or prefer. Humans are "out of the loop" for many cyberactions, but only because they created the loop in the first place. And so the complexities of human cognition matter for the understanding, assessment, and interpretation of even those cyberactions that appear to bypass human actors.

9.4 Soldiers, Saboteurs, Snipers, and Spies in Cyberspace

Rules of law, such as international humanitarian law, have been established for, and govern much of, kinetic warfare. The rules for traditional espionage are

---

[34] For references to this finding and other types of self-projection bias, see Loewenstein, O'Donoghue, and Rabin, "Projection Bias in Predicting Future Utility."

[35] Danks and Danks, "Moral Permissibility of Automated Responses during Cyberwarfare."

[36] Rosen, "Culpability and Ignorance," 61–84.

[37] Guerrero, "Don't Know, Don't Kill," 59–97.

less well codified, but nations nonetheless recognize certain common practices and guidelines.[38] The question is whether these same rules of law for warfare and uncodified rules for espionage can be easily extended to cyberwarfare and cyberespionage, particularly since the distinction between cyberwarfare and cyberespionage is much fuzzier than between warfare and espionage in the traditional kinetic case. The *Tallinn Manual*,[39] as part of Rule 66—Cyber Espionage, holds that:

> [C]yber espionage is defined narrowly as any act undertaken clandestinely or under false pretences that uses cyber capabilities to gather (or attempt to gather) information with the intention of communicating it to the opposing party.... Cyber espionage must be distinguished from computer network exploitation (CNE), which is a doctrinal, as distinct from an international law, concept. CNE often occurs from beyond enemy territory, using remote access operations. (159)

The distinction between cyberwarfare and cyberespionage also is codified within the United States as Title 10 (cyberwarfare) and Title 50 (cyberespionage),[40] but the laws and rules of other countries do not distinguish the two activities so sharply. In the kinetic domains, the two activities are easier to separate based typically on uniformed military versus civilian spies. In cyberspace, in contrast, identifying the actor is not so clear. The same cyberintrusion could have a damaging effect on the target computer network but also be more benign surveillance.

Normally, the side being attacked can recognize the attacking military by their uniforms and insignia.[41] In cyberattacks, targets must instead look for other indicators of who the attackers are (i.e., the attribution problem) and whether they are military or civilians. In the Estonian DDoS attack, it remains uncertain as to whether the attack was driven by the Russian military, dissident Estonian civilians of Russian background, or some other force. More generally, perfidy is normally not acceptable in warfare lest it become espionage, but perfidy seems inherent in both cyberwarfare and cyberespionage. Unlike in the kinetic domain, recognizing "cyboteurs" is not at all obvious.[42]

The difficulty in distinguishing between cyberespionage, cyberattack, and cyberdefense, but also the importance of doing so, is starkly evident in the

---

[38] Rowe, "Perfidy in Cyberwarfare," 394–404; Roff, "Cyber Perfidy, Ruse, and Deception."

[39] Schmitt, *Tallinn Manual*.

[40] Wall, "Demystifying the Title 10–Title 50 Debate."

[41] But not always, as shown by the confusion in eastern Ukraine during 2014 about whether "rebels" were Ukrainian or Russian, as discussed in, e.g., Roth and Tavernise, "Russians Revealed among Ukraine Fighters."

[42] Arquilla and Ronfeldt, "Advent of Netwar (Revisited)," 1–25.

increasingly heated accusations between China and the United States about various cyberintrusions.[43] The Mandiant Report attributed numerous cyberattacks to China's People's Liberation Army (PLA), 2nd Bureau, 3rd General Staff Department Unit 61398 in Shanghai, and presented substantial evidence supporting this attribution.[44] Over many years, the active exploits (that Mandiant attributes to the Shanghai group) have become legion, involving the exfiltration of massive amounts of intellectual property from US businesses, including defense contractors and high-profile security firms such as RSA Security LLC.[45] The attribution of the exploitation to Unit 61398 helped lead to the indictments of five members of the PLA—Wang Dong, Sun Kailiang, Wen Xinyu, Huang Zhenyu, and Gu Chunhui, all officers in Unit 61398, all pictured in uniforms of the PLA—by a grand jury in Pittsburgh.[46] These five were accused of hacking into computer networks operated by several US businesses operating critical infrastructure in western Pennsylvania, such as Westinghouse, U.S. Steel, SolarWorld, Alcoa, and United Steel Workers.

One key, largely unanswered question is whether the indicted officers of Unit 61398 should be viewed as combatants or as spies. Is it relevant that the indictments included photographs of the accused in PLA uniforms? Standard international laws of war would contend that it is conceivably relevant, as the agreements governing soldiers are very different than the national laws applicable to spies operating within a country, but that presupposes that simply wearing a uniform matters in cyberspace. More generally, these diverse, but linked, roles—noncombatant versus military versus support versus spy—are partly culturally constructed and constituted, and so our analysis of them must include information about the cognitively bounded humans involved. Even if the officers were moonlighting in their off-duty hours as civilians and outside their employment in the military, their activities are just as elusive and difficult to prove in a court of law, either national or international.

Moreover, the Chinese government vigorously denied this attribution and attacked the motives of the US government in pursuing the indictments.[47] The US government and military's hands are arguably far from clean, as the Chinese and other countries are wont to point out: "China is a victim of severe US cyber theft, wiretapping and surveillance activities. Large amounts of publicly disclosed

---

[43] Sanger, Barboza, and Perlroth, "Chinese Army Unit Is Seen as Tied to Hacking against U.S."; Sanger and Perlroth, "Hackers from China Resume Attacks on U.S. Targets"; Sanger, "With Spy Charges, U.S. Draws a Line That Few Others Recognize."

[44] See citations earlier in section 9.3.

[45] RSA Security LLC is an American computer and network security firm known for the RSA public-key cryptography algorithm.

[46] Office of Public Affairs, US Department of Justice, "U.S. Charges Five Chinese Military Hackers for Cyber Espionage Against U.S. Corporations and a Labor Organization for Commercial Advantage"; Schmidt and Sanger, "5 in China Army Face U.S. Charges of Cyberattacks."

[47] Ministry of Foreign Affairs of the People's Republic of China, "China Reacts Strongly to U.S. Announcement of Indictment against Chinese Personnel."

information show that relevant US institutions have been conducting cyber intrusion, wiretapping and surveillance activities against Chinese government departments, institutions, companies, universities and individuals."[48]

A hypothetical case can help to draw out some of the complexities that can arise in this fuzzy area between cyberwarfare and cyberespionage. Suppose an agency of the fictional Yahere government infiltrated firms in the country of Nowhere; let ChatMor be one such company, a leading computer communications firm with a worldwide reach. Further suppose that the country of Nowhere is considered to be an ideological, political, economic, and military challenger to Yahere. Successful access to ChatMor's computer networks to exfiltrate its intellectual property would also allow Yahere's government agencies to spy not only on Nowhere's departments and institutions but also on organizations in other countries that purchased and installed ChatMor's products. Whether such actions were taken by Yahere's military or civilian personnel is largely irrelevant when the actions and end use is for espionage and reconnaissance. But when the knowledge of ChatMor's network operations might be exploited for cyberattacks, then the legal status arguably changes under international law.

These observations about the fuzzy boundary between cyberwarfare and cyberespionage all draw on the importance of perceptions, whether about the cyberaction, the effects of the action, or the roles of the people involved. These kinds of perceptions depend in key ways on the cultural frames and biases that we bring to bear. Whether a particular cyberaction is defense or stealing is in the eye of the actor or the target. For example, the Chinese and Americans have quite different views of cyberespionage, at least when motivated by defense or economic exploitation. There is no international law governing economic espionage.[49] The Chinese position, as illustrated in the previous quotation from the PRC Ministry of Foreign Affairs, conflates activities of governments, especially actions related to national defense, with economic activities of private enterprise. Defending the nation and defending the economy are literally one and the same. In contrast, the United States distinguishes quite sharply between government and private enterprise. As US Attorney General Eric Holder asserted in announcing the indictments:

> The range of trade secrets and other sensitive business information stolen in this case is significant and demands an aggressive response. Success in the global market place should be based solely on a company's ability to innovate and compete, not on a sponsor government's ability to spy and steal business secrets.[50]

---

[48] Ibid.
[49] Fidler, "Economic Cyber Espionage and International Law."
[50] Office of Public Affairs, US Department of Justice, "U.S. Charges Five Chinese Military Hackers for Cyber Espionage against U.S. Corporations and a Labor Organization for Commercial Advantage."

These sorts of alleged activities by both US and Chinese governments, whether by uniformed military or civilians, blur the lines between cyberwarfare, cyberespionage, and cyberdefense.[51] Technical specifications of the activities are not at issue. Rather, the human motives, perceptions, and social, political, and economic perspectives are the issue.

Human motives are neither simple nor pure. One might initially attribute cyberactivities of military personnel to nationalistic or patriotic motives, or even ideology, but hackers are also motivated by money, power, or success. Economic gain is a frequently attributed extrinsic motivator since it can be observed and measured. But hackers' motives, like those of all of us, are multiply determined. The need to gain power over an individual, organization, or target can be a powerful motivator for the psychological satisfaction it yields. Hackers by their practice are good at solving problems, especially the technical problems of computers and computer networks, and this general problem-solving orientation can lead them to use that skill to achieve power and control over computer networks. Hackers are social beings as well and can be motivated by the need to affiliate with fellow hackers or to be loyal to their organizations or country. They also may desire recognition from their peers. The prevalence of "black hat" organizations attests to the strength of several of these motives.[52]

The Stuxnet virus that attacked the uranium centrifuges in Iran is another instance of the blurring of the line between cyberwarfare and cyberespionage, as there were both kinetic and cybereffects. The Stuxnet worm has been commonly attributed to the United States and Israel although neither government has publicly acknowledged responsibility for it. The physical damage to the centrifuges has been well publicized, but what is less well known is the cybereffect of the Stuxnet virus that was intended to hide the physical attack. Specifically, Stuxnet altered the information feedback to the operators so that the centrifuges appeared to be operating normally. As a result, the operators were unaware that the centrifuges were speeding up and out of control.[53]

A possible Iranian response to Stuxnet has recently been revealed by iSightPartners.[54] A social networking campaign, dubbed NEWSCASTER by iSightPartners, targeted US military officers and diplomats as well as both US and Israeli defense contractor personnel. The hackers invented numerous online personae in various social networking sites such as Facebook and LinkedIn, and then sent friendly messages with known contacts of their targets, including links to a fake

---

[51] Sanger, "With Spy Charges, U.S. Draws a Line That Few Others Recognize."
[52] Denning, "Cyber Conflict as an Emergent Social Phenomenon," 170–86; Rogers, "Psyche of Cybercriminals: A Psychosocial Perspective."
[53] Kushner, "Real Story of Stuxnet"; Rid, *Cyber War Will Not Take Place*.
[54] iSightPartners, *NEWSCASTER*.

site, NewsOnAir.org, a technique known as spearphishing.⁵⁵ (NewsOnAir.com is a legitimate English-language news site in India, so the ruse was plausible.) Once the target's friends and contacts had been compromised, the targets themselves could be spearphished. Once the primary target had been compromised, data and information could be exfiltrated for a variety of adverse uses. NEWSCASTER is a well-defined social engineering campaign that relies on human attackers and spies being able to take advantage of the cognitive and emotional biases and weaknesses of the human targets.

Successful cyberattacks and cyberespionage, and even cyberdefense, require a full understanding of the human target, as well as implicit self-knowledge of the attacking operator. An analysis of specific instances of cyberattacks and cyberespionage reveals this human-centric character, but cybersecurity experts rarely, if ever, talk about the human element, except for campaigns to get computer users to improve their security activities. Effective cyberattacks and cyberespionage frequently depend on social engineering, including manipulating people psychologically to get them to take actions they might not otherwise take, especially with regard to confidential information. These manipulations are based on understanding the biases and weaknesses of human decision-making; these cognitive foibles might be called "bugs" in the human wetware, but are inevitable and perhaps even necessary. Understanding the cognitive, behavioral, and social processes of the humans in the loop is crucial when exploring the distinctions between cyberwarfare and cyberespionage across a range of cultural contexts.

9.5 Keeping Humans in the Picture

Humans—not computers, networks, or technology writ large—are the agents who can be held responsible for ethical decisions. Key issues in the ethics of cyberwarfare depend critically on recognizing and understanding these agents as cognitively real humans, rather than idealized fictions. Humans must be kept "in the loop" to decide: who is the aggressor; whether an action discriminates between effects on combatants and civilians; if an action is a proportional response; whether combatants are clearly identified as legitimate targets; or if there can even be perfidy or treacherous deceit in cyberspace.⁵⁶ Technical information about a cyberattack is

---

⁵⁵ Phishing is a social engineering technique to obtain private information. The attacker sends an email message that appears to be from a trusted source requesting private information, such as passwords and bank account numbers. Phishing attacks are sent to a large number of targets without personal links to the individual target. Spearphishing attacks also appear to come from a trusted source, often from within the recipient's own organization or a website the target uses frequently, and are targeted to particular individuals using personal targeting information.

⁵⁶ Lin, Allhoff, and Rowe, "War 2.0: Cyberweapons and Ethics," 24–26.

insufficient to answer these questions, as was demonstrated in the DDoS attacks on Estonia. In criminal activity, the key elements in identifying a perpetrator are whether he has the motivation, opportunity, and capability to commit the crime. The same is true for attributing a cyberattack or cyberespionage to an agent: we must determine which individual, organization, or nation has the technical capability to actually effect the cyberattack or cyberespionage, and just as importantly, the motivation to carry out a cyberattack and the opportunity to do so. Technical information is helpful for these determinations, but not sufficient; we also must think about the actual humans behind the cyberactions.

The speed and velocity at which cyberactions take place is a fundamental challenge for human cyberoperators. Automatic responses leave humans outside of the real-time loop. Humans can be held responsible for developing and programming the automatic responses, but it is unclear under what conditions the responsibility for the actual event transfers to those agents. If humans were rational in their planning and decisions, then assigning responsibility to originating agents might be reasonable, but neither people nor groups exhibit the necessary rationality or self-knowledge. In particular, people are frequently unable to accurately predict their future preferences, goals, and values. Moreover, preferences for a particular course of action often are modified by changes in the physical situation as well as the person's internal mental state.

In addition to not recognizing their own (future) motivations, people frequently misjudge others' motivations, a finding often called the "fundamental attribution error."[57] Our own motives are commonly attributed to transitory situational factors at the moment when the decision is made and action taken, a stance that is sometimes called "situationism." The motivations of others, however, are typically attributed to relatively enduring (personality) traits, that is, a stance of "dispositionism." These persistent personality traits are more unidimensional, and so attributing actions to them can result in simplistic, and often erroneous, interpretations of another's actions, especially when that person is a potential adversary. Moreover, the fundamental attribution error is culturally bound. It is not found as frequently in groups from so-called collectivistic cultures (e.g., East Asian) as opposed to groups in individualistic cultures (e.g., Western European and American).[58] This cultural differentiation results from East Asians holding a more situational perspective than westerners: "East Asians believe dispositions to be more malleable and have a more holistic conception of the person as being situated in a broad social context."[59] If we

---

[57] Jones and Harris, "Attribution of Attitudes," 1–24; Nisbett and Ross, *Person and the Situation: Perspectives of Social Psychology*.

[58] Miller, "Culture and the Development of Everyday Social Explanation," 961–78.

[59] Choi, Nisbett, and Norenzayan, "Causal Attribution Across Cultures: Variation and Universality," 49; see also Norenzayan, Choi, and Nisbett, "Cultural Similarities and Differences in Social Inference," 109–20.

instead recognize other humans as complex cognitive agents, then we can consider their situational context as well as their enduring personality traits, and thereby recognize and conceive of multiple motives.

All of these factors point to the need for a human-centric analysis of cyberactors and cyberevents. Only with a full understanding of the human in the loop can we begin to assign and understand ethical responsibility for these actions and, more generally, explicate the complex moral and ethical dimensions of cyberwarfare, cyberespionage, and other actions in the cyberdomain.

Bibliography

Albright, David, Paul Brannan, and Christina Walrond. *Stuxnet Malware and Natanz: Update of ISIS December 22, 2010 Report*, Institute for Science and International Security, February 15, 2011. http://isis-online.org/uploads/isis-reports/documents/stuxnet_update_15Feb2011.pdf.

Andress, Jason, and Steve Winterfeld. *Cyber Warfare: Techniques, Tactics and Tools for Security Practitioners*, Waltham, MA: Syngress, 2011.

APT1: Exposing One of China's Cyber Espionage Units, Mandiant Intelligence Center, February 18, 2013. http://intelreport.mandiant.com.

Ariely, Dan. *Predictably Irrational: The Hidden Forces That Shape Our Decisions*. New York: Harper Collins, 2008.

Arquilla, John, and David Ronfeldt. "The Advent of Netwar (Revisited)." In *Networks and Netwars: The Future of Terror, Crime, and Militancy*, ed. John Arquilla and David Ronfeldt, 1–25. Santa Monica, CA: RAND Corporation, 2001. http://www.rand.org/pubs/monograph_reports/MR1382.html.

Broad, William J., and David E. Sanger. "Worm Was Perfect for Sabotaging Centrifuges." *New York Times*, November 18, 2010.

Campbell, John P. "Individual versus Group Problem Solving in an Industrial Sample." *Journal of Applied Psychology* 52, no. 3 (1968): 205–10.

Choi, Incheol, Richard E. Nisbett, and Ara Norenzayan. "Causal Attribution across Cultures: Variation and Universality." *Psychological Bulletin* 125 (1999): 47–63.

Clark, David D., and Susan Landau. "Untangling Attribution." *Harvard National Security Journal* 2, no. 2 (2011). http://harvardnsj.org/2011/03/untangling-attribution-2/.

Cornish, Paul, David Livingstone, Dave Clemente, and Claire Yorke. *On Cyber Warfare*. London: Chatham House, 2010.

Danks, David. *Unifying the Mind: Cognitive Representations as Graphical Models*. Cambridge, MA: MIT Press, 2014.

Danks, David, and Joseph H. Danks. "The Moral Permissibility of Automated Responses during Cyberwarfare." *Journal of Military Ethics* 12, no. 1 (2013): 18–33.

Denning, Dorothy E. "Cyber Conflict as an Emergent Social Phenomenon." In *Corporate Hacking and Technology-Driven Crime: Social Dynamics and Implications*, ed. T. J. Holt and B. H. Schell, 170–86. Hershey, NY: Information Science Reference, 2011.

Dipert, Randall R. "Preventive War and the Epistemological Dimension of the Morality of War." *Journal of Military Ethics* 5, no. 1 (2006): 32–54.

Dipert, Randall R. "The Ethics of Cyberwarfare." *Journal of Military Ethics* 9, no. 4 (2010): 384–410.

Fidler, David P. "Economic Cyber Espionage and International Law: Controversies Involving Government Acquisition of Trade Secrets through Cyber Technologies." *ASIL Insights* 17, no. 10 (March 20, 2013). http://www.asil.org/insights/volume/17/issue/10/economic-cyber-espionage-and-international-law-controversies-involving.

Guerrero, Alexander A. "Don't Know, Don't Kill: Moral Ignorance, Culpability, and Caution." *Philosophical Studies* 136, no. 1 (2007): 59–97.

Heller, Kevin Jon. "'One Hell of a Killing Machine': Signature Strikes and International Law." *Journal of International Criminal Justice* 11, no. 1 (2013): 89–119.
Herzog, Stephen. "Revisiting the Estonian Cyber Attacks: Digital Threats and Multinational Responses." *Journal of Strategic Security* 4, no. 2 (2011): 49–60.
Hunker, Jeffrey, Bob Hutchinson, and Jonathan Margulies. *Role and Challenges for Sufficient Cyber-Attack Attribution*. Institute for Information Infrastructure Protection, January 2008. http://www.thei3p.org/docs/publications/350.pdf.
iSightPartners. *NEWSCASTER: An Iranian Threat within Social Networks*, May 28, 2014.
Jones, Edward E., and Victor A. Harris. "The Attribution of Attitudes." *Journal of Experimental Social Psychology* 3, no. 1 (1967): 1–24.
Kahneman, Daniel. *Thinking, Fast and Slow*. New York: Farrar, Straus & Giroux, 2011.
Kushner, David. "The Real Story of Stuxnet." *IEEE Spectrum*, February 26, 2013. http://spectrum.ieee.org/telecom/security/the-real-story-of-stuxnet.
Langner, Ralph. "Stuxnet's Secret Twin." *Foreign Policy*, November 19, 2013.
Lin, Patrick, Fritz Allhoff, and Neil C. Rowe. "War 2.0: Cyberweapons and Ethics." *Communications of the ACM* 55, no. 3 (March 2012): 24–26.
Loewenstein, George, Ted O'Donoghue, and Matthew Rabin. "Projection Bias in Predicting Future Utility." *Quarterly Journal of Economics* 118, no. 4 (2003): 1209–48.
Miller, J. G. "Culture and the Development of Everyday Social Explanation." *Journal of Personality and Social Psychology* 46, no. 5 (1984): 961–978.
Ministry of Foreign Affairs of the People's Republic of China. "China Reacts Strongly to U.S. Announcement of Indictment against Chinese Personnel," May 19, 2014. http://www.fmprc.gov.cn/mfa_eng/xwfw_665399/s2510_665401/t1157487.shtml.
Nakashima, Ellen. "Iranian Hackers Are Targeting U.S. Officials through Social Networks, Report Says." *Washington Post*, May 29, 2014.
Nisbett, Richard E., and Lee Ross. *The Person and the Situation: Perspectives of Social Psychology*. New York: McGraw-Hill, 1991.
Nisbett, Richard E., and Timothy DeCamp Wilson. "Telling More Than We Can Know: Verbal Reports on Mental Processes." *Psychological Review* 84, no. 3 (May 1977): 231–59.
Norenzayan, Ara, Incheol Choi, and Richard E. Nisbett. "Cultural Similarities and Differences in Social Inference: Evidence from Behavioral Predictions and Lay Theories of Behavior." *Personality and Social Psychology Bulletin* 28 (2002): 109–20.
Office of Public Affairs, U S Department of Justice. "U.S. Charges Five Chinese Military Hackers for Cyber Espionage against U.S. Corporations and a Labor Organization for Commercial Advantage." May 19, 2014. http://www.justice.gov/opa/pr/2014/May/14-ag-528.html.
Popper, Nathaniel. "Flood of Errant Trades Is a Black Eye for Wall Street." *New York Times*, August 1, 2012.
Rid, Thomas. *Cyber War Will Not Take Place*. Oxford: Oxford University Press, 2013.
Roff, Heather M. "Cyber Perfidy, Ruse, and Deception," this volume.
Rogers, M. K. "The Psyche of Cybercriminals: a Psychosocial Perspective." In *Cybercrimes: A Multidisciplinary Analysis*, ed. Sumit Ghosh and Elliott Turini, 217–35. Springer eBooks, 2011. http://link.springer.com/book/10.1007/978-3-642-13547-7.
Rosen, Gideon. "Culpability and Ignorance." *Proceedings of the Aristotelian Society* 103, no. 1 (2003): 61–84.
Rosenzweig, Paul. *Cyber Warfare: How Conflicts in Cyberspace Are Challenging America and Changing the World*. Santa Barbara, CA: Praeger, 2013.
Roth, Andrew, and Sabrina Tavernise. "Russians Revealed among Ukraine Fighters." *New York Times*, May 27, 2014.
Rowe, Neil C. "Perfidy in Cyberwarfare." In *Routledge Handbook of Ethics and War*, ed. Fritz Allhoff, Nicholas G. Evans, and Adam Henschke, 394–404. New York: Routledge, 2013.
Rowe, Neil C. "The Ethics of Cyberweapons in Warfare." *International Journal of Cyberethics* 1, no. 1 (2009): 20–31.

Sanger, David E. *Confront and Conceal: Obama's Secret Wars and Surprising Use of American Power*. New York: Random House, 2012.

Sanger, David E. "Iran Fights Malware Attacking Computers." *New York Times*, September 25, 2010.

Sanger, David E. "With Spy Charges, U.S. Draws a Line That Few Others Recognize." *New York Times*, May 19, 2014.

Sanger, David E., and Nicole Perlroth. "Hackers from China Resume Attacks on U.S. Targets." *New York Times*, May 19, 2013.

Sanger, David E., David Barboza, and Nicole Perlroth. "Chinese Army Unit Is Seen as Tied to Hacking against U.S." *New York Times*, February 19, 2013.

Schmidt, Michael S., and David E. Sanger. "5 in China Army Face U.S. Charges of Cyberattacks." *New York Times*, May 19, 2014.

Tallinn Manual on the International Law Applicable to Cyber Warfare. Ed. Michael N. Schmitt. Cambridge: Cambridge University Press, 2013.

Shackelford, Scott J., and Richard B. Andres. "State Responsibility for Cyber Attacks: Competing Standards for a Growing Problem." *Georgetown Journal of International Law* 42, no. 4 (2011): 971–1016.

Sparrow, Robert. "Killer Robots." *Journal of Applied Philosophy* 24, no. 1 (February 2007): 62–77.

Wall, Andru E. "Demystifying the Title 10–Title 50 Debate: Distinguishing Military Operations, Intelligence Activities and Covert Action." *Harvard National Security Journal* 3, no. 1 (2011). http://harvardnsj.org/2011/12/demystifying-the-title-10-title-50-debate-distinguishing-military-operations-intelligence-activities-covert-action/.

Waxman, Matthew C. "Cyber-Attacks and the Use of Force: Back to the Future of Article 2(4)." *Yale Journal of International Law* 36, no. 2 (2011): 421–59.

Woolley, Anita Williams, Margaret E. Gerbasi, Christopher F. Chabris, Stephen M. Kosslyn, and J. Richard Hackman. "Bringing in the Experts: How Team Composition and Work Strategy Jointly Shape Analytic Effectiveness." *Small Group Research* 39, no. 3 (2008): 352–71.

Zsambok, Caroline E., and Gary Klein. *Naturalistic Decision Making*. New York: Psychology Press, 1997.

Part IV

CYBERWARFARE, DECEPTION, AND PRIVACY

10

Cyber Perfidy, Ruse, and Deception

HEATHER M. ROFF

> Perfidy and ruses of war share a common ground: both categories stem from deception and stratagem. Yet, perfidy is largely a violation of LOIAC [laws of international armed conflict], whereas ruses of war are perfectly legitimate. The question is how to tell them apart, and *the answer depends on a modicum of mutual trust which must exist even between enemies*, if LOIAC is to be fully complied with.

The opening epigraph provides us with a helpful insight into warfare: that if we are to forbear from perfidious actions, we are obligated, to some degree, to trust our enemies.[1] Yet trust between enemies seems such a strange sentiment. My enemy is, by definition, one who is opposed to me, and this opposition is usually hostile in nature. Therefore, it would seem prudent not to trust my enemy, for it is in my enemy's interest to deceive me in such a way that he gains the upper hand. In warfare, this hostile relationship increases the incentive to deceive, lie, and mislead, as any advantage one gains may ensure not only that one saves lives but also that one is victorious. This is why, four thousand years later, Sun Tzu's dictum that "all warfare is based on deception" still rings true.[2] The prudent commander ought to "hold out baits to entice the enemy" and "feign disorder" to confuse and mislead his enemy in order to win.[3]

Today's warfighters continue to utilize deceptive strategies, from mock combat operations to psychological operations aimed at changing the enemy's and the enemy population's perceptions.[4] More recently, however, cyberoperations have surfaced as a new theater for deceptive tactics, for they not only "promise coercion on the cheap"[5] but also exploit information, disseminate misinformation, and cause

---

[1] Dinstein, *Conduct of Hostilities*, 198.
[2] Tzu, *Art of Warfare*.
[3] Ibid.
[4] Latimer, *Deception in War*; Hart, *Strategy*.
[5] Feaver, "Blowback: Information Warfare and the Dynamics of Coercion," 100.

physical destruction and harm in ways more subtle than other traditional domains. Cyberoperations also promise advantage without having to endanger one's own troops or assets, as they work silently, from a distance, "creeping in on little cat feet."[6]

Deceptive strategies, and cyberoperations in particular, present us with a bit of a puzzle: we are obligated to trust our enemies, but we are also permitted—to some extent—to deceive them, and we should also expect their reciprocal deception of us. The effect seems to be that mutual trust in this environment is impossible, unless one places trust in a minimal acceptance of and compliance with a publicly recognized rule. In warfare, this rule is the prohibition on "perfidy" or "treacherous killing."[7] The question before us, then, is to what extent cyberoperations challenge or support the prohibition on perfidy. As discussed later in the chapter, cyberoperations are deceptive, but the degree and extent to which they are perfidious is not clear. Do cyberoperations fall on the side of permissible ruses, or do they amount to perfidy and erode the minimal levels of mutual trust between enemies?

I argue that if we are to hold fast to the internationally legally recognized and customary rule prohibiting perfidy, then *any* use of a cyberweapon that results in the killing, wounding, or capture of an adversary is impermissible and amounts to perfidy. Other cyberoperations that do not result in these outcomes are still wrongful but do not rise to the graver wrong of perfidy. Given this conclusion, I suggest that the long-term effects of cyberoperations may undermine or erode the minimal trust necessary between belligerents, as well as in international law or institutions. In particular, the (over)reliance on cyberoperations may threaten the ability to negotiate peaceful settlements and adversely affect international stability. The argument proceeds in three sections.

The first section outlines the laws of armed conflict (LOAC) regarding perfidy and ruse, and offers a purposive reading of the prohibition on perfidy. While LOAC is not tantamount to the just war tradition, they are co-constitutive. Moreover, since LOAC provides the clearest account of rules regarding deception, whereas the just war tradition only indirectly addresses this topic, I rely on international legal rules to argue analogously against the moral permissibility of deceptive cyberstrategies. The second section argues that existing scholarly work on cyber perfidy misses the importance of the interaction between the rules prohibiting perfidy and the technological reality of cyberoperations. I draw from publicly available information to discuss the current technological capacities of cyberweapons. I then apply a purposive account of perfidy to cyberoperations and find that cyberoperations are permissible only when carried out as espionage, or when such operations are not intended to establish any technical or human "confidence" nexus to then produce death, injury, or capture. The third section examines the concept of trust and argues

---

[6] Arquilla, Interview with *Frontline*.
[7] Protocol Additional to the Geneva Conventions of 12 August 1949.

that deceptive cyberstrategies will ultimately undermine the fragile trust between adversaries and have negative impacts for peace settlements.

10.1 Ruse and Perfidy

The laws of armed conflict—based on norms, conventions, and treaty law—attempts to regulate the behavior of belligerents during hostilities. It is a prohibitive body of law that "forbids rather than authorizes certain manifestations of force."[8] While often overridden by leaders, governments, and individual warfighters, these laws are necessary bulwarks against the potential for gross atrocities. The just war tradition informs much of these laws and, in turn, is informed by them.[9] These rules require, at bare minimum, a belief that one's enemy will abide by the protections and prohibitions laid down by them. As Augustine first noted, and Lassa Oppenheim later reiterated, "*fides etiam hosti servanda*": faith must be kept even to the enemy, otherwise peace is impossible.[10]

While war is a rule-governed activity, the negative character of these rules provides leeway for belligerents to pursue their military objectives. What is not prohibited is permitted, and as such, some practices of deception are licit. Ruses, or those acts "intended to mislead the enemy, or to induce him to act recklessly," do not invite an enemy's confidence.[11] Ruses do not form a bond (or nexus) of trust that generates a right or duty based on this confidence. Thus, if there is no "confidence," there can be no breach. Examples of permissible ruses include: camouflage; the use of (mis)information, such as images, data, or communications with the intent to mislead or to cause rebellion, mutiny, or desert (though not surrender); the use of dummy constructions and weapons; and the use of false codes.[12] Transmitting and broadcasting propaganda over radio waves (and, by extension, through websites) is likewise permitted, as are, it would seem, attempts at Internet "spamming" or using pop-up ads. However, manipulation of images or communications to induce one's adversary to surrender is prohibited.[13]

---

[8] Baxter, "So-Called 'Unprivileged Belligerency': Spies, Guerrillas, and Saboteurs," 324.

[9] Bellamy, "Responsibilities of Victory," 601–25.

[10] Oppenheim, *International Law: A Treatise*, 226.

[11] Article 24 of the Hague Regulations and Article 37(2) of Protocol I. "Ruses of war are not prohibited. Such ruses are acts which are intended to mislead an adversary or to induce him to act recklessly but which infringe no rule of international law applicable in armed conflict and which are not perfidious because they do not invite the confidence of an adversary with respect to protection under that law. The following are examples of such ruses: the use of camouflage, decoys, mock operations and misinformation." Dinstein, *Conduct of Hostilities*, 206.

[12] Dinstein, *Conduct of Hostilities*, 206–s7.

[13] Department of Defense Office of General Counsel, "An Assessment of International Legal Issues in Information Operations."

In the context of naval warfare, which is governed by a more permissive body of laws, ships may fly false flags up until the moment they engage in hostile activities. But a belligerent warship must still show its "true colors" before firing.[14] Warships may also use deceptive lighting techniques, where they disguise themselves to look like civilian ships. The use of "dummy ships and other armament, decoys, simulated forces, feigned attacks and withdrawals, ambushes, false intelligence, electronic deceptions, and use of enemy codes, passwords and countersigns" is likewise permissible.[15] However, in any domain, including the international waters, emitting a distress signal to do anything other than indicate distress is clearly prohibited. Ruses, then, permit a belligerent to engage in any type of deceptive activity that does not violate a law or use and pervert the law in such a way to endanger the protections provided by LOAC. The nexus of confidence, or "trust," here is in the reciprocal recognition—and protection accorded by law and ethics—that some persons and things are sacrosanct and will not be used treacherously.

Perfidy, by contrast, constitutes an act whereby a belligerent uses the laws of armed conflict to make its adversary believe he has been accorded some protection with the intent to breach that trust through killing, wounding, or capturing.

Article 37 of Additional Protocol I to the Geneva Conventions states that:

1. It is prohibited to kill, injure or capture an adversary by resort to perfidy. Acts inviting the confidence of an adversary to lead him to believe that he is entitled to, or is obliged in accord, protection under the rules of international law applicable in armed conflict, with intent to betray that confidence, shall constitute perfidy. The following acts are examples of perfidy:

    (a) the feigning of an intent to negotiate under a flag of truce or of a surrender;
    (b) the feigning of an incapacitation by wounds or sickness;
    (c) the feigning of civilian, non-combatant status, and
    (d) the feigning of protected status by the use of signs, emblems or uniforms of the United Nations or of neutral or other States not Parties to the conflict.[16]

Other articles in Additional Protocol I reiterate the impermissibility of using protected emblems, such as the Red Cross, Red Lion and Sun, the United Nations, particular signs or signals (like distress or surrender), as well the wearing of enemy or neutral uniforms, and improperly using flags to gain battle advantages.[17]

---

[14] San Remo Manual on International Law Applicable to Armed Conflict at Sea.
[15] US Navy, "Commander's Handbook."
[16] Protocol Additional to the Geneva Conventions of 12 August 1949.
[17] See Articles 38 and 39. There is further support for these principles in the Rome Statutes, the 1923 Hague Rules of Air Warfare; the 1923 Hague Rules for the Control of Radio in Time of War, and Additional Protocol II.

Notably absent from the discussion regarding perfidy is damage to property. Damaging property does not rise to the level of a perfidious action. This is not to say that one can damage property to any extent in warfare, as rules governing necessity and due care, as well as proportionality and excessive harm, still apply. Rather, the point is that if one feigns a protected status to damage property, more than likely such an act amounts to sabotage. For instance, donning an enemy's uniform to gain proximity to a tank for the sake of making the tank inoperable would not be classified as perfidy. Rather, it is an impermissible use of a uniform. Sabotage, as Thomas Rid notes, "is not ultimately focused on the human body as a vehicle to the human mind—instead, sabotage, first and foremost, attempts to impair a technical or commercial system and to achieve a particular effect by means of damaging that system."[18] Perfidy, on the contrary, uses killing, wounding, or capture to breed widespread distrust in the minds of the enemy and enemy population. The thinking goes that assuming a nonthreatening appearance, such as a civilian, a medic, or a representative of the United Nations, to inflict harm or death makes vulnerable all persons. Abusing protected emblems or assuming protected status provides one's adversary with incentive and excuse to begin targeting all classes of people and things, thereby increasing the amount of suffering, violence, and destruction during the conflict. The principles of proportionality, distinction, and necessity then collapse along with the protections.

Rules notwithstanding, perfidious action is not always easy to identify. The limitless nature of human creativity provides a vast array of deceptive techniques with which to wage war. Following Mike Madden, I urge that we should view the laws prohibiting perfidy through a purposive lens "by importing concepts of causation commonly used within domestic legal systems."[19] For the purposive test can work to uphold the clear cases—waving a white flag only to turn and fire—but it can also illuminate the more ambiguous ones, such as cyberoperations. The test asks, "if but for this use [of X]" would the death, injury, or capture have resulted? If we answer affirmatively, then the act would constitute perfidy; if it is unclear, we might conjecture that the use is wrongful, though not perfidious. Ambiguous cases require further examination of the circumstances ruling at the time. The purposive account traces the causation of a particular act and allows for the assigning of, at a minimum, liability in cases where there are multiple or concurrent causes. Ultimately, the point of adopting a purposive interpretation for the perfidy rule is to trace the intentions of perpetrators, the foreseeability of harm or risk of a particular act, as well as identifying the proximate cause of the death, injury, or capture.

Madden notes that there are three types of perfidy at work in international jurisprudence. First are those acts that are clearly perfidious and "constitute criminal

---

[18] Rid, *Cyber War Will Not Take Place*, 57.
[19] Madden, "Of Wolves and Sheep," 440.

conduct amounting to a war crime," such as the white flag example above.[20] Second are acts that are perfidious but do not rise to the level of a war crime, such as acts that might resort to treachery to capture or wound but do not result in "serious bodily injury" or killing.[21] Third are ostensibly perfidious acts—acts that are deceptive in intent, but presumably do not constitute a criminal act of any kind. An example of this last category would be the deceptive lighting techniques used by belligerent ships at sea.[22] That the law of the sea does not explicitly forbid such acts, and that they did not result in the injury, killing, or capture of enemy combatants, does not necessarily make such acts morally permissible, though they are legally so.[23] This third category is most likely where cyberoperations will take place, as there are no laws prohibiting cyberoperations at present, and no operations to date have resulted in the killing, injury, or capture of combatants.

While we are not concerned with strict jurisprudence, Madden's formulation is helpful in illuminating the spectrum of impermissible uses of deception. On one end, we have permissible ruses, such as camouflage or stealth technology, and on the other, we have obviously perfidious acts resulting in the death, injury, or capture of a combatant by abusing a protected status. Between these two positions, we have *at least* two additional positions: using deception to induce one's adversary to believe they are accorded some protections under LOAC (but where harm is not as grave as killing), and those gray areas that attempt to induce a "fog" wherein a belligerent may or may not believe he is accorded some protection or is facing a nonthreatening party.

To date, scholars' treatment of deception and cyberoperations, particularly in reference to the issue of perfidy, is sparse. Three sources, however, are germane to my arguments here: the *Tallinn Manual*, Neil Rowe, and William Boothby. I will argue that each, however, is flawed in their arguments concerning cyber perfidy and their applications of the rule in international law. These flaws perpetuate a nonproblematic reading of cyberoperations, whereas, as I will argue, cyberoperations are anything but. To see why this is so, we must fully unpack their arguments alongside the technological aspects of cyberoperations. Once we understand how cyberoperations function, we are faced with a serious question: whether we are prepared to identify *any* killing, wounding, or capture via cybermeans as perfidious—and

---

[20] Ibid., 444.

[21] Ibid. Madden also notes that the wording of Additional Protocol I implies that perfidious actions are not illegal tout court, but only those perfidious acts leading to death, injury, or capture are. There is some ambiguity, admittedly, about how one might draw lines between death and injury and capture.

[22] Ibid.

[23] This last category is suspect, though, a fact of law. This is due more to the tensions between the laws of the sea and the laws regulating combat in other domains. It makes little sense to say that something is perfidious, but neither illegal nor criminal, as the very category of perfidy means it is *at least* illegal. Unfortunately it is outside the scope of this chapter to engage in the tensions inherent in the laws of the sea and the LOAC.

thus uphold the black letter of international law—or whether we want to abandon the customary moral and legal rules of prohibiting perfidy to make permissible the killing, wounding, or capturing by cybermeans. If one believes that the laws pertaining to perfidy must be upheld, and thus side with the former position, then we must be prepared to accept that any killing, wounding, or capturing through cyberweapons is a war crime. If we side with the latter, and claim that cyberoperations resulting in death, injury, or capture do not rise to the types of harm envisioned by the likes of Augustine, Oppenheim, and the Additional Protocols, then we must construct some sort of justification to use cyberoperations as a prima facie proximate cause of ostensibly perfidious acts that are not, however, in fact perfidious.

10.2 Cyberweapons and Cyber Perfidy

As the scholarship on cyber perfidy is limited, it is in our interests to work through it.[24] In particular, we ought to consider how the authors' arguments fail to consider the convergence of technical *and* legal aspects of cyberoperations. The first set of arguments, those presented in the *Tallinn Manual*, developed by Michael Schmitt and authored by a group of international legal experts, is rapidly becoming an authoritative work on issues of cyberconflict and LOAC. While the manual can also be read as supporting a similarly purposive account of perfidy—it states that "in order to breach the prohibition against perfidy, the perfidious act must be the proximate cause of death or injury," where "proximate" does not mean temporal proximity, but rather *purposeful* causal connection—it does not offer a coherent picture of cyberdeception, perfidy, or ruse.[25] To be fair, this conclusion is more than likely the result of the authors being "divided" on a number of issues. For instance, the authors were divided as to what (or who) must be the object of deception. A majority found that deception of a cybersystem, and not merely deception of a human being, qualified as perfidy if the action results in killing, wounding, or capture.

---

[24] I will not engage with Neil Rowe's arguments beyond stating that his position errs in its understanding and application in international law. He argues that "cyber perfidy can be defined as the feigning of civilian computer software, hardware or data as *a step towards doing harm to an adversary."* This reading, however, is incorrect. At best, it amounts to a prohibition on cybersabotage; at worst, it prohibits any action defined as "harm," like the changing of data or communications. Altering communications or data on a network to induce an adversary to take risks is not, however, perfidy, and is in fact neither wrongful nor unlawful. While feigning an "ordinary" or perhaps more concretely "civilian" status is prohibited, damage to property through this use is likewise not perfidious. This is not to say it is permissible, as such damage may be prohibited through other laws (such as laws regulating distinction or the protection of works and installations containing dangerous forces). See Rowe, "Perfidy in Cyberwar," 394–404.

[25] *Tallinn Manual*, 181 (italics added).

Yet the route by which a cybersystem becomes deceived is not as clear. In particular, how or to what extent one must feign a protected status or use a protected emblem to "deceive" the system (or person) is not evident. The authors do not explicitly address this, and are divided, again, as to whether an email sent to an adversary with a protected status domain name (such as "icrc.org") could be a wrongful act or not. Some viewed that it would not be wrongful if merely limited to the domain name and did not utilize the emblem of the International Committee of the Red Cross.[26]

Using enemy insignia, uniforms, or flags is also a contentious point from the cyberoperations perspective. The rule the manual puts forth states, "it is prohibited to make use of the flags, military emblems, insignia or uniforms of the enemy *while visible to the enemy* during an attack, including a cyber attack."[27] Adding the clause "while visible to the enemy" sidesteps more difficult questions. For instance, the authors posit that it is only when "the attacker's use *is apparent* to the enemy that the act benefits the attacker or places its opponent at a disadvantage"; however, the experts fail to consider whether "visible" or "apparent" entails not merely human eyesight but also cybernetwork perception. Here is the rub: if the experts grant that killing, wounding, or capturing by "deceiving" a cybersystem is perfidious, they must answer the antecedent question of whether a cybersystem can "see" or "perceive." To be deceived through the "apparent" use of a protected status or emblem necessarily entails that: (i) one knows what and who is protected, and (ii) one can identify through some perceptive capacity that the email, person, or credential is protected. One cannot be deceived if one cannot perceive. The manual, however, fails to make this connection.

Second, we ought to consider William Boothby's discussion of deception and cyber perfidy on its own terms. Boothby partook in the editorial committee of the *Tallinn Manual*, but did not contribute. He has, however, written separately and in greater detail on cyber perfidy and espionage, offering a series of arguments and vignettes to aid in the identification of perfidious actions. Two of those vignettes deserve our attention. First is a cyberoperation that

> deceives the targeted computerized perimeter security system to believe that the enemy personnel are in fact friendly forces. The enemy personnel then enter the closed military facility protected by the security system and wreck the facility, capture its personnel and kill the commander. If the attackers enter in uniform, the operation would not be prohibited perfidy. If they enter in civilian clothes, it likely would be.[28]

---

[26] Ibid., 187–88.
[27] Ibid. (italics added).
[28] Boothby, "Cyber Deception and Autonomous Attack," 16.

The second scenario involves similar cyberdeception of a closed military information technology system. In the second scenario, the system accepts an email from what it "believes" to be a "non-threatening civilian source," and when the someone opens the attachment to the email (presumably a human operator), some sort of unspecified cyberattack causes a server to shut down, thereby denying access to all users "with the result that the targeted unit's water purification system instantly malfunctions causing death and injury through disease/infections."[29] Such a conclusion, he finds, would amount to cyber perfidy.

Unfortunately, an inconsistency arises between the two examples, as well as with existing international law. In the first scenario, it is unclear why Boothby requires the wearing of civilian clothing by the commandos for this to count as perfidious action. If we take a purposeful account as our starting point, or what the manual identifies as a "proximate cause," the permitting condition for the enemy commandos' ability to kill is the deceptive use of a cyberweapon, not their wearing of civilian clothing. Only if the commandos' entrance was *dependent upon their wearing civilian clothing* would that be considered the perfidious proximate cause.[30] We are not concerned with the intermediate cause of the commando's wearing (or not) civilian clothing, but rather the "deception" of the network system, and how that system was deceived. We ask, if "but for" the illegitimate use of a protected symbol or status via cyberoperations did the deaths, injury, or capture result? If the answer is affirmative—and it is—then that act (or set of acts) is perfidious. The way in which the system was deceived matters.

Second, consider the rather burdened and unlikely example of the water purification system.[31] Boothby finds that these actions *would* amount to perfidy because the proximate cause was the deception of a closed military network system (and surprisingly not the deception of the human who opened the email), whereby the unspecified cyberattack triggers a shutdown of a water purification system. While this scenario specifies that the deception owes to the "civilian" and "nonthreatening" status of the email, this description only tidies up the case. First, this case has multiple intervening causes, but Boothby finds that the proximate cause of perfidy is the deception of the cybersystem, unlike his conclusions in commando example. Second, that Boothby grants that the deaths are perfidious necessarily entails that the harm inflicted (deaths due to waterborne illness) was reasonably foreseeable and predicted. While he does not explicitly address this, as he constructed the case to show it was the "nonthreatening" source of the email as the determining factor,

---

[29] Ibid., 15.

[30] Baxter, "So-Called 'Unprivileged Belligerency': Spies, Guerrillas, and Saboteurs."

[31] It is quite unlikely that the malfunction of a water purification system would "instantly" cause death and disease. Only if one poisoned the water system would this occur. While the length of time between the malfunction and death does not matter to a purposeful account of perfidy, the inclusion of "instantly" only confuses the example.

in reality the example is rather telling because the very notion of a proximate cause means that one must have reasonable foreseeability that one's actions will result in a particular event.

Granting, then, the notion of proximate cause, we ought to reread the commando raid example to also amount to perfidy, for it is the deception of the system that permitted the intervening cause (the commandos) to enter and kill, as well as the fact that it is more than reasonably foreseeable that the deception of the system allows the commandos to enter and kill. Moreover, that the perimeter security system is "deceived" also means that it was more than likely breached on similar grounds as the closed military technology system. Both systems are, presumably, closed military systems and thus configured only to admit "friendly" signals, communications, and the like. One must possess a trust certificate, be a protected person/entity, or possess military credentials (similar to a uniform) to gain access to the network. In the first case we are left to wonder *how* the cyberattack deceived the perimeter defense system; in the second, the fact that a human soldier opened the email from icrc.org is merely one more intervening cause. The initial deception was that the email was permitted to enter the network.[32] Both examples amount to perfidy, and, moreover, *any* similar use of cyberweapons that results in the death, injury, or capture is perfidious.

Admittedly, much of my argument hinges upon how a cybersystem or cybernetwork is "deceived" because we must be able to properly trace the proximate cause of killing, wounding, or capturing by treacherous (cyber) means. This requires, in turn, a rudimentary understanding of how cyberoperations take place. Such means include, but are not limited to, intrusions and infiltrations.[33] Intrusions, such as trapdoors or Trojans, are unauthorized software additions that enable an adversary to gain access to a network or software program. Trapdoors, unlike Trojans, do not require a human user to execute or "implement" the file, and are capable of predated activation commands. Both intrusion methods would be considered a "deception" because they enter a network or software program through seeming "friendly" or authorized means, but with deceptive and potentially malicious intent. When it comes to military systems, which would seem to be the primary targets for

---

[32] If the virus, worm, or what-have-you was not designed to take out the water purification system, but was rather an unforeseen and unintended effect of, say, a spearphishing attack, then this would not be considered as perfidy. However, this is not how Boothby's example is constructed. Presumably, the intent is to cause death through the cyberweapon. "Spearphishing" is when an attacker targets not merely a network, but a particular person inside the network, and such targeted attacks "require prior intelligence gathering to figure out how to trick a particular person and are mostly reserved for prime targets." Singer and Freidman, *Cybersecurity and Cyberwar*, 41.

[33] I am not concerned with denial of service attacks or distributed denial of service attacks. These types of attacks merely prevent access to a website, and while they may rely on some level of deception in order to gain access to a network of computers (botnets) to flood a server, the server merely goes offline. Death, injury, or capture are not at issue, nor presumably are any moral rules regarding combat.

perfidious actions, such friendly status may be most easily accomplished through the feigning of a civilian status (either through software, administrators, or maintenance). Infiltrations, while different in means of attack, are still deceptive in intent. Clarke and Knake identify five methods of infiltration attack: logic bombs, viruses, worms, packet sniffers, and keystroke logging.[34]

The multilayered, and generally more deceptive, tactic of "advanced persistent threat" (APT) utilizes many of the previously identified methods of attack, but does so in a targeted and often autonomous manner. Most APTs are used for data exfiltration, using "multiple phases to break into a network, avoid detection, and harvest valuable information over the long term,"[35] rather than attack; however, this is not to say that APTs cannot be used as an attack vector. Stuxnet, the only verifiable use of a cyberweapon to cause physical damage, used this as one of its approaches.[36]

Less cited means of attack also include cookie hacks and router hacks. Recent leaks by former National Security Agency (NSA) contractor Edward Snowden explain how the NSA and Britain's Government Communications Headquarters (GCHQ) "are using the small tracking files or 'cookies' that advertising networks place on computers to identify people browsing the Internet."[37] The identification links the numeric codes in the website to a person's browser; once the person is identified, malicious software or "malware" is sent out to hack the person's computer. That the NSA and GCHQ merely claim to have exfiltrated information does not mean that the same tactic could not also be used to upload malware for destructive purposes as part of a larger intrusion or APT operation. Cyberespionage (computer network exploitation) uses the same techniques as cyberconflict (computer network attacks).

Router hacks enable attackers to commandeer a particular router, and then use a network of commandeered routers to deliver an attack by hopping, in a lily pad fashion, from router to router to beat the legitimate network provider's directed traffic. Jacob Appelbaum reports that this technique, termed "QUANTUMINSERT" by the NSA and GCHQ, is a sophisticated version of a "honeypot."[38] A particular

---

[34] Clarke and Knake, *Cyber War*. As Valeriano and Maness explain, "(a) logic bombs are programs that cause a system or network to shut down/or erase all data within that system or network; (b) viruses are programs which need help by a hacker to propagate and can be attached to existing programs in a network or as stand-alone programs. . . . (c) worms are essentially the same as viruses, except they have the ability to propagate themselves; (d) packet sniffers are software designed to capture information flowing across the web; (e) keystroke logging is the process of tracking the keys being used on a computer so that the input can be replicated in order for a hacker to infiltrate secure parts of a network." Valeriano and Maness, "Dynamics of Cyber Conflict between Rival Antagonists," 347–60.

[35] Symantec, "Advanced Persistent Threats."

[36] Lindsay, "Stuxnet and the Limits of Cyber Warfare," 365–404.

[37] Soltani, Peterson, and Gellman, "NSA Uses Google Cookies to Pinpoint Targets for Hacking."

[38] A "honeypot" is "a trap set to detect, deflect or in some manner counteract attempts at unauthorized use of information systems": http://ethics.csc.ncsu.edu/abuse/hacking/honeypots/study.php, accessed March 15, 2014.

user will attempt to log onto what s/he believes is a legitimate website/server, like Yahoo!, LinkedIn, or CNN; once detected, QUANTUMINSERT attempts to redirect the user's traffic, not to the legitimate site, but to an NSA-hosted mirror site that then injects malicious code into the user's system. The malicious code, then, allows the attacker full access to any information on the system, as well as installing a trapdoor for full remote access.[39] This process is largely undertaken without any action or interruption from a human operator, as it looks for particular preidentified targets. If any data packet linked to an identified target passes through a cable or router monitored by the NSA, then the system sounds the alarm and generates an automated attack.[40]

Undoubtedly, the array of attacks is expanding, and there are many more that are not publicly known. Attacks undertaken by governments or their militaries are, moreover, quite sophisticated and designed to be "silent" or virtually undetectable. Consequently, any successful military cyberoperation must be able to pass system security, such as firewalls or even "air gaps."[41] To do this, it must either feign a "friendly" or civilian status like Yahoo! and wait for a human being to unknowingly upload malware, or it must exploit a fault in the software code, like finding a hole in a fence, or pass itself off as a friendly data packet or authorized user. Since many of the sophisticated attacks rely on multilayered or multidirectional tactics, it is difficult to determine whether any one particular method is being used, though given what is known about Stuxnet and other attacks, it is likely that all of the means mentioned here are operative.[42]

This brings us full circle to the question posed at the beginning of the chapter: are cyberattacks permissible ruses or wrongful cases of perfidy? I contend that any death, injury, or capture of an enemy through cybermeans is perfidious. Unlike other weapons systems, the very nature of a cyberweapon is deceitful, and deployment will more than likely rely on the use of a protected status (such as a civilian source). One might object here and claim that stealth technology is also inherently deceptive, as its purpose is to catch an adversary unawares or off-guard. While this

---

[39] *Der Spiegel*. "Inside TAO."

[40] Ibid. It should be noted that *Der Speigel* reports that the NSA has successfully spied on the "SEA-ME-WE-4" underwater cable system that links Europe, North Africa, the Gulf States, Pakistan, India, Malaysia, and Thailand. In other words, it is likely that any traffic originating in any of these countries or any countries that rely on this cable network will pass through the NSA's automated data collection system, and open any potential targets to attack.

[41] An air gap is a system not connected to the Internet.

[42] One might cite the recent "heartbleed" bug as an example of a software problem without malicious intent. The bug is a vulnerability in the open-source OpenSSL cryptography library, which is used to implement the Internet's Transport Layer Security. It basically allows an attacker to gain access to sensitive data, particularly trust keys used for encryption. Any individual who accesses a website that is affected by the bug risks their data and passwords. See http://www.symantec.com/outbreak/?id=heartbleed, accessed June 14, 2014.

is certainly true, a stealth bomber does not paint itself as a civilian airliner to do so, and this is the fundamental difference between the two.

One might still object and claim that even if the deception relies on a protected status, any death, injury, or capture must be reasonably foreseeable on my account to count as perfidy. If the deaths or injuries (doubtful for capture) are merely accidental, they are not perfidious. Take as an example a cyberoperation directed against a SCADA (Supervisory Control And Data Acquisition) system at a chemical munitions plant. While the attack's goal is to shut down the system to prevent further chemical weapons production, the cybercommanders are unaware that a plant operator is performing maintenance near a steam valve when the attack commences. The attack opens the valve as the means of shutting down the chemical reactor, and the plant operator, not expecting the valve to open, is severely burned. The operator's injury may not have been foreseeable, and thus, we might rule this was merely unfortunate, but not perfidious. However, such a case confirms rather than undermines my case. If an attack's purpose is to manipulate an object, and that manipulation has the foreseeable potential to result in harm—such as severe steam burning—then it does not matter how that harm comes about. One could hit the valve with a hammer, open it by hand, or send a cyberweapon. That is only one aspect of the attack that proves its foreseeability. The second aspect of the attack that matters for perfidy is how one gained access to open that valve, and here, we would claim that the probability is that the "weapon" relied on assuming a protected status to deceive the SCADA operating system.

Cyberoperations that merely use signs and emblems of internationally recognized, protected entities for the purpose of deception during an attack but do not result in death, injury, or capture are also wrongful, but following Madden, are not perfidious. In cyberoperations such uses might look like email phishing, such as receiving an email from *user.x@icrc.org*, or more likely, will come from protected networks and the use of trust/root certificates associated with them, or the creation of mirror sites, such as those in the QUANTUMINSERT operation. Attempts to mislead that are based on a protection offered by a legally recognizable symbol, entity, or class are impermissible; however, the graver wrong is to use cyberweapons to purposefully (and/or) foreseeably induce the loss of life, wounding, or capture.

Contrarily, cyberoperations may be deemed permissible when the operation is one of espionage or does not intend to establish a "confidence" nexus. For instance, if the means (cyberweapons) and intent to deceive (exploiting trust certificates or feigned credentials) do not actually result in death, injury, or capture, the act might be wrongful—given the type and degree of unauthorized credentialed use—but not perfidious. Or, the act may be perfectly legitimate if the intent is to exfiltrate information or emplace disinformation. These acts would be deemed ruses or espionage.

Additionally, if a belligerent uses cyberweapons with the intent to cause property damage or corruption of data to slow or make inoperable a military target/objective, then the manipulation of information, communications systems, networks, or virtual or physical infrastructure may also be permissible.[43] These acts would be more than likely be considered as sabotage. While intent to deceive is present, given the likely use of a trusted certificate, software exploit, or unauthorized credential, as long as the attacker limits the damage to the property, data, or information of the adversary, then it is likely a morally (and legally) permissible attack. The difficulty remains, however, in knowing whether any given attack is a ruse, espionage, or a failed or ongoing effort at perfidy. Unlike warheads with blast radii, we cannot easily calculate the damage of or infer the intent of a cyberattack.

If my arguments are correct, and one is committed to upholding the letter and spirit of international law, then one must deem any killing, wounding, or capture from a cyberoperation as perfidy and potentially a war crime. Thus when the *Tallinn Manual* finds that causing the death of an adversary commander via cybermanipulation of his pacemaker is perfidious, we must also find that *any* death to combatants or noncombatants by similar weapons would likewise be perfidious.[44] The feigning of a "friendly signal" designed to disrupt the pacemaker and kill the commander is no different than the cyberattack designed to take down a computerized defensive perimeter to allow commandos to enter or to take out water purification systems. Moreover, the likelihood that a cyberattack utilizes some protected status, such as a civilian software, a trust certificate, or some other credentialed material, entails that the intent is clearly malicious and prohibited.

This conclusion has sweeping implications for the law and ethics of armed conflict. First and legally, it entails that the *Tallinn Manual*'s definition of cyberattack necessitates that all attacks on persons are perfidious. The manual states that a cyberattack "is a cyber-operation, whether offensive or defensive, that is reasonably expected to cause injury or death to persons or damage or destruction to objects."[45] Yet as I have argued here, any cyberoperation that is reasonably expected—that is foreseen and intended—to cause injury or death is morally and legally prohibited. Some might object to this conclusion as too strong, but it is the logical conclusion

---

[43] This is termed a "cybereffect" in the language of the US Department of Defense. A cybereffect is "the manipulation, disruption, denial, degradation, or destruction of computers, information or communications systems, networks, physical or virtual infrastructure controlled by computers or information systems or information resident thereon." Obama, "Presidential Policy Directive 20," 2. The leaked top-secret directive also notes that a "significant cyber effect" is one that results in the "loss of life, significant responsive actions against the United States, significant damage to property, serious adverse U.S. foreign policy consequences, or serious economic impact on the United States." Ibid., 3. As long as a "cybereffect" did not violate another law of LOAC, then it would be permissible; "significant cybereffects," however, as I argue would be perfidious if it results in the loss of life.

[44] *Tallinn Manual*, 182. The manual specifies that the signal is a medical one, thereby tidying up the case. However, my arguments show that if one relies on any civilian credential, like a trust/root certificate, would equally be impermissible, and would result in perfidious actions.

[45] Ibid., 106. Obviously, if the attack results in property damage, as I have argued, then it would not amount to perfidy.

to reach. However, if we desire to permit cyberattacks resulting in the death, injury, or capture of adversaries, then we must either amend the law of armed conflict to justify the use of cyberweapons that are ostensibly perfidious, or we must ignore the law. It is not enough to allow an exception for cyberoperations under the prohibition on perfidy. For such an exception would contradict all other principles and law related to perfidy, and open up a cyber-minefield for jurisprudence. We would require instead a full justification of why these particular weapons, and their deceptive uses, do not undermine the minimal levels of trust established by protected statuses and symbols[46] If we want to uphold LOAC, then, we must reconsider our intuitive acceptance of such cyberattacks as "coercion on the cheap."

Second, the ethics of war cannot permit an exception for the use of perfidious cyberoperations. While the law may never be fully consistent, morality ought to be so. To admit of an exception for perfidious cyberoperations means that we are comfortable with the indirect and duplicitous means of cyberkilling. In principle, if one uses a protected symbol, say a white flag, to lure one's enemy into a vulnerable position to kill him, this is no different than using a civilian status, a root certificate, or an authorized credential to gain access to a system to alter or damage that system so as to result in those same violations. The only difference between the two is that one must be present on a battlefield to wave a white flag, and one's waving of the flag is a direct and temporal cause of the enemy's demise. Cyberoperations merely do not require the presence or the direct causation.

10.3 Strategies of Deception and Trust

> And yet, in other things [like deceptive activities] those who avail themselves of the aid of bad men against an enemy are thought to sin before God, but not before men; that is, they are thought not to commit wrong against the law of nations, because in such cases—Custom has brought law beneath its sway; and "to deceive" as Pliny says, "in light of the practices of the age, is prudence."[47]

It is no secret that war permits actions otherwise considered impermissible during times of peace. The extent to which we permit deceptive tactics, however, is not boundless, as I have argued here. Perfidious action, in cyber and other domains, remains impermissible. However, that other forms of deception are considered "permissible" is a result of Pliny's observation: the short-term calculation of belligerents dictates that they engage in deceptive ruses and subterfuge, and by doing so breeds a collective authorization of deceptive tactics. However, while Pliny observes that

---

[46] Unfortunately, it is outside of the scope of this essay to offer such an argument.
[47] Grotius, "On the Right of Killing Enemies in a Public War,"

"bad men" must answer to God, I would push this and argue that we will ultimately answer to ourselves.

The increased intrigue of cyberoperations, the increasing comfort with indirect forms of violence or destruction, and the belief in the very necessity of deception will have wide-ranging effects on the efficacy and authority of international institutions (such as the laws of war, the United Nations, or the ICRC), as well as for peace and stability. In making this claim, however, I am moving beyond the use of cyberoperations during hostilities to the (ab)use of them in peacetime as well. Strategies of deception do not remain bound to wartime; rather, in Kant's terms, they "carry over" into peacetime activities.[48]

For deception to work, though, one must believe it is the truth, and one only believes it is the truth when one trusts the deceiver. I began this chapter with the idea that trust is a necessary feature for the moral and legal governance of war; even if that trust is the minimal belief that one's opponent will forbear from abusing protected statuses and symbols to kill, wound, or capture. If this trust is lost, war becomes utterly barbarous.

Now, however, I would like to take a long view of trust and deception. The remainder of this essay will briefly unpack the concept of trust and argue that deceptive peacetime activities, such as espionage and sabotage, weaken the social and political bonds of trust between states in international society. The breakdown of these relationships, in turn, threatens international peace and stability. I will then turn to how deceptive cyberoperations undermine the opportunity for peaceful settlements after war. Using findings from empirical political science, I suggest that the inability to trust one's opponent increases the likelihood that belligerents will choose to fight to surrender or destruction, thereby increasing the cost and destructiveness of war.

Trust is a complex concept.[49] It is far beyond the scope of this essay to do it justice, so I limit my understanding of trust to be an actor's belief and attitude that "at worst, others will not knowingly do him harm, and at best, that they will act in his interests."[50] As Newton observes, many synonyms such as "mutuality, empathy, reciprocity, civility, respect, solidarity, empathy, toleration, and fraternity" are used in place of trust, but for our purposes, I take the broad definition

---

[48] Kant, *To Perpetual Peace*, 5.

[49] For instance, there is much work done on social trust and social capital as distinct from political trust and political capital. See Newton, "Trust, Social Capital, Civil Society, and Democracy," 201–14. Critical IR scholars look to the unobservable and emotive characteristics of trust. See Rathbun, "Before Hegemony Generalized Trust and the Creation and Design of International Security Organizations," 243–73; Booth and Wheeler, *Security Dilemma*; Michel, "Time to Get Emotional," 869–90. Psychologists and economists also attempt to explain how or why trust exists by positing explanations such as evolutionary requirements and norms, and philosophers attempt to explain trusting behavior as a desire for the good opinion of others. For an excellent explanation, see Petit, "Cunning of Trust," 202–25.

[50] Newton, "Trust, Social Capital, Civil Society, and Democracy," 202.

above. Trust can be generated through a variety of mechanisms, such as nonrational or emotive reasons like shared histories, common understandings, norms, or feelings, or through more rational and instrumental decision-making processes based on data, patterns of reliability, verification mechanisms, and the like.[51] Trust may also be generated by both nonrational and rational reasons simultaneously. In general, trust is a necessary feature of interpersonal, societal, and international relations and, as such, can take a myriad of forms in a variety of degrees.

Empirically, we know that trust works at the personal and the societal level circuitously.[52] Survey work establishes that how much people "trust" is not based on their dispositions or characters per se, but how they view the world in which they live.[53] If one tends to be in a decent spot, surrounded by others similarly situated, then trust is an easier belief to establish.[54] If poverty, despair, cheating, and lying abound, then trust is not easily established or widely held. Much of the work on trust assumes, however, that one is a member of a community with some level of shared interests and purposes.[55] These societies can be thought of along the lines of "purposive associations" that consist of agents who "get together to further some particular ends and who, if they adopt rules, adopt them as instruments of that pursuit."[56] This purposive end could be minimalist, such as the mere maintenance of the society or association, or it could entail more robust goals, such as the pursuit of justice of a particular sort, or some more deep communal value or belief. The difficulty for our purposes, however, is that we are concerned with establishing and maintaining trust between states in the international community, a community quite unlike a purposive association.

---

[51] Michel, "Time to Get Emotional," argues that more rational and instrumental types of reasons are not considered "trust" but "reliability." However, those like Rathbun (2012) do not classify them as different concepts, but different classes, such as "moralistic" and "strategic" trust. It is outside the scope of this essay to give an ontological argument for/or about trust. I merely claim that both nonrational and rationalist grounds may generate reasons for trusting.

[52] Scholarly work demarcates these two types of trust into "social" trust and "political" trust. Social trust is trust created between individual citizens of a community, whereas political trust is trust in one's governmental institutions. While both types of trust are explained by different variables, they both hold in common that they are virtuous or vicious circles. If one begins in a trusting position, then that trust can further grow; if one is in a distrusting position, than it is very difficult to establish. See Newton (2001) for further explanations regarding social and political trust and social and political capital.

[53] Newton, "Trust, Social Capital, Civil Society, and Democracy."

[54] In an analogous manner, one might claim that trust, like virtue, requires the right kind of society, the right kinds of actions, and the right dispositions.

[55] Seligman, *Problem of Trust*;
Scholz and Lubell, "Trust and Taxpaying: Testing the Heuristic Approach to Collective Action," 398–417; Hardin, "Do We Want Trust in Government?" 22–41; Levi, "Social and Unsocial Capital," 45–55; Putnam, "Tuning in, Tuning out," 664–83; Putnam, *Bowling Alone*.

[56] Nardin, *Law Morality and Relations of States*, 4.

International society is, rather, best seen as a "practical association" whereby agents "are related in terms of constraints that all are expected to observe whatever their individual purposes may be." As Nardin further explains, a "practical association is a relationship among those who are engaged in the pursuit of different and possibly incompatible purposes, and who are associated with one another, if at all, only in respecting certain restrictions on how each may pursue his own purposes."[57] States may be said to cooperate, but they do so out of the recognition that they must observe some sort of authoritative rule to pursue their own self-interests free from the hindrance of others.

Establishing trust between rational egoists in a practical association is, however, difficult. Maintaining a level of trust in a practical association when the members might be at war with one another is even more challenging. The cornerstone for establishing trust in a practical association is that all members uphold the common authoritative practices, forms, and procedures when interacting with one another. In international society, this body of common authoritative practices, forms, and procedures is the law of nations (*genitum*). When states disregard it, they undermine and corrode its authoritative status, and moreover, they weaken bonds of trust between states and between states and international institutions. It is, therefore, the product of centuries of opinion, thought, norms, and state practice, and is, on this basis, the way the society of states orders itself. To disregard *genitum* is to put oneself below or above it—in Aristotle's words, to it is to abandon society and become a beast or a god.[58]

As I argued in the previous section, cyberoperations, both inside and outside of wartime, rely on deception to succeed. They rely on a network, a software program, or a person trusting that the information seen is true. Some might view that cyberoperations do not degrade or undermine trust between states because there was never trust there to begin with. If one is the subject of a cyberattack, this only confirms the view of other states. Yet such a pessimistic view is to ignore the very structure of the society of states. While the international system might be considered "anarchical" in that there is no single authoritative sovereign, this is not to say that there is no authoritative body of rules and practices that enjoins states to act a certain way and trust that others will do so too. Customary law, by very definition, means that all states accept the law as custom and will act accordingly.

What is more, allies in the international system rely on each other to greater extents than other relations between states. Alliances provide assurance and even greater levels of trust than nonallied states.[59] Military alliances, such as the North

---

[57] Ibid., 9.

[58] "He who is without a country on account of nature rather than chance is either . . . a beast or a god." What Aristotle meant was that a person outside of society (polis) was denying or denied a necessary feature of being a human being. Thus if one was not a human, and thus a citizen of some country, one was either an animal or divine. Aristotle, 1253a3–29.

[59] Russett, O'Neal, and Davis, "Third Leg of the Kantian Tripod for Peace," 441–67.

Atlantic Treaty Organization, for instance, require that all member states come to the military aid of any member if or when that member state is attacked.[60] An attack on one is treated as an attack on all. Yet such a response would require a state to believe that its allies mean what they say and will act when required; it requires that it trusts other states with its interests. Deceptive activities between allies, however, may throw such a relationship into doubt, deepen mistrust, and have negative impacts on the continuation of stable alliances.

One need only point to the rather contentious relationship between US President Barack Obama and German Chancellor Angela Merkel after evidence came to light that the United States spied on its ally Germany. Merkel explicitly claimed, "we need trust" for "spying among friends is never acceptable."[61] More recently, this relationship was again shaken when allegations of a double agent working within the German security services came to light. Merkel again noted that "it would be for me a clear contradiction as to what I consider to be trusting cooperation between agencies and partners," as well as a "serious breach of trust."[62] While there is no law prohibiting the United States from spying on Germany, there is an authoritative practice, or at least a common understanding, that allied nations, like the United States and Germany, do not engage in this level of deceit. Cooperation requires faith. Yet the ease of cyberespionage and cybersabotage promises to increase distrust in international society, entrenching a vicious rather than a virtuous circle of trust in international society. The long-term consequences of such actions may threaten international stability if these alliances degrade or fall apart.

To see how this is so, we can look to the process of ending a war. Arguably, the ending of a war is most likely the most difficult and delicate of all situations during conflict. Both sides must agree to disarm, to make concessions or compromises, and, in some instances, to create new political constitutions. Sometimes territory or resources must be shared or relinquished, and in all cases, the social fabric of wartorn societies must be rewoven. In cases of civil war, these problems are more acute. Peaceful settlements, therefore, are about more than a mere signing of a treaty. However, to even get to the negotiating table, the parties must have some confidence, *at minimum*, that one's opponent is negotiating in good faith.[63] In other words, they require a minimal level of trust.

---

[60] See Article 5 of the Charter of the North American Treaty Organization.

[61] Smith-Spark, "Germany's Angela Merkel." While this comment was made in reference to listening to Chancellor Merkel's cell phone conversations, the project involves widespread levels of cyber-spying and espionage.

[62] Reuters, "It's a Serious Breach of Trust,"

[63] Fearon, "Commitment Problems and the Spread of Ethnic Conflict," 107–26; Fortuna, "Scraps of Paper? Agreements and the Durability of Peace," 337–72; Fortuna, *Peace Time*; Posen, "Security Dilemma and Ethnic Conflict," 103–24; Smith and Stam, "Bargaining and the Nature of War," 783–813; Walter, "Critical Barrier to Civil War Settlement," 335–64; Werner and Yuen, "Making and Keeping Peace," 261–92.

The use of cyberoperations during a conflict, however, portends to make peaceful settlement between belligerents more difficult. This is so for two reasons. First, the use of deceptive tactics, particularly of a perfidious kind, will erode the trust necessary to reach the negotiating table. We can call this a "precommitment problem," for one side has no incentive to believe that a particularly duplicitous adversary will actually live up to its word.[64] While admittedly this claim is about an unobservable event, we can look to the evidence of failed cease-fires and peace settlements—that is, instances where negotiations already took place—to generate insights.[65] In particular, in cases where there is no settlement, parties to a dispute fight to either surrender or total defeat.

We are faced with an additional difficulty that the only publicly known cyberoperation to cause destruction is Stuxnet. We have not witnessed a stand-alone cyberoperation that rises to the level of perfidy, and while there is evidence of cyberoperations used as adjuncts to traditional military operations, their details remain obscure.[66] Since I am concerned with the level of trust in one's enemy as a precondition for peaceful settlement negotiation, I will, therefore, have to argue analogously to make my claims pertaining to peaceful settlements. The closest analogue we have to the type of cyberoperations envisioned here would be aerial bombardment.

Aerial bombardment might seem, at first glance, disanalogous to cyberoperations. However, this is not entirely so. Aerial bombardment strategies can take two forms: denial and punishment.[67] Denial operations seek to disrupt "the military capabilities of the defender, including forces on the battlefield, military production facilities and supply lines."[68] Denial is about taking out the military's ability to function by targeting military objects and assets.

---

[64] In international relations literature, there is a related problem known as the "commitment problem." The commitment problem basically states that states must commit to "certain courses of action in the future"; however, "these promises or threats are not credible because when the contingency arises the actors have no incentives to abide by them." The commitment problem is about actually abiding by the agreements in the future, while the precommitment problem is about even negotiating with one's adversary about the future.

[65] Greig and Regan do, however, look to the conditions under which low-intensity civil conflicts are likely to receive mediations. Literature on peaceful settlements, however, usually assumes that for whatever reason both sides are at the table and then scholars look to whether or if settlement or peace is successful. See Greig and Regan, "When Do They Say Yes? An Analysis of the Willingness to Offer and Accept Mediation in Civil Wars," 759–81.

[66] "Operation Orchard" is said to have begun with Israeli cyberattacks against Syrian air defense networks. After deceiving the air defense networks, seven Isreali F-151 fighter jets entered Syrian air space and leveled the Kibar complex with ordinance. See Singer and Friedman, *Cybersecurity and Cyberwar*, 127.

[67] See Horowitz and Reiter, "When Does Aerial Bombing Work?" 147–73; Pape, *Bombing to Win*; Snyder, *Deterrence and Defense*.

[68] Horowitz and Reiter, "When Does Aerial Bombing Work?" 152.

Punishment bombing targets civilian-industrial assets, "including electricity grids, water supplies, and other important underpinnings of industrial society."[69] The aim of punishment bombing is to inflict enough suffering on the civilian population that it rises up and overthrows its political leadership. As Horowitz and Reiter explain, there are two models of punishment bombing: The Douhet model and the industrial web model. The Douhet model "holds that the punishment bombings can destroy the morale of the target population, causing the government of the defender to be blamed for air strikes by the attacker," with the result being a popular revolt or political concessions.[70] The industrial web model, however, "holds that bombardment must attack those choke points or bottlenecks that are crucial to modern industrial economies."[71] The rationale here is destroying the entire economy will lead to widespread suffering and thus "destroy the will of the population to resist." The only difference between the two models seems to be in their temporal effects. The Douhet model seeks to target the civilian population directly, while the industrial web model seeks to target the civilian population indirectly and secondarily.

Cyberoperations will, for the foreseeable future, look more like the industrial web model of aerial bombardment, even if militaries seek to create discriminate and proportionate cyberweapons aimed at denial strategies. Notoriously, cyberweapons are difficult to control once released. Indeed, we have evidence that in the Iraq war "US military officers were very excited by the prospects of taking down *an* enemy computer network facilitating suicide bombings;" however, that operation could not be contained and ultimately "took down over 300 servers in the wider Middle East, Europe, and the United States."[72] Even Stuxnet, claimed to be the most discriminate worm of all, affected 100,000 computers inside and outside of Iran.[73] If states did not take years, and perhaps decades, to develop such discriminate weapons, but rather engaged in cyberoperations more generally, the result promises to look more like those of the Iraq war than the attack against Natanz.

Moreover, there is evidence to support the view that denial strategies are actually tantamount to industrial web targeting. For example, US military targeting doctrine includes the targeting of objects which "by their nature, location, purpose, or use, effectively contribute to the enemy's *warfighting or war-sustaining capability*."[74] While this doctrine departs significantly from international legal standards, it would not only permit cyber spillover effects in the civilian population, it would require targeting civilian infrastructure.

---

[69] Ibid., 151.
[70] Ibid., 151.
[71] Ibid., 152.
[72] Singer and Friedman, *Cybersecurity and Cyberwar*, 132.
[73] Zetter, "Report: Obama Ordered Stuxnet to Continue."
[74] United States, Navy/Marine/Coast Guard, *Commanders Handbook on the Law of Naval Operations*.

This is, of course, the industrial web targeting strategy. While the terms look to uphold a division between military and civilian, targeting a "war sustaining capability" includes everything from roads to electric grids, water treatment facilities, and any other infrastructure that allows a state to "sustain" a war. Thus, if we are unlucky enough to witness or be part of the widespread use of cyberweapons during hostilities, targeting doctrine and state practice, of at least the United States—a major cyberplayer—supports the industrial web bombardment strategy and, as such, would more than likely support cyberweapons to be used against such targets as well. This means, then, that we should look not merely to the type of weapon (bombs or binary bullets) but also to the targeting practice and strategies of states. Aerial bombardment is the favored strategy of high-tech military powers because it offers air dominance with relatively little human cost. Cyberweapons also contain this promise.

The problem, however, with strategies of aerial bombardment (and the portended problem with cyberstrategies following or mirroring industrial web targeting) is that they do not work. Pape and Horowitz and Reiter find that targeting the civilian population through punishment strategies fails to achieve the attacking party's objectives. In fact, not only do such strategies fail to weaken resolve and demoralize a population, they have the opposite effect: they stiffen resolve and keep a target population's leadership in power. This only lengthens the war and tends to drive both sides to view the conflict as zero-sum (defeat/surrender or win). It drives the belligerents to more extreme versions of conflict because neither side trusts the other, and each would rather fight barbarously to the end.

The second potential problem with cyberoperations and peaceful settlements concerns the ability to reach some sort of convergence of expectations about the military consequences of continued fighting.[75] Such a convergence is seen as the very basis for negotiations, for otherwise, states will choose to continue fighting to receive a better settlement or payoff. Unfortunately this convergence is not easily achieved, especially if there are factors that increase uncertainty. As Werner and Yuen note, external third-party pressure and variance in military dominance increases the likelihood of settlement failure.[76] Belligerents may not want to lay

---

[75] Werner and Yuen, "Making and Keeping Peace," 262. Wolford, Reiter, and Carruba, however, find that in some cases convergence of beliefs can actually prolong conflict rather than terminate it. This is so when one side believes they have an "unappeasable opponent." Wolford, Scott, Dan Reiter and Clifford J. Carrubba. "Information, Commitment, and War," 556–79. See also Powell, "Bargaining and Learning While Fighting," 344–61. Slantchev, "How Initiators End Their Wars,." 813–29. Stanley and Sawyer, "Equifinality of War Termination," 651–76. Smith and Stam, "Bargaining and the Nature of War."

[76] Werner and Yuen, "Making and Keeping Peace." Outside pressure could increase the belief that the third party is required for enforcement, or that the third party might weigh in on one side. Uncertainty as to military dominance, or when states trade victories, means that each side has a rational belief that they might actually win should they keep fighting, as the evidence suggests that they have at least an equal chance.

down their arms, or they may believe that they have an equal probability of winning if they just keep fighting. Deceptive cyberoperations may ultimately increase uncertainty by sending mixed signals to those fighting or, as previously noted, may increase resolve because of the imposed suffering.

One might object here and claim that cyberoperations will not affect war termination and settlement any more than any other tactic or strategy, and the notion that trust between must be present the parties is unimportant. Wars cease and states settle all the time, so we ought not to worry too much about deceptive cyberoperations.[77] Admittedly, I am engaging in a bit of hypothetical argumentation concerning cyberoperations; however, this is not to say that cyberoperations do not change the preconditions for negotiation, nor does it correctly characterize the little we do know about peace settlements and the endings of wars. First, the fact that major cyberplayers, such as the United States, are now expressly showing restraint in the use of cyberoperations, as well as urging others to do the same, signals that at least it views these tactics as nontraditional and potentially game-changing.[78] Fighting with binary bullets may have unintended or unforeseen consequences, and at least the United States is saying that it is treading carefully. Second, as I have attempted to argue here, empirical evidence suggests that how one fights matters to how a war ends, and moreover, when there are deep-seated tensions, hatreds, and distrust, such as in civil conflict, fighting the other side to the end is the predominant strategy. Perfidious action in warfare heightens distrust and enmity, and it follows that if cyberoperations are considered perfidious, then peaceful settlements will also be more difficult to achieve, for they will have further undermined trust.

Finally, in terms of how cyberoperations will likely affect the ability for the *just* settlement between belligerents, we ought look to what David Rodin terms "*jus terminatio*."[79] Rodin is concerned with initial proportionality calculations made to justify going to war and how those calculations ought to include a wider range of costs imposed during the course of a war. He argues that *jus terminatio* is tied to *jus ad bellum*. I would like to add to Rodin's framework and note that extensive deceptive

---

[77] It is outside the scope of this paper to discuss all the varieties of conflict. However peace settlement does seem to differ in terms of type and intensity of conflict. In interstate conflict, Slantchev identifies that 67 out of 104 interstate wars ended in negotiated settlement between 1816-1991. Civil or intrastate conflict, however, is far less. Slantchev, "Principle of Convergence in Wartime Negotiations," 621–32. Zartman finds that two-thirds of all civil wars end with either surrender or destruction, and Greig finds that intrastate low-intensity conflicts are less likely to see mediation than civil wars. Zartman, *Elusive Peace*. Greig, "Nipping Them in the Bud."

[78] "US Promises 'Restraint' in Cyberoperations"; Sanger, "Syria War Stirs New U.S. Debate on Cyberattacks"; United States Department of Defense, "Department of Defense for Operating in Cyberspace."

[79] Rodin, "Ending War," 359–67. Rodin notes that "*Jus terminatio* governs the transition from a state of war back into a state of peace" (360). He extends this argument in a forthcoming piece, "The War Trap," in *Ethics* (2014).

strategies during war will adversely affect the potential for belligerents to terminate hostilities. That is, *jus terminatio* may need to be extended to forbid certain practices *during* war if those practices adversely affect the probability of peace. Thus I would contend that *jus terminatio* is also tied to *jus in bello*. How and by what means a war is fought may change whether it is morally permissible to continue fighting it. Belligerents who cannot trust their adversaries to fight fairly will not seek peace. *Fides etiam hosti servanda*.

10.4 Conclusion

This essay argued that cyberoperations that kill, wound, or capture ought to be considered perfidious. Acts of perfidy erode the trust necessary between belligerents with two unfortunate consequences. The first is that war becomes more indiscriminate because no one trusts that a protected status or person is truly nonthreatening. The second is that the breakdown in trust between belligerent parties significantly affects the likelihood of peaceful settlements. Moreover, *jus in bello* affects the possibility of *jus post bellum* (or, in Rodin's terms, *jus terminatio*).[80]

We ought, then, to tread lightly and not be overly accepting of cyberoperations. While new technologies are always seductive in their promise of advantage and superiority, they do not always portend a more stable and peaceful world. That cyberoperations appear to be less costly and more discriminate is mainly due to how little we know about them. Our population of known cyberoperations is small, and all have mainly been undertaken by states with an eye toward high levels of discrimination and low levels of detection. It is uncertain, therefore, how we might see cyberattacks in the future that do not comport with these two principles. Yet regardless of how the future plays out, any deaths, injury, or capture by cybermeans during hostilities will be perfidious. For if we hold fast to the internationally recognized standards and prohibitions on perfidious action, we cannot find otherwise. Granting this, we should be aware that their deployment might then have serious consequences in the long term, the least of which is the threatening of a delicate balance of trust between warring parties.

Bibliography

Annex to the Hague Convention, October 18, 1907. http://www.icrc.org/applic/ihl/ihl.nsf/Article.xsp?action=openDocument&documentId=090BE405E194CECBC12563CD005167C8. Accessed May 28, 2014.
Aristotle. *Nicomachean Ethics*, ed. Roger Crisp. Cambridge: Cambridge University Press, 2000.
Arquilla, John. Interview with *Frontline*, March 4, 2003. http://www.pbs.org/wgbh/pages/frontline/shows/cyberwar/interviews/arquilla.html. Accessed May 28, 2014.

---

[80] Rodin, "Ending War."

Baxter, Richard. "So-Called 'Unprivileged Belligerency': Spies, Guerrillas, and Saboteurs." 28 British Year Book of International Law. 323 (1951): 323–45.

Bellamy, Alex. "The Responsibilities of Victory: Jus Post Bellum and the Just War." *Review of International Studies* 34 (2008): 601–25.

Booth, Kenneth, and Nicholas J. Wheeler. *The Security Dilemma: Fear, Cooperation and Trust in World Politics*. Houndsmills: Palgrave Macmillan, 2008.

Boothby, William. "Cyber Deception and Autonomous Attack—Is There a Legal Problem?" 5th International Conference on Cyber Conflict, ed. K. Podins, J. Stinnissen, and M. Maybaum (2013): 1–17. © NATO CCD COE Publications, Tallinn.

Charter of the North American Treaty Organization. April 4, 1949. http://www.nato.int/cps/en/natolive/official_texts_17120.htm. Accessed July 6, 2014.

Clarke Richard A., and Robert K. Knake. *Cyber War: The Next Threat to National Security and What to Do About It*. New York: HarperCollins, 2010.

Der Spiegel. "Inside TAO: Documents Reveal Top NSA Hacking Unit." http://www.spiegel.de/international/world/the-nsa-uses-powerful-toolbox-in-effort-to-spy-on-global-networks-a-940969-2.html. Accessed January 3, 2014.

Dinstein, Yoram. *The Conduct of Hostilities under the Law of International Armed Conflict*. Cambridge: Cambridge University Press, 2004.

Feaver, Peter. "Blowback: Information Warfare and the Dynamics of Coercion." *Security Studies* 7, no. 4 (1998): 88–120.

Fearon, James. "Commitment Problems and the Spread of Ethnic Conflict." In *The International Spread of Ethnic Conflict: Fear, Diffusion, and Escalation*, ed. David Lake and Donald Rothchild. Princeton: Princeton University Press, 1998, 107–26.

Fortuna, Virginia Page. "Scraps of Paper? Agreements and the Durability of Peace." *International Organization* 57, no. 2 (2003): 337–72.

Fortuna, Virginia Page. *Peace Time: Cease-fire Agreements and the Durability of Peace* (Princeton: Princeton University Press, 2004).

Gartzke, Erik, and Jon Lindsay "Weaving Tangled Webs: Offense, Defense and Deception in the Cyber Age." Unpublished essay, 2013.

Glozman, Edy, Netta Barak-Corren, and Ilan Yaniv. "False Negotiations: The Art and Science of Not Reaching an Agreement." *Journal of Conflict Resolution* 59, no. 4 (2014): 671–97.

Greig, Michael. "Nipping them in the Bud: The Onset of Mediation in Low-Intensity Civil Conflicts" *Journal of Conflict Resolution*, 2013.

Greig, Michael J., and Patrick M. Regan. "When Do They Say Yes? An Analysis of the Willingness to Offer and Accept Mediation in Civil Wars." *International Studies Quarterly* 52, no. 4 (2008): 759–81.

Grotius, Hugo. "On the Right of Killing Enemies in a Public War, and on Other Violence Against the Person." In *The Rights of War and Peace*, Book III, Chapter 4 (1625). http://oll.libertyfund.org/titles/grotius-the-rights-of-war-and-peace-1901-ed. Accessed April 27, 2014.

Hardin, R. "Do We Want Trust in Government?" In *Democracy and Trust*, ed. M. E. Warren. Cambridge: Cambridge University Press, 1999, 22–41.

Hart, B. H. Liddell. *Strategy*. New York: Meridian, 1991.

Horowitz, Michael, and Dan Reiter. "When Does Aerial Bombing Work? Quantitative Empirical Tests, 1917–1999." *Journal of Conflict Resolution* 45, no. 2 (2001): 147–73.

Kant, Immanuel. *To Perpetual Peace: A Philosophical Sketch*, tran. Ted Humphrey. Indianapolis: Hackett, 2003.

Latimer, Jon. *Deception in War: The Art of the Bluff, the Value of Deceit, and the Most Thrilling Episodes of Cunning in Military History from the Trojan Horse to the Gulf War*. New York: Overlook Press, 2003.

Levi, M. "Social and Unsocial Capital: A Review Essay of Robert Putnam's Making Democracy Work." *Politics and Society* 24 (1996): 45–55.

Lindsay, Jon R. "Stuxnet and the Limits of Cyber Warfare." *Security Studies* 22 (2013): 365–404.

Madden, Mike. "Of Wolves and Sheep: A Purposive Analysis of Perfidy Prohibitions in International Humanitarian Law." *Journal of Conflict and Security Law* 17, no. 3 (2012): 439–63.

Melzer, Nils. "Cyber-Operations and Jus in Bello." *United Nations Institute for Disarmament Research* (2011): 12. http://www.isn.ethz.ch/Digital-Library/Publications/Detail/?lng=en&id=143275. Accessed May 28, 2014.

Michel, Thorsten. "Time to Get Emotional: Phronetic Reflections on the Concept of Trust in International Relations." *European Journal of International Relations* 19, no. 4 (2013): 869–90.

Nardin, Terry. *Law Morality and Relations of States*. Princeton: Princeton University Press, 1983.

Newton, Kenneth. "Trust, Social Capital, Civil Society, and Democracy." *International Political Science Review* 22, no. 2 (2001): 201–14.

Obama, Barack. "Presidential Policy Directive 20: U.S. Cyber-Operations Policy" (2012). http://www.fas.org/irp/offdocs/ppd/ppd-20.pdf. Accessed May 28, 2014.

O'Brien, Gregory J. CDR. "Information Operations and the Law of Perfidy." Master's thesis, 2001. Available at: http://handle.dtic.mil/100.2/ADA395074. Accessed May 27, 2014.

Oppenheim, Lassa. *International Law: A Treatise*. New York and Bombay: Longmans, Green., 1906.

Pape, Robert. *Bombing to Win: Air Power and Coercion in War*. Ithaca: Cornell University Press, 1996.

Petit, Philip. "The Cunning of Trust." *Philosophy and Public Affairs* 24, no. 3 (1995): 202–25.

Posen, Barry. "The Security Dilemma and Ethnic Conflict." In *Ethnic Conflict and International Security*, ed. Michael E. Brown. Princeton: Princeton University Press, 1993,: 103–24.

Powell, Robert. "Bargaining and Learning While Fighting." *American Journal of Political Science* 48, no. 2 (2004): 344–61.

Protocol Additional to the Geneva Conventions of 12 August 1949, and relating to the Protection of Victims of International Armed Conflicts (Protocol I), 8 June 1977. http://www.icrc.org/ihl/WebART/470-750046?OpenDocument. Accessed May 14, 2014.

Putnam, Robert. "Tuning in, Tuning out: The Strange Disappearance of Social Capital in America." *PS: Politics and Political Science* 28, no. 4 (1995): 664–83.

Putnam, Robert. *Bowling Alone: The Collapse and Revival of American Community*. New York: Simon & Schuster, 2000.

Rathbun, Brian. "Before Hegemony Generalized Trust and the Creation and Design of International Security Organizations." *International Organization* 65, no. 2 (2011): 243–73.

Rathbun, Brian. *Trust In International Cooperation: International Security Institutions, Domestic Politics and American Multilateralism*. New York: Cambridge University Press, 2012.

Reuters. "It's a Serious Breach of Trust: Angela Merkel Condemns US Spying Allegations that German 'Double Agent' Was Working for America" *Daily Mail*, July 7, 2014. http://www.dailymail.co.uk/news/article-2683601/Angela-Merkel-calls-Germanys-arrest-US-double-agent-breach-trust-turns-spying-Americans.html. Accessed July 7, 2014.

Rid, Thomas. *Cyber War Will Not Take Place*. Oxford: Oxford University Press, 2013.

Rodin, David. "Ending War: Reply to Miller." *Ethics and International Affairs* 25, no. 3 (2011): 359–67.

Rodin, David. "The War Trap: Dilemmas of *jus terminatio*." *Ethics* 125, no. 3 (2015): 674–95.

Rowe, Neil. "Perfidy in Cyberwar." In *Routledge Handbook for Ethics and War: Just War Theory in the Twenty-First Century*, ed. Fritz Allhoff, Nicholas J. Evans, and Adam Henschke. New York: Routledge, 2013, 394–404.

Russett, Bruce, John R. O'Neal, and David R. Davis. "The Third Leg of the Kantian Tripod for Peace: International Organizations and Militarized Disputes, 1950–85." *International Organization* 52, no. 3 (1998) 441–67.

Sanger, David. "Syria War Stirs New U.S. Debate on Cyberattacks." *New York Times*, February 24, 2014. http://www.nytimes.com/2014/02/25/world/middleeast/obama-worried-about-effects-of-waging-cyberwar-in-syria.html?_r=0. Accessed July 7, 2014.

San Remo Manual on International Law Applicable to Armed Conflict at Sea, June 12, 1994. http://www.icrc.org/applic/ihl/ihl.nsf/Treaty.xsp?documentId=5B310CC97F166BE3C12563F6005E3E09&action=openDocument. Accessed May 29, 2014.

Schmitt, M. N. "Wired Warfare: Computer Network Attack and the *Jus in Bello*." *International Review of the Red Cross* 84, no. 846 (2002): 365–99. http://www.icrc.org/eng/assets/files/other/365_400_schmitt.pdf. Accessed May 12, 2014.

Scholz, J. T., and M. Lubell. "Trust and Taxpaying: Testing the Heuristic Approach to Collective Action." *American Journal of Political Science* 42, no. 2 (1998): 398–417.

Seligman, A. B. *The Problem of Trust*. Princeton: Princeton University Press, 1997.

Singer, Peter W., and Allan Freidman. *Cybersecurity and Cyberwar: What Everyone Needs to Know*. Oxford: Oxford University Press, 2014.

Slantchev, Branislav. "The Principle of Convergence in Wartime Negotiations." *American Political Science Review* 47 (2003): 621–32.

Slantchev, Branislav. "How Initiators End Their Wars: The Duration of Warfare and the Terms of Peace." *American Journal of Political Science* 48, no. 4 (2004): 813–29.

Smith, Alastair, and Allan C. Stam III. "Bargaining and the Nature of War" *Journal of Conflict Resolution* 48, no. 6 (2004): 783–813.

Smith-Spark, Lauren. "Germany's Angela Merkel: Relations with U.S. 'Severely Shaken' over Spying Claims." CNN World. http://www.cnn.com/2013/10/24/world/europe/europe-summit-nsa-surveillance/. Accessed July 5, 2014.

Snyder, Glenn H. *Deterrence and Defense: Toward a Theory of National Security*. Princeton: Princeton University Press, 1961.

Soltani, Ashkan, Andrea Peterson, and Barton Gellman. "NSA Uses Google Cookies to Pinpoint Targets for Hacking." *Washington Post*, December 10, 2013. http://www.washingtonpost.com/blogs/the-switch/wp/2013/12/10/nsa-uses-google-cookies-to-pinpoint-targets-for-hacking/. Accessed February 16, 2014.

Stanley, E. A., and J. P. Sawyer. 2009. "The Equifinality of War Termination: Multiple Paths to Ending War." *Journal of Conflict Resolution* 53, no. 5 (2009): 651–76.

Sun Tzu. *The Art of Warfare*. http://classics.mit.edu/Tzu/artwar.html. Accessed April 14, 2014.

Sydney Morning Herald. "US Promises 'Restraint' in Cyberoperations." http://www.smh.com.au/world/us-promises-restraint-in-cyber-operations-20140329-zqof7.html. Accessed July 7, 2014.

Symantec. "Advanced Persistent Threats." http://www.symantec.com/theme.jsp?themeid=apt-infographic-1. Accessed March 3, 2014.

Tallinn Manual on the International Law Applicable to Cyber Warfare, ed. Michael Schmitt. Cambridge: Cambridge University Press, 2013.

United States Department of Defense. "Department of Defense for Operating in Cyberspace" (2011). http://www.defense.gov/news/d20110714cyber.pdf. Accessed July 7, 2014.

United States Air Force. JP 3-60, *Joint Targeting* (2007). https://www.aclu.org/files/dronefoia/dod/drone_dod_jp3_60.pdf. Accessed July 7, 2014.

United States, Navy/Marine/Coast Guard. *The Commanders Handbook on the Law of Naval Operations*, NWP 1-14M, MCWP 5.21, COMDTPUB P5800.7 (1995).

Valeriano, Brandon, and Ryan Maness. "The Dynamics of Cyber Conflict between Rival Antagonists, 2001–2011." *Journal of Peace Research* 51, no. 3 (2014): 347–60.

Walter, Barbara. "The Critical Barrier to Civil War Settlement." *International Organization* 51, no. 3 (1997): 335–64.

Werner, Suzanne, and Amy Yuen. "Making and Keeping Peace." *International Organization* 59 (2005): 261–92.

Wolford, Scott, Dan Reiter, and Clifford J. Carrubba. "Information, Commitment, and War." *Journal of Conflict Resolution* 55, no. 4 (2011): 556–79.

Zartman, I. W. *Elusive Peace: Negotiating an End to Civil Wars*. Washington, DC: Brookings Institution Press, 1993.

Zetter, Kim. "Report: Obama Ordered Stuxnet to Continue after Bug Caused It to Spread Wildly." *Wired Magazine*, June 1, 2012. http://www.wired.com/2012/06/obama-ordered-stuxnet-continued/. Accessed July 7, 2014.

11

Cyberattacks and "Dirty Hands"

Cyberwar, Cybercrime, or Covert Political Action?

SEUMAS MILLER

Cyberwar is a new form of conflict. Contemporary nation-states and, for that matter, nonstate actors such as corporations, now suffer and inflict ongoing cyberattacks on a large scale, although whether all or any of these attacks constitute war rather than conflict short of war,[1] or are mere breaches of security (criminal or otherwise), is not always entirely clear. In this paper I distinguish between cyberwar, cyberterrorism, cybercrime, cyberespionage, and what I will refer to as "covert political cyberaction"—a species of covert political action. I argue that many, if not most, cyberattacks perpetrated by nation-states on other nation-states for political reasons are best understood neither as acts of war (not even acts of economic war) nor as crimes (counting acts of terrorism as crimes) but rather as a new form of covert political action—that is, covert political cyberaction.

Covert political cyberaction does not include purely defensive measures such as building firewalls, layers of password protection, and the like. Rather it is offensive action, including offensive action undertaken in response to a present or future attack. Covert political cyberaction is multifarious and includes covert political cyberattacks as well as certain forms of cyberespionage.

In the final section of this chapter (11.3), I provide a preliminary ethical analysis of covert political cyberaction. More specifically, I argue that much covert political cyberaction is best understood as a species of "dirty hands" action—harmful and unlawful action undertaken to achieve an (alleged) greater good. This being so, it turns out that the two currently available, well-developed moral frameworks—namely, the law enforcement framework (for criminal activity) and the just war theory framework (for war) —are not straightforwardly applicable to many covert political cyberactions, although both are relevant, both qua frameworks and by virtue of

---

[1] See Ford, "*Jus Ad Vim* and the Just Use of Lethal Force-Short-of-War," 63–75.

some of their constituent principles, such as the principle of proportionality. On the other hand, there are two moral principles of reciprocity that are directly applicable to covert political cyberactions, or so I argue. The first of these principles is retrospective in form, the second prospective.

11.1 Some Definitions: Cyberwar, Cyberterrorism, Cybercrime, and Cyber espionage

Recent high-profile cyberattacks and/or acts of cyberespionage include the following:[2]

(1) The denial of service cyberattack on Estonian banks, media, and government websites in 2007 perpetrated (it is presumed) by Russia.
(2) The Stuxnet malware attack—the software worm, Stuxnet, was used to disrupt Iran's nuclear enrichment ICT (information and communication technology) infrastructure in the context of a joint US and Israeli operation (Olympic Games) established to disrupt Iran's nuclear program.[3]
(3) Operation Orchard—the Israeli bombing of a Syrian nuclear facility after they had penetrated Syrian computer networks and "turned off" Syrian air defense systems.
(4) Mandiant[4]—the US computer-security firm that has documented ongoing Chinese cybertheft and disruption of the websites and other ICT infrastructure of US corporations and government agencies.
(5) Snowden[5]—the release by US National Security Agency (NSA) contractor Edward Snowden of a large amount of confidential NSA data to the international press, including in relation to Verizon and PRISM incidents.
(6) Verizon—the collection by the NSA of the metadata[6] from calls made within the United States, and between the United States and any foreign country, of millions of customers of Verizon and other telecommunication providers.
(7) PRISM—the agreements between the NSA and various US-based Internet companies (Google, Facebook, Skype, etc.) to enable the NSA to monitor the online communications of non–US citizens based overseas.

---

[2] See Singer and Friedman, *Cybersecurity and Cyberwar*, for an outline of these various cyberattacks. On the Stuxnet and Estonia cases, see also Rid, *Cyber War Will Not Take Place*, 32–34.
[3] Sanger, *Confront and Conceal*.
[4] See Mandiant Intelligence Centre, *APT1: Exposing One of China's Cyber Espionage Units*.
[5] For a sympathetic account see Harding, *Snowden Files*.
[6] Metadata is the unique phone/email numbers of the caller and recipient, the time and duration of the call, and the location of the caller and the recipient, but not the content of the communication.

A further point to be noted here is that whereas cyberattacks by terrorists have not been common—due, presumably, to the lack of relevant technical expertise in the context of sophisticated state-based cyberdefense systems—recent international terrorist attacks, nevertheless, have relied heavily on ICT. For terrorist attacks are a lethal means of terrorizing members of some social or political group to achieve the terrorists' political purpose.[7] Accordingly, terrorism relies on the violence receiving a high degree of publicity—a degree of publicity necessary to engender widespread fear in the target political or social group.

Take the 9/11 attacks on the Twin Towers in New York. These were not cyberattacks. However, qua terrorist attacks, the Twin Towers attacks were a huge success. They were a huge success from the terrorists' perspective, not only because they killed almost 3,000 people, destroyed an iconic building, and disrupted global financial flows, but also because they received an extraordinarily high level of publicity. Consider, for example, the endless repetition by global media outlets of the images and video footage of the hijacked planes crashing into the Twin Towers buildings.

Crucially, for our purposes here, much of the disruption, and certainly the extraordinary level of global visibility, was only enabled by ICT-based, densely interconnected international media networks, global financial systems, and so on. Moreover, Osama bin Laden created his own text, images, and video footage that was also circulated globally, initially by means of traditional media but then also via services such as YouTube. Accordingly, in discussing cyberattacks we should distinguish, but also keep in mind, noncyberattacks, which, nevertheless, rely in very important ways on ICT.

In the case of the cyberattack on Estonia, it is important to note that Estonia is a member of NATO and that Russia was the chief suspect. However, there were no deaths or destruction of property, and computer technicians unblocked the networks relatively quickly thereby ensuring the disruption was minimal. Moreover, Russia denied responsibility for the attack and NATO did not declare war. It seems, therefore, that this cyberattack did not constitute an act of war but rather something short of war.[8]

By contrast, Operation Orchard was presumably an act of war since it involved the Israeli bombing of a Syrian nuclear facility immediately after an Israeli cyberattack on Syrian air defense systems. More specifically, the cyberattack itself was undertaken in the context of an act of war (the bombing), and only undertaken for the specific purpose of enabling the bombing.[9]

---

[7] See Miller, *Terrorism and Counter-Terrorism*, 30–59, for a full definition.

[8] Rid, *Cyberwar Will Not Take Place*, 30–32; BBC News, "Estonia Hit by 'Moscow Cyber War' "; Finn, "Cyber Assaults on Estonia Typify a New Battle Tactic."

[9] Rid, *Cyberwar Will Not Take Place*, 42–43; Follath and Stark, "Story of 'Operation Orchard': How Israel Destroyed Syria's Al Kibar Nuclear Reactor."

Other cases are more difficult to classify. Presumably a cyberattack in which both the aggressor and the victim were nation-states, and which not only disabled ICT infrastructure but destroyed it and caused many deaths, could well count as an act of war. But in such cases it is the destruction of *physical* infrastructure and, especially, the loss of *human life* that elevates these cyberattacks to cyberwar.[10] What to make of a denial of service cyberattack that did not destroy physical infrastructure or immediately cause any loss of life, but that did cause, and was intended to cause, a prolonged period in which government-funded and administered welfare and other services were unable to be provided, leading to severe hardship among large sections of the population, albeit not to loss of life? And what are we to make of so-called collateral damage by contagion? Should those nation-states severely affected, albeit unintentionally, regard themselves as at war? Stuxnet, for example, while targeted at Iranian ICT infrastructure also caused collateral damage by contagion; it infected and shut down computers and computer networks in places such as Indonesia and India.[11]

Whether or not collateral damage by contagion constitutes war partly depends on the nature and extent of the collateral damage in question, and on whether it was foreseen or reasonably foreseeable. Presumably, neither India nor Indonesia ought to have regarded themselves as being at war with the United States or Israel as a consequence of Stuxnet. On the other hand, if a nation-state hell-bent on prosecuting war with an enemy nation-state released a highly virulent form of malware knowing that it would disable and destroy key components of the ICT infrastructure—including the components of life-supporting medical facilities—of various neutral nation-states leading to substantial loss of life, then those erstwhile neutral states might well, and justifiably, regard themselves as being in a state of war with the aggressive state in question.

An important question to be addressed at this point is the nature and extent of the harm culpably caused. Presumably, thresholds of harm can be delineated, or at least described, to serve as benchmarks in determining what is, and what is not, an act of war. The authors of the *Tallinn Manual* and some others[12] have argued that such thresholds have to be specified, at least legally and, presumably, also morally, in terms of the nature and/or extent of the injury, loss of human life, and/or physical destruction caused. Apparently the idea informing such proposals is that cyberattacks cannot in and of themselves constitute war, properly understood. Rather a cyberattack could only count as an act of war if it had consequential effects

---

[10] This definition of cyberwar as necessarily involving destruction of physical property and/or loss of human life appears to be the one favored by the authors of the *Tallinn Manual*. See Schmitt, *Tallinn Manual*.

[11] Jared Anwer, "India Caught in Crossfire of Global Cyber War"; Bachrach, "Stuxnet Worm Turns."

[12] See, e.g., Eberle, "Just War and Cyberwar," 54–67.

in terms of human injury, loss of life, and/or substantial damage to physical objects (buildings, etc.).[13]

An important point to be stressed here is that whatever international and domestic law might have to say on this matter, the moral justification for any specific threshold setting is very much context-dependent. In the context of the possibility of nuclear war between superpowers, such as the United States and Soviet Russia during the Cold War, or between the United States and China now, the threshold setting at which an aggressive act on the part of one of these powers ought to count as an act of war must be set very high indeed (perhaps so high it can never be met in practice).[14] This has important implications for the practice of covert political action, as I argue in the next section.

At any rate, here we need to distinguish four kinds of harm or damage. First, there is harm (physical or psychological)[15] done to human beings per se. Second, there is damage done to buildings, ICT hardware, and other human artifacts (as well as to the natural environment insofar as it supports individual and collective human life). Third, there is, as Randall Dipert notes,[16] cyberharm, for example damage to software and data (as opposed to the physical ICT hardware itself). Fourth, there is institutional harm; that is, the undermining of institutional processes and purposes, for example major breaches of confidentiality in a security agency or loss of institutional control of territory.[17]

The point to be made here is that the third and fourth kinds of harm (cyberharm and institutional harm) might have thresholds at which war might be justified, independently of the level of the first two kinds of harm caused (i.e., the level of physical or psychological harm caused to humans per se and the level of destruction of physical property and the like).[18] More importantly, for our purposes in this paper, the third and fourth kinds of harm might have thresholds at which a seriously *harmful* response short of war is morally, and perhaps

---

[13] Territorial integrity is also in play. However, whether or not a loss of territorial integrity can be specified independently of control of a geographical area (including the human activities undertaken in that area) and, therefore, independently of what I refer to as institutional harm, in particular, is questionable.

[14] I am assuming that these nuclear powers would not engage in an actual war with one another and yet in doing so refrain from using nuclear weapons in this war.

[15] Violations of individual rights to various aspects of autonomy should be included as psychological harms, albeit they are also acts of wrongdoing—all violations of human rights being acts of (pro tanto) wrongdoing. For more on expanding the understanding of cyberharms, see the chapter by Daphna Canetti, Michael L. Gross, and Israel Waismel-Manor in this volume.

[16] Dipert, "Ethics of Cyberwarfare," 400.

[17] The latter may well also be a violation of the individual moral right to autonomy, depending on the nature of the political institutions in place; perhaps an invasion of an authoritarian state might not be a violation of the autonomy rights of the citizens, but rather the reverse.

[18] Or at least that cyberharm and/or institutional harm could conceivably reach such a threshold independently *to some extent* of the first two kinds of harm.

legally, justified. Such harmful responses might include economic sanctions and the like; but they might also include various forms of covert political action, notably covert political cyberattacks. I return to this question in detail below in section 11.2.

In this context it is important to distinguish cybertheft from other forms of cyberattack. As with theft in general—and fraud, defined as theft by means of deception—cybertheft does not necessarily involve damaging any human person or artifact. Nor does it necessarily include cyberharm as such. Conversely, other forms of cyber-attack cause cyberharm but do not necessarily involve theft. Thus a so-called logic bomb might destroy data and algorithms and do so without damaging the actual physical hardware, and in a manner that does not enable the attacker himself to possess the data or algorithms in question.

Of course, in the case of cybertheft the "item" stolen is typically intellectual property, for example, data algorithms. Being theft of intellectual property, cybertheft does not necessarily deprive the owner of the use of the property, although the owner may well be deprived of many of the rights and benefits of ownership, such as *exclusive* use and the economic benefits that flow from exclusive access.

Cybertheft needs to be distinguished from cyberespionage. The latter refers to the theft by some computer-based means (as opposed to, for example, by physical removal of paper-based documents): (i) of data or other intellectual property stored in an ICT system; (ii) that is reasonably regarded as *confidential from a national security perspective*; (iii) in order to realize some *political or military purpose*.

Here the Snowden case is salient.[19] Edward Snowden was a low-level private contractor to the NSA who breached legal and moral confidentiality obligations by engaging in unauthorized accessing, retrieving, and/or releasing of a large volume of confidential data from the NSA to the international press. Snowden's activities are a major, indeed stunning, breach of institutional confidentiality and were enabled by ICT and, specifically, the existence of vast amounts of communicable, searchable, analyzable, stored data on a computer linked to a network. Given the importance of compliance with confidentiality requirements to the integrity of security agencies, and given the large volume of confidential data released, Snowden's actions surely did considerable institutional damage to the NSA, in particular.

Nevertheless, perhaps release of some of this data to the press was morally justified by the public's right to know, for example, the public's right to know that the NSA was engaged in an extremely large-scale collection process of the metadata of US and other citizens (see the brief accounts of Verizon and PRISM above). Certainly, Verizon and PRISM raise important privacy concerns pertaining both to security agencies' collecting and analyzing metadata on their own citizens (Verizon)

---

[19] As mentioned earlier, for more on this case, see Harding, *Snowden Files*.

and their interception of the content of communications between their citizens and foreign citizens, and between foreign citizens (PRISM). I note that metadata enables the construction of a detailed profile of a person (e.g., of the person's associates and activities), especially when combined with financial and other data, and enables, also, the tracking of a person's movements. Accordingly, it is not necessarily innocuous from a right to privacy perspective.[20]

Insofar as such metadata and content collection and analysis has targeted the confidential data and communications of the personnel of foreign governments and their security agencies for US national security purposes, it is perhaps best understood as cyberespionage. To the extent that the target has been the data and communications of terrorists, it is perhaps best thought of as cyber-based law enforcement, since terrorism is a crime (including in the context of armed conflict). Insofar as such metadata and content collection and analysis has targeted the private data and communication of ordinary citizens (both domestic and foreign), it constitutes an infringement (and in some cases, evidently, a violation) of their privacy rights.

What if it turns out, as some have speculated, that Snowden has released confidential data to foreign powers, such as China or Russia? Arguably this would be a form of cyberespionage with the potential to undermine legitimate security purposes and processes and/or to put security personnel and others in harm's way. Indeed, some such as the US Defense Intelligence Agency[21] have suggested that the confidential data actually released by Snowden to the international media has compromised security processes and put security personnel in harm's way. If so, then Snowden's action of releasing the data was morally wrong *pro tanto*. Whether or not it was morally wrong, *all things considered*, depends on the countervailing moral weight to be attached to Snowden's enabling the exercise of the public's right to know about the NSA's metadata collection and related activities.

On this admittedly somewhat stipulative definition, there would also be a distinction between cyberespionage and what might be referred to as industrial cyberespionage. The latter refers to the theft by some computer-based means of: (i) data or other intellectual property stored in an ICT system; (ii) that is reasonably regarded as confidential from a commercial perspective; (iii) in order to realize some commercial purpose.

The above-mentioned investigations by the US computer-security firm, Mandiant, indicate that China is a major cyberthief.[22] For there are multiple acts of cybertheft originating from the headquarters of the China's People's Liberation Army Unit 61398. Indeed, according to Mandiant most cyberattacks on US corporations, US infrastructure (e.g., power grids), and US government agencies

---

[20] Lucas, "Press Corps Full of Snowdenistas."; Henschke, "Morality of Metadata."
[21] Leopold, "Pentagon Report."
[22] Mandiant Intelligence Centre, *APT1*.

originate from China, and China's large-scale cybertheft comprises hundreds of terabytes from 140 countries.

Much of this stolen information is apparently commercial in character. However, much of it is politically and military sensitive, at least potentially. For example, such cybertheft might enable the Chinese state to manipulate power grids, air-traffic control operations, financial systems, and so on; accordingly, cybertheft can be a major threat to critical ICT infrastructure. Accordingly, much of this information might simultaneously be both politically and militarily sensitive information, as well as being confidential commercial information. As such, the acts of cybertheft in question might be both acts of cyberespionage and acts of industrial cyberespionage. Consider, for example, the design and performance details of a fighter aircraft being developed by a commercial company for the exclusive use of the air forces of a particular nation-state and its allies, as in the case of the US F-35 Joint Strike Fighter plane[23] (evidently the victim of a Chinese cyber-theft operation).[24]

11.2 Covert Political Action, Covert Political Cyberattacks, and the "Problem" of Attribution

A key problem in relation to cyberattacks by nation-states on other nation-states is the so-called problem of attribution.[25] Unlike most attacks in conventional wars or, for that matter, conventional crimes of assault or theft, there is a major *epistemic* problem in cybersecurity: the problem of *reliably* attributing responsibility and, conversely, the *credibility of denial* of responsibility on the part of culpable aggressors. Because harmful cyberactivity is difficult to distinguish from benign cyberactivity, and because actors in the cyberworld are densely interconnected by indirect pathways, it is often extremely difficult to pinpoint the source of a cyberattack, or even to know that an attack rather than, say, a malfunction has taken place.

Moreover, the attribution problem is not simply a technical issue; it is not simply a matter of, so to speak, technical computer forensics. As with the determining of culpability for crimes in general, or ascribing responsibility for covert acts of aggression in wartime, there is a complex mix of rational and evidential considerations in play.[26] These include: (i) elements of the framework of rationality, such as motive, ability, and opportunity; (ii) physical evidence, for example as a basis for computer forensics; and (iii) testimony. There is also the question of weighing the different

---

[23] See Singer and Friedman *Cybersecurity and Cyberwar*, 93.
[24] Mullins, "China F-35."
[25] Lucas, "*Jus in Silico*," 371; Rowe, "Perfidy in Cyberwarfare," 401.
[26] See Miller and Gordon, *Investigative Ethics*.

kinds of evidence in play and the internal coherence of the overall narrative attributing responsibility to this or that actor.[27]

However, a key problem in the case of cyberattacks emanating from foreign nation-states, as opposed to from within a domestic jurisdiction, is the problem of access. It is not possible, for example, for the United States to send a team of investigators, replete with computer forensics specialists, to China to the People's Liberation Army building, from which Mandiant claims cyberattacks have emanated, for the purpose of interviewing relevant personnel, removing the computers for forensic scrutiny, and so on. China can both deny responsibility for the crimes in question and (on grounds of national sovereignty) deny access to investigators; yet without access to such evidence, criminal responsibility may be extremely difficult to prove and, as a consequence, denial of criminal responsibility may well be credible.

At any rate, the existence of the "problem" of attribution and, as a consequence, the credibility of denial makes cyberattacks an extremely useful tactic for nation-states seeking to avoid outright war but, nevertheless, engaged in the age-old strategy of covert political operations against other nation-states they regard as enemies but with whom they are not actually at war. Historically, the tactics deployed in covert political operations have included assassination of the political leaders of such "enemy" states, the financing of coups d'état and other insurrectionary movements, and destabilizing "enemy" states by spreading disinformation and propaganda, deploying agents provocateurs, and so on.[28]

Some covert political operations, if they were done overtly, may well constitute acts of war and be taken as such. Assassinations of foreign leaders and orchestrations of coups d'état are cases in point. On the other hand, some covert political operations, such as political espionage during peacetime, would probably not be regarded as acts of war, even if acknowledged by the offending nation-state.[29] Consider, for example, the recent revelations of US spying on the chancellor of Germany, Angela Merkel. While it has certainly soured US-German relations it is not even close to

---

[27] Some of these points are also covered in the chapter by Joseph Danks and David Danks in this collection.

[28] Perry, *Partly Cloudy*.

[29] The *Tallinn Manual* restricts espionage and cyberespionage, in particular, to activity conducted during war behind enemy lines (Schmitt, *Tallinn Manual*, 158). Therefore, it holds that covert *remote* cyberinformation gathering in war is not cyberespionage but rather computer network exploitation (CNE). This definitional move seems somewhat artificial. Moreover, the term "exploitation" that is frequently used in these contexts seems to me to be unhelpful. Why should not, for example, A exploit B's vulnerability by attacking B rather than merely stealing from B? State A can exploit state B's internal dissension and invade B or exploit B's lack of cyberdefense by destroying B's physical ICT infrastructure. Perry excludes espionage from his definition of covert political actions (Perry, *Partly Cloudy*, 163). However, it is not entirely clear to me why he does so when other covert politically motivated, *nonviolent* actions, for example, telling lies, spreading disinformation or propaganda, and so on are included.

triggering war between the United States and Germany.[30] Moreover, espionage has not typically been regarded as a casus belli from a legal perspective.

Covert political operations are typically, but not necessarily, unlawful. This is one reason why they are not conducted openly, albeit not, I submit, the main reason. Covert political operations, while they may involve killings and the destruction of property are designed to stop short of war; the whole point of such *covert* political operations is to weaken an enemy state, or defend oneself from being weakened, while plausibly denying that one is doing so, thereby averting outright war. It is, therefore, no accident that during the Cold War in the shadow of nuclear war, the covert political operation was a favored tactic of both the Soviet Union and the United States.

Thus with the assistance and under the influence of Moscow, if not under its direction, communist-controlled labor unions orchestrated a series of violent strikes in the late 1940s in key industries in France and Italy in an attempt to destabilize the democratically elected governments. For their part, the CIA responded with various covert operations, including financing and otherwise supporting anti-communist groups in these countries.[31]

In South America in the 1960s and 1970s, notably in Chile, the CIA went much further and actively supported the overthrow of democratically elected President Salvador Allende, presumably on the grounds of his links with Cuba and with the Soviet Union. A further interesting case mentioned by Perry involved the British duping the United States into believing Nazi Germany had a secret plan to attack the United States.[32] The British did so by allegedly discovering a secret Nazi map. However, the "map" was a forgery by British intelligence; the forgery was made with a view to influencing the United States to go to war against Germany.

While covert political operations are, and always have been, used by nation-states against nonstate actors, notably terrorist groups, they are not so often used *by* terrorist groups, in particular, for reasons provided in the opening section. Terrorist groups are primarily interested in drawing attention to their crimes; publicity is, as they say, the oxygen of terrorism. Accordingly, I suggest that many, if not most, cyberattacks by state actors against other state actors, especially cyberattacks by nuclear powers against other nuclear powers or their allies, can typically be appropriately regarded, not as acts of war, but rather as covert political operations—specifically, covert political cyberattacks—which stop short of war. I suggest that many of the cyberattacks emanating from China against the United States (Mandiant) and emanating from the United States against, for example, Iran (Stuxnet), or from Russia against Estonia and other European states, can be so regarded.

---

[30] BBC News, "Angela Merkel."
[31] Perry, *Partly Cloudy*, 167f.
[32] Ibid., 166.

This is, as we have seen, not to say that there are not cyberattacks that are, in fact, acts of war, such as Operation Orchard; I am not denying the possibility of what would be quite literally cyberwar. Nor am I settling the difficult questions broached above concerning the threshold settings for war. Moreover, there are many cyberattacks that are neither acts of war nor plausibly characterized as covert political operations. For in some cases cyberattacks have no political or military purpose and are neither conducted by nation-states (or their security agencies or proxies thereof) nor directed at nation-states (or at individuals or organizations qua members of a nation-state). For example, many cyberattacks are simply crimes directed at corporations and carried out by criminals or criminal organizations for financial gain.

It might be argued that whereas I am correct in thinking that cyberattacks typically take place in conflicts short of war, nevertheless, there are other frameworks or models available that are preferable to my favored one of covert political action. Two such models immediately come to mind, namely, economic war and armed interventions short of war. Let me briefly mention my reasons for preferring the model of covert political action to these.

Economic warfare consists of blockades, sanctions, and the like. Admittedly, these kinds of action are similar to cyberattacks in that they are not, at least in the first instance, lethal (or physically injurious) actions (although they may well ultimately lead to death, e.g., sanctions preventing importation of medicine, as happened in the case of US sanctions against Iraq under Saddam Hussein).[33] However, economic warfare is not typically *covert* whereas cyberattacks, as we have seen, usually are. Moreover, if and when aggressive economic measures are covert—and conducted to serve political purposes—then they may well count as covert political action rather than as acts of war per se. Consider in this connection the sabotage during peacetime of some economically important installations of an "enemy" state by a state actor who subsequently denies responsibility. Since, unlike sanctions, this is manifestly a violation of national sovereignty and may well be a casus belli, it is likely to be a covert political operation. Moreover, in this respect it is very similar to cyberattacks of economic sabotage.

Armed interventions short of war, such as UN peacekeeping operations in Bosnia, East Timor, and so on, while they involve, at least potentially, the use of lethal force, are akin to cyberattacks insofar as they are actions short of war. However, unlike cyberattacks per se, armed interventions rely on the use, or threat of the use, of lethal force. Moreover, such armed interventions are not typically covert and if they are covert, as in the case of the CIA's air support for Guatemalan antigovernment forces in 1954, they are plausibly thought of as covert political actions.

[33] Coates, *Ethics of War*, 222–27.

11.3 The Morality of Covert Political Cyberaction: "Dirty Hands"

The actions that constitute the core of covert political actions are multifarious. As already mentioned, they include assassination, support for coups d'état, sabotage, theft, spreading of disinformation, use of agents provocateurs, espionage, and so on. Aside from their political motivation they have another thing in common: they are harmful actions normally regarded as immoral. Moreover, as stated above, covert political action and, therefore, covert political cyberaction are typically illegal,[34] either in terms of international or domestic law (or both).

In short, covert political actions and, therefore, covert political cyberactions, are morally justified, if at all, by the greater good that they serve—specifically, the greater good that consists of the realization of their motivating political purposes. Naturally, the political purposes served by covert political actions do not necessarily morally justify these actions and, indeed, in many cases the political purposes themselves are not morally acceptable, for example covert operations conducted to further the political interests of the Soviet Union under Stalin.

However, the most appropriate moral category, or general description in the philosophical tradition, under which to file most[35] covert political actions and, therefore, many, if not most, covert political cyberactions is, I suggest, that of so-called dirty hands.[36] Covert political action is typically a paradigm of dirty hands: doing what is wrong in order to achieve some (allegedly) greater good.

---

[34] Perhaps some kinds of covert political cyberattacks are not crimes merely because the law has yet to catch up. Hence recent initiatives such as the *Tallinn Manual* and its successor. In relation to cybertheft see Green, *Thirteen Ways to Steal a Bicycle*. Perhaps some kinds of covert political actions are longstanding but there is a reluctance to criminalize them, even though they ought to be criminalized, for example some kinds of assassination in some jurisdictions at different times. Moreover, the United States and Israel, in particular, have in recent times sought to legalize *within their respective domestic jurisdictions* covert political activity that many *within those jurisdictions* regard as actions that ought to be unlawful, e.g., torture (or at least so-called torture lite), drone attacks outside declared theaters of war, targeted killings in civilian settings, and so on. Further, espionage engaged in by nation-state A against nation-state B might be lawful within A's domestic jurisdiction but not within B's (foreign spies are jailed) and vice versa. However, the point to be stressed here is that whether or not such actions are in fact unlawful within those domestic jurisdictions, they are typically regarded as unlawful *outside those domestic jurisdictions*, notably by the nation-states on whose soil they take place but also under international law or, at the very least, they are highly problematic from a legal perspective, both within domestic jurisdictions and under international law. In short, even when covert political action—including covert political cyberaction—is not unlawful, there is an extreme tension between such action and the law.

[35] Albeit not all; not, for example, the 1981 US covert operation to rescue the US diplomats and other US citizens held hostage by Iran—its breach of Iranian sovereignty notwithstanding.

[36] There is a voluminous literature on this topic. For an influential treatment see Walzer, "Political Action: The Problem of Dirty Hands," 160–80. See also Miller, "Noble Cause Corruption in Politics," 92–112. The problem of dirty hands is sometimes cast in terms of the distinction between deontological

Here is it important to distinguish dirty hands actions from lawful and morally justifiable but, nevertheless, harmful actions.[37] Presumably, the lethal and other harmful actions of soldiers in wartime, insofar as they comply with just war theory (both the *jus ad bellum* and the *jus in bello*) are not instances of dirty hands actions. Nor are the harmful actions of police officers (e.g., the use of coercive force to effect an arrest) instances of dirty hands insofar as they comply with legally enshrined moral principles.

If this is correct then covert political action and, therefore, covert political cyberaction pose particular challenges, both for the law enforcement model and just war theory. On the one hand, covert political cyberaction is (more or less) by definition action short of war; its whole raison d'être is typically to harm an "enemy" state without triggering war and, especially, in the case of nuclear powers, to avoid triggering nuclear war. So the application of just war theory is, at least for the most part, inappropriate; it largely misses the mark.

On the other hand, covert political cyberaction is (more or less) by definition unlawful. Accordingly, there is a strong moral presumption against its use. Yet, for reasons elaborated below, it does seem morally justified on some occasions and in some areas, for example cyberespionage. So the application of the law enforcement model leaves the problem largely untouched—the problem being the apparent moral justifiability of many instances of covert political action and, therefore, of covert political cyberaction, notwithstanding their unlawfulness.

Although admittedly the distinction is not clear-cut let us, nevertheless, distinguish between two species of covert political cyberaction, namely, covert political cyberattacks and cyberespionage. As mentioned above, cyberattacks do not include purely defensive measures such as firewalls and password protection. Again, as mentioned above, cyberattacks, if successful, are harmful (directly or indirectly) in one or more of the following ways: (i) physical or psychological harm to human beings per se; (ii) physical destruction; (iii) cyberharm, for example data destruction; (iv) institutional harm.

As already stated, covert political cyberattacks are, in the paradigm cases, covert unlawful, harmful actions short of war undertaken by one nation-state against another nation-state (or nonstate political actor) for political purposes. Since such actions are typically unlawful, an immediate response might be as follows: (i) one's own government ought not authorize covert political cyberattacks and one's own security agencies ought to cease to carry out such attacks; (ii) foreign governments who authorize covert political cyberattacks and their security agencies who carry them out ought to be investigated and, if appropriate, prosecuted and punished

---

and consequentialist perspectives. However, both the "dirty" action and the (allegedly) greater good are intended or otherwise aimed at. So a better way to frame the issue is in terms of means and ends (realized ends not being mere consequences).

[37] Miller and Blackler, *Ethical Issues in Policing*, 5–30.

in accordance with (presumably) international law. In short, the law enforcement model ought to be relied on to deal with this problem.

Unfortunately, as argued above, in the case of covert political cyberactions this law enforcement approach is not practicable, given the attribution problem and the current state of the international criminal justice system. This is not to say that it is not worth striving to bring into existence a more effective international criminal justice system in respect of cyberattacks in general and covert political cyberattacks in particular; quite the contrary, as in fact I suggest below. However, to reiterate, it is to say that full-blown application of the law enforcement model to covert political cyberactions is not practicable at this stage in the development of the international order.

So the question to be addressed is: Can our own covert political cyberattacks be morally justified in an overall context in which other nation-states are routinely engaging in such attacks on us and on one another? In short, can covert political cyberattacks be morally justified in what is in effect a state of nature—a cyberstate of nature (if this is not a contradiction in terms)?

The existence of this cyberstate of nature notwithstanding, covert political cyberattacks do need to be morally justified; I am not advocating a so-called realist view of the international order.[38] In particular, they need to be justified, at least in the first instance, by recourse to some morally weighty political purpose. For example, it was not morally justifiable for Russia to launch a covert cyberattack on, say, Estonia's ICT infrastructure merely because it judged it to be in its political interest to do so. On the other hand, if the United States finds itself under frequent and ongoing covert cyberattack from, say, China, and these attacks threaten to destroy or seriously disrupt key US ICT infrastructure, then the United States may well be morally justified on self-defense grounds in responding in kind.

So, on the one hand, we confront a cyberstate of nature and, on the other, we are not absolved from the need to provide moral justifications for our own covert political cyberattacks. I suggest that a number of familiar moral principles remain in play albeit, as I argue below, in a somewhat different form. The principles in question exist in both criminal law (and are, therefore, in part constitutive of the law enforcement model) and in just war theory, albeit in somewhat different forms. First, there is the principle of self-defense, for example defense of a national infrastructure asset. Second, there is the principle of necessity; a cyberattack might be morally justified if diplomatic means, for example, have been or would be ineffective. Third, there is the principle of proportionality; the United States might not be entitled, for example, to destroy China's ICT infrastructure if China has only been engaged in disruption of US infrastructure. Fourth, there is the principle of discrimination; it is prohibited to intentionally harm innocent third parties.

---

[38] Assuming realism is the view that conflict between nation-states is somehow outside morality.

However, as is typically the case in states of nature, there is another moral principle in play, namely, the principle of reciprocity. I note that the principle of reciprocity is not normally taken to be constitutive of just war theory nor is it typically invoked by proponents of the law enforcement model.[39] Moreover, as mentioned above, the various moral principles constitutive of just war theory and of the law enforcement model take on a different form, or at least must be differently applied, in contexts of covert political action. (And, of course, they are differently applied in war than in criminal justice contexts in peacetime.) The principle of discrimination, in particular, is evidently in need of adjustment in the context of (otherwise morally justified) covert political cyberattacks, as I argue below.

Accordingly, I suggest the following: (1) The principle of reciprocity, at least in its retrospective form (of which more below), applies to covert political cyberattacks and its effect is to render some such attacks morally permissible, notwithstanding that these same attacks may well not be permitted under the more stringent conditions imposed by just war theory (supposing it were to be applied to them) and certainly not under the law enforcement model (at least as it typically applies to criminal justice contexts in liberal democracies in peacetime). (2) A principle of discrimination is applicable to covert political cyberattacks but it is more stringent than the one constitutive of just war theory and yet less stringent than the one that is generally applicable under the law enforcement model. Let me turn first to the principle of reciprocity.

On one rendering of the principle of reciprocity, it is essentially retrospective in form and takes its inspiration from the ancient prescription, "an eye for an eye and a tooth for a tooth."[40] On this version of the principle, if one is unjustifiably attacked, or otherwise unjustifiably harmed or wronged, then one is morally entitled to respond in kind, irrespective of whether it is necessary for the specific purpose of, say, self-defense. On the other hand, one is not entitled to do more harm to an attacker than the attacker did to oneself, whether by mounting a single more harmful counterattack or by mounting a series of counterattacks that in aggregate are more harmful.

As we have seen, covert political cyberattacks are typically dirty hands actions and, therefore, difficult to morally justify. However, dirty hands actions are evidently justified in many situations of conflict by the retrospective principle of reciprocity; one can dirty one's hands as long as the other guy is doing so. Yet the prescription "an eye for an eye and a tooth for a tooth" is too permissive; it would license reciprocal attacks on others for any purpose whatsoever, just so long as this attack was not more harmful than the one it was in response to.

---

[39] Although it is apparently a principle of international law. See Osiel, *End of Reciprocity*. Michael Skerker's chapter in this collection also invokes a notion of reciprocity.

[40] See also Rule 9 in Schmitt, *Tallinn Manual*, 36–41.

Accordingly, we need to place a restriction on the principle: a restriction with respect to the purposes it is to serve.[41] More specifically, a morally acceptable version of this retrospective principle would justify nation-state A, engaging in covert political cyberattacks against nation-state (or nonstate actor) B, in circumstances in which B had engaged, or was engaging, in unjustifiable cyberattacks on A, but only if A's attacks were in the service of A's morally justifiable political purposes.[42]

So far so good. However, there is another version of the reciprocity principle that is salient; this version is prospective in form. It is a tit-for-tat principle in the service of bringing about a morally desirable future state of affairs.[43] The state of affairs in question is an equilibrium state among nation-states; more specifically, a morally justifiable equilibrium under the rule of international law. This is not tit-for-tat in the service of the very general purpose of doing whatever is in one's political self-interest, legitimate or otherwise (in the manner of rational choice theories); nor is it tit-for-tat measures short of war in the service of the narrow purpose of averting a future large-scale lethal attack that would constitute war (as might be justified under some extension of just war theory). Of course, in this equilibrium state of affairs there would be no covert political cyberattacks or, at least, they would be few and far between. So this principle does not justify dirty hands actions in the manner of its sister retrospective principle; rather it has as its purpose to eliminate, or at least greatly reduce, dirty hands actions and, in this case, covert political cyberattacks and, thereby, move the international order out of its current cyberstate of nature and into a cybersocial contract (so to speak). However, the equilibrium that is its raison d'être is at best a long-term goal; it is unlikely to be achieved anytime soon.[44]

---

[41] Perhaps compliance with the "eye for an eye and a tooth for a tooth" principle has as one purpose that of getting even, that is, revenge. But the point is that the principle does not rule out other additional purposes. Note that on the revised version of the "eye for an eye and a tooth for a tooth" principle (the retrospective reciprocity principle), the "eye for an eye and a tooth for a tooth" principle becomes a necessary part of some sufficient condition that justifies engaging in covert political cyberattacks. Note also that the retrospective reciprocity principle is not purely retrospective in character since it consists in part in serving some (typically forward-looking) morally legitimate political purpose.

[42] In order to simplify matters I will exclude from consideration here the possibility of responding with a cyberattack to a noncyber-, presumably kinetic, attack or vice versa.

[43] So this is not the same as the tit-for-tat principle deployed in rational choice theory understood in terms of rational self-interested actors. Dipert, "Ethics of Cyberwarfare," suggests in passing the application of rational choice theory and its tit-for-tat principle to cyberconflict. In my view rational choice theory is useful up to a point as a descriptive theory but not as a normative theory. Moreover, the practical reasoning required to move to the social contract presupposes joint action at some point among at least some of the main actors. Elsewhere I have argued against the adequacy of rational choice–based modes of practical reasoning in joint action. Miller, "Rationalising Conventions," 23–41.

[44] What I am calling the prospective reciprocity principle has a retrospective aspect insofar as its application is only triggered by a past (or present) attack.

I suggest that these two contrasting principles of reciprocity, one retrospective and the other prospective, one relatively permissive (though less permissive than the prescription, "an eye for an eye and a tooth for a tooth") and the other much less so, both may well be applicable to covert political cyberattacks.[45] If so, then there are moral justifications for covert political cyberattacks other than that of self-defense against present or future attack.

The retrospective principle of reciprocity justifies the pursuit of one's morally legitimate political interests by means of dirty hands actions, including covert political cyberattacks, given the other side is pursuing their political interests by such means. So it is relatively permissive and might encourage reciprocal attacks that would not otherwise be justified; on the other hand, at other times, it may have a deterrent effect and discourage initiating attacks. At any rate, its application is unlikely to lead to a large reduction, let alone the elimination, of covert political actions in general or of covert political cyberattacks in particular.

The contrasting prospective principle of reciprocity justifies tit-for-tat covert political cyberattacks in the cyberstate of nature if they are undertaken in the pursuit of the cybersocial contract, that is, a future morally justifiable equilibrium state under the rule of international law in which dirty hands actions are eliminated or greatly reduced. So it is far more restrictive than its sister principle, although it does permit present covert political cyberattacks if they deter future ones and are likely to lead to the cybersocial contract.

Let me now turn to the principle of discrimination. The application of this principle in criminal justice contexts is quite different from its application in war (or at least its application to war according to just war theory). Combatants are entitled to shoot enemy combatants in circumstances in which they put the lives of innocent civilians at serious risk; indeed, they can do so, even when knowing that innocent lives will probably be lost, albeit unintentionally so. By contrast, a police officer cannot, for example, shoot an armed offender in circumstances in which if he does so he will, thereby, unintentionally kill an innocent bystander, even if this is the only means to prevent the armed offender from escaping. Rather, he must simply allow the armed offender to escape.[46] In short, the application of the principle of discrimination in war is far more permissive than in ordinary law enforcement.

What of the application of the principle of discrimination in relation to covert political cyberattacks? As noted above, there are complications here arising from the nature and quantum of the harm (e.g., institutional harm and cyberharm as opposed to physical harm) potentially done to innocent bystanders as a result of a cyberattack. There is also the problem of the ubiquity of so-called dual-use ICT

---

[45] They may also conflict with one another. However, I don't see one consistently dominating the other in cases where they conflict.

[46] Miller and Blackler, *Ethical Issues in Policing*, chap. 3.

infrastructure: infrastructure that is used both by governments and their security agencies as well as by ordinary citizens. A further problem arises from the "unpredictable, unquantifiable and diffuse effects of cyberattacks";[47] in this respect they are somewhat different from, for example, bomb blasts.

If covert political cyberattacks can be morally justified in accordance with, for example, the retrospective principle of reciprocity, then the principle of discrimination as it operates in ordinary criminal justice contexts will need to be relaxed in its application to covert political cyberattacks. For these attacks take place in conflicts in which there is no, or limited, recourse to highly restrictive enforceable laws, such as exist in well-ordered domestic jurisdictions and apply to kinetic conflict between individual citizens. In this respect, cyberconflict is akin to war. Can we then simply apply the principle of discrimination constitutive of just war theory in these cases? I think not; this version of the principle, or at least its application in contexts of war, is too permissive and for three reasons.

The first reason is that there is less at stake or, at least, there is less immediately at stake in the contexts of conflict in question: contexts short of war. Therefore, the grounds for putting the well-being of innocent citizens at risk in order to win such conflicts are weaker. The second reason is that, as is the case with counterterrorist activities, it is often a more difficult matter in cyberconflicts to distinguish between aggressors and innocent civilian nonaggressors than it is in conventional warfare. The third reason is that, as mentioned above, in general it is considerably more difficult to contain collateral damage in the case of cyberattacks than in the case of kinetic attacks. Therefore, other things being equal, the risks to innocent civilians are greater.[48]

The fourth reason pertains to the prospective principle of reciprocity. As is the case with its sister principle, this principle is here being applied to conflict composed of frequent and ongoing cyberattacks on and by an "enemy" nation-state in the overall context of the cyberstate of nature. However, the prospective principle has a different purpose from the retrospective principle, namely, to bring about the cybersocial contract: a future morally justifiable equilibrium state of the international order under the rule of international law. Therefore, it is of great importance to avoid making an enemy of the citizenry of "enemy" nation-states. At the very least, the foreign citizenry must be able to differentiate between attacks on their government and its security agencies, on the one hand, and attacks on themselves as citizens, on the other. In this respect, covert political cyberattacks conducted under the prospective principle of reciprocity are more akin to counterterrorist operations

---

[47] Lin, "Overview of Relevant IHL Rules and Principles That May be Challenged by Cyberwar."

[48] Of course things might not be equal. For example, the quantum of harm done to innocent third parties by a cyberattack might be considerably less than in the case of a kinetic attack launched to achieve the same military or political purpose.

than conventional warfare between nation-states. As with counterterrorism, winning hearts and minds is of paramount importance.

In light of these notions of a cyberstate of nature and a cybersocial contract, and of the role of the retrospective and prospective principles of reciprocity, let me now turn to a particular, but currently very prominent, species of covert political cyberaction, namely, cyberespionage.

As we saw above, Verizon and PRISM have raised legitimate privacy concerns, both for US citizens and for foreigners, for example in relation to metadata collection and analysis. Regarding metadata collection and analysis in the context of domestic law enforcement, the solution, at least in general terms, is evidently at hand: extend the existing principles of probable cause (or, outside the United States, reasonable suspicion), and the existing relevant accountability requirements, for example, the system of judicial warrants.[49]

However, some of these privacy concerns pertain only to foreign citizens. Consider the FISA (Foreign Intelligence Surveillance Act) Amendments Act of 2008. It mandates the monitoring of, and data gathering from, foreigners who are outside the United States by the NSA. Moreover, data gathered but found not to be relevant to the foreign intelligence gathering purpose of, say, counterterrorism is not allowed to be retained. Importantly, however, there is no probable cause (or reasonable suspicion) requirement unless the person in question is a US citizen.

This is problematic insofar as privacy is regarded as a *human* right and, therefore, a right of all persons, US citizens or not. Moreover, these inconsistencies between the treatment of US citizens and foreigners are perhaps even more acute or obvious when it comes to the infringement of the rights to privacy and, for that matter, confidentiality of non-US citizens in liberal democratic states allied with the United States, for example EU citizens.[50]

Intelligence gathering, surveillance, and so on of citizens by domestic law enforcement agencies is reasonably well defined and regulated, for example in accordance with probable cause/reasonable suspicion principles and requirements for warrants; hence the feasibility of simply extending the law enforcement model to metadata collection within domestic jurisdictions. However, this domestic law enforcement model is too restrictive, and not practicable, in relation to intelligence gathering from, for example, hostile foreign states during peacetime, let alone wartime.

The privacy rights of the members of the citizenry during wartime are curtailed under emergency powers, and the privacy and confidentiality rights of enemy citizens are almost entirely suspended. Military intelligence-gathering during

---

[49] This is, of course, a simplification, however I do not have the space to go into details here. I have done so in my unpublished manuscript, Seumas Miller, "Cyber-security, Privacy and Confidentiality."

[50] See, e.g., Kleinig et al., *Security and Privacy*.

wartime has few privacy constraints and, given what is at stake in all-out wars, such as World War II, this may well be justified. However, these are extreme circumstances and the suspension of privacy rights is only until the cessation of hostilities. Accordingly, this military model of intelligence gathering is too permissive in relation to covert intelligence gathering from, for example, fellow liberal democracies during peacetime.

As with cyberattacks, the above-mentioned intelligence-gathering activities, notably cyberespionage, of the NSA do not fit neatly into the law enforcement model or just war theory. At any rate the question arises as to what is to be done in relation to cyberespionage in particular. First, I invoke the retrospective and the prospective principles of reciprocity utilized above. On the one hand, the United States and its allies cannot be expected to defend their legitimate national interests with their hands tied behind their backs. So their recourse to cyberespionage seems justified and the retrospective principle of reciprocity provides a specific moral justification for this. On the other hand, understood as a prospective tit-for-tat procedure in the service of bringing about a cybersocial contract, the principle of reciprocity requires the moral renovation of cyberespionage as it is currently conducted. Second, I make a couple of suggestions: (i) the clustering of nation-states and; (ii) a demarcation between government and security personnel on the one hand, and ordinary citizens on the other.

Under existing arrangements, the United States, United Kingdom, Canada, Australia, and New Zealand—the so-called "Five Eyes"—share information gathered from other states. These nation-states are, so to speak, allies in espionage, notably cyberespionage; for example, they share intelligence.[51] They are the members of my first cluster. There are, of course, other liberal democratic states outside the Five Eyes, such as various EU countries, which have "shared core liberal democratic values" with one another and with the Five Eyes and, specifically, a commitment to privacy rights. This is a second cluster.

The members of these two clusters ought to make good on their claims to respect privacy rights by developing privacy-respecting protocols governing their intelligence-gathering activities in relation to one another. Of course, determining the precise content of such protocols is no easy matter given, for example, that there are often competing national political interests in play, even between liberal democracies with shared values and many common political interests. But there does not appear to be any in-principle reason why such protocols could not be developed; and the fact that this might be difficult is no objection to attempting to do so. Since adherence to the protocols in question would consist, insofar as it is practicable, in

---

[51] Documents released about the UKUSA Agreement, which established the "Five Eyes," are available from the US National Security Agency, at http://www.nsa.gov/public_info/declass/ukusa.shtml. For a brief overview see Privacy International, "Five Eyes Fact Sheet."

ensuring compliance with some of the standard moral principles protecting privacy and confidentiality rights, such as probable cause/reasonable suspicion and use of judicial warrants, these two clusters would essentially consist of an extension of the law enforcement model to cyberespionage conducted within and between these countries.

Further, such a process of clustering of liberal democratic states would be in accordance with the prospective principle of reciprocity. Each of these nation-states would need to agree to, and actually comply with, the privacy-respecting protocols in question but each might be deterred from not doing so by the tit-for-tat procedure of the prospective principle.[52] What of authoritarian states known to be supporting international terrorism and/or engaging in hostile covert political operations, including cyberespionage, for example China and North Korea?

In respect of authoritarian states of this kind, the retrospective principle of reciprocity reigns. Accordingly, there are few if any constraints on intelligence gathering and analysis, including cyberespionage, if it is done in the service of a legitimate political interest such as national security.[53] Nevertheless, it is important to demarcate within such an authoritarian state between the government and its security agencies, on the one hand, and private citizens, on the other. Notwithstanding the applicability of the retrospective reciprocity principle, the need to respect the privacy rights of private citizens in authoritarian states remains; perhaps all the more so given these rights (and, for that matter, human rights in general) are routinely violated by their own governments.

So a stringent principle of discrimination ought to govern cyberespionage directed at authoritarian states. At the very least, the citizens of these states ought to be able to differentiate between morally justified infringements of the privacy and confidentiality rights of members of their government and its security agencies, on the one hand, and violations of their own privacy and confidentiality rights, on the other, and be justified in believing that whereas the former might be routine the latter are few and far between.

---

[52] There are, of course, considerable difficulties here in relation to democracies outside the "Five Eyes" and NATO. For example, in part for historical reasons some of the South American democracies do not necessarily trust the United States and have different and often competing political interests from the United States. Moreover, within the "Five Eyes" there are power-imbalances, for example United States versus Australia, which might render the tit-for-tat procedure ineffectual. However, there is the possibility of smaller powers forming a collective and, thereby, reducing the power imbalance to the extent necessary to enable an effective tit-for-tat procedure.

[53] There are important questions here concerning what counts as a legitimate purpose, particularly in the context of the blurring of the distinction between a political interest and an economic interest, for example China's cybertheft operations. For reasons of space I cannot pursue these here.

Bibliography

Anwer, Jared. "India Caught in Crossfire of Global Cyber War." *Times of India*, August 20, 2012. Accessed June 24, 2014. http://timesofindia.indiatimes.com/tech/it-services/India-caught-in-crossfire-of-global-cyber-war/articleshow/15567180.cms.

Bachrach, Judy. "The Stuxnet Worm Turns." *World Affairs Journal*, January 30, 2013. Accessed June 24, 2014. http://www.worldaffairsjournal.org/blog/judy-bachrach/stuxnet-worm-turns.

BBC News. "Angela Merkel: Spy Claims Test US Ties with Germany." *BBC News*, November 18, 2013. Accessed June 24, 2014. http://www.bbc.com/news/world-europe-24992485.

BBC News. "Estonia Hit by 'Moscow cyber war.'" *BBC News*, May 17, 2007. Accessed June 24, 2014. http://news.bbc.co.uk/2/hi/europe/6665145.stm.

Coates, A. J. *The Ethics of War*. Manchester: Manchester University Press, 1997.

Dipert, Randall R. "Ethics of Cyberwarfare." *Journal of Military Ethics* 9 (2010): 384–410.

Eberle, Christopher J. "Just War and Cyberwar." *Journal of Military Ethics* 12 (2013): 54–67.

Finn, Peter. "Cyber Assaults on Estonia Typify a New Battle Tactic." *Washington Post*, May 19, 2007. Accessed June 24, 2014. http://www.washingtonpost.com/wp-dyn/content/article/2007/05/18/AR2007051802122.html

Follath, Erich, and Holger Stark. "The Story of 'Operation Orchard': How Israel Destroyed Syria's Al Kibar Nuclear Reactor." *Spiegel Online International*, November 2, 2009. Accessed June 24, 2014. http://www.spiegel.de/international/world/the-story-of-operation-orchard-how-israel-destroyed-syria-s-al-kibar-nuclear-reactor-a-658663.html.

Ford, S. B. "*Jus Ad Vim* and the Just Use of Lethal Force-Short-of-War." In *Routledge Handbook of Ethics and War: Just War in the 21st Century*, ed. F. Allhoff, N. G. Evans, and A. Henschke, 63–75. Abingdon: Routledge, 2013.

Green, Stuart P. *Thirteen Ways to Steal a Bicycle: Theft Law in the Information Age*. Boston, MA: Harvard University Press, 2012.

Harding, Luke. *The Snowden Files: The Inside Story of the World's Most Wanted Man*. London: Guardian Books, 2014.

Henschke, Adam. "The Morality of Metadata: Not Just Innocuous Adornment." *The Conversation*, December 13, 2013. Accessed June 24, 2014. http://theconversation.com/the-morality-of-metadata-not-just-innocuous-adornment-21160.

Kleinig, John, Peter Mameli, Seumas Miller, Douglas Salane, and Adina Schwartz. *Security and Privacy*. Canberra: ANU ePress, 2011.

Lin, Herb. "Overview of Relevant IHL Rules and Principles That May be Challenged by Cyberwar." Paper presented at the *Cyberwarfare, Ethics and International Humanitarian Law* workshop, Geneva, May 21–22, 2014.

Lucas, Edward. "A Press Corps Full of Snowdenistas." *Wall Street Journal*, January 29, 2014. Accessed June 24, 2014. http://online.wsj.com/news/articles/SB10001424052702303519404579350663554949356.

Lucas, George R. "*Jus in Silico*: Moral Restrictions on the Use of Cyberwarfare." In *Routledge Handbook of Ethics and War: Just War in the 21st Century*, ed. F. Allhoff, N. G. Evans, and A. Henschke, 367–81. Abingdon: Routledge, 2013.

Mandiant Intelligence Centre. *APT1: Exposing One of China's Cyber Espionage Units*. Mandiant Intelligence Centre, 2013. Accessed June 23, 2014, http://intelreport.mandiant.com/Mandiant_APT1_Report.pdf.

Miller, Seumas. "Rationalising Conventions." *Synthese* 84 (1990): 23–41.

Miller, Seumas. "Noble Cause Corruption in Politics." In *Politics and Morality*, ed. I. Primoratz, 92–112. Basingstoke: Palgrave Macmillan, 2007.

Miller, Seumas. *Terrorism and Counter-terrorism: Ethics and Liberal Democracy*. Oxford: Blackwell, 2009.

Miller, Seumas. "Cyber-security, Privacy and Confidentiality." Unpublished manuscript.

Miller, Seumas, and John Blackler. *Ethical Issues in Policing*. Aldershot: Ashgate, 2005.

Miller, Seumas, and Ian Gordon. *Investigative Ethics*. Oxford: Wiley-Blackwell, 2014.

Mullins, Michael. "China F-35: Secrets Stolen from US Show Up in Its Stealth Fighter." *The Wire*, March 14, 2014. Accessed June 24, 2014. http://www.newsmax.com/TheWire/china-f-35-secrets-stolen/2014/03/14/id/559556/.

Osiel, Mark. *The End of Reciprocity: Terror, Torture and the Law of War*. Cambridge: Cambridge University Press, 2009.

Perry, David L. *Partly Cloudy: Ethics in War, Espionage, Covert Action and Interrogation*. Lanham, MD: Scarecrow Press, 2009.

Privacy International. "The Five Eyes Fact Sheet." Last modified November 27, 2013. Accessed June 24, 2014. https://www.privacyinternational.org/blog/the-five-eyes-fact-sheet.

Rid, Thomas. *Cyber War Will Not Take Place*. New York: Oxford University Press, 2013.

Rowe, Neil C. "Perfidy in Cyberwarfare." In *Routledge Handbook of Ethics and War: Just War in the 21st Century*, ed. F. Allhoff, N. G. Evans, and A. Henschke, 394–404. Abingdon: Routledge, 2013.

Sanger, David E. *Confront and Conceal: Obama's Secret Wars and Surprising Use of American Power*. New York: Broadway Books, 2013.

Schmitt, Michael M., ed. *Tallinn Manual on the International Law Applicable to Cyber Warfare*. Cambridge: Cambridge University Press, 2013.

Singer, Peter W., and Allan Friedman. *Cybersecurity and Cyberwar: What Everyone Needs to Know*. Oxford: Oxford University Press, 2014.

Walzer, Michael. "Political Action: The Problem of Dirty Hands." *Philosophy and Public Affairs* 2 (1973): 160–80.

12

Moral Concerns with Cyberespionage

Automated Keyword Searches and Data Mining

MICHAEL SKERKER

States engage in espionage, including cyberespionage, as part of a continuum of actions to pursue their national security. This chapter will address the moral permissibility of two types of remarkable electronic intelligence collection that former National Security Agency (NSA) contractor Edward Snowden charged the NSA and the Government Communications Headquarters (GCHQ) with undertaking: keyword searches, in which automated collectors record electronic communications anywhere in a targeted region containing phrases, words, or names of intelligence interest; and metadata analysis, in which the pattern of communications in a particular region is mapped. I develop a standard for assessing coercive government action based on respect for the autonomy of inhabitants of liberal states and argue that both types of signals intercepts (SIGINT) described by Snowden can potentially meet this standard. That said, the collection of metadata creates the conditions for some unsavory government behavior and so judgments about the trustworthiness and competence of the government engaging in the collection, as well as the threat level the state faces, must be made in order to decide whether metadata collection is justified in a particular state. Further, the moral standard I propose has a reflexive element justifying adversary states' intelligence collection against one another. Therefore, high-tech forms of SIGINT can only be justified at the cost of justifying cruder versions of signals intelligence collection practiced by some technologically advanced states' less-advanced adversaries.

12.1 Politically Legitimate Forms of Coercive State Action

The novel methods of espionage made possible by cybertechnology resist easy comparison with more traditional methods of spying. We therefore need to be clear about the moral foundations of espionage and military operations (discussed in this section) as well as the moral right(s) potentially violated by keyword searches and data mining (see section 12.2). Section 12.3 will distinguish two morally different types of information gathering helpful to refine our argument about the SIGINT techniques discussed in section 12.4.

I will stipulate certain starting assumptions relevant to a moral foundation for espionage.[1] In doing so, I will be assuming a liberal political system as the background for the relevant domestic and international intelligence operations. I will also inquire as to what sort of intelligence operations are "politically legitimate" in these states—that is, what government actions are just, given the government's liberal underpinnings.[2] Politically legitimate actions do not in principle violate the rights of inhabitants of the state even if these actions are coercive in nature (I use this term broadly to refer to actions or laws that compel people to do things they do not otherwise want to do, e.g., tax collection, business regulation, arrest, prosecution, etc.). The criterion for politically legitimate state actions is consent-worthiness by the inhabitants of that state.[3] To judge whether some policy is consent-worthy, the theorist first conceives of an abstract consenter (with a particular moral constitution), then judges whether consent to the policy is logically necessary to protect the consenter's moral constitution; or, to put it another way, whether it would be self-contradictory to dissent to the policy given that the policy contributes to the consenter's protection. Since we are concerned here with national security actions, and thus chiefly with securing the negative freedom of a state's inhabitants (i.e., freedom from rights violations), we can operate with a fairly simple version of autonomy characterizing the abstract consenter. Since many versions of autonomy assume that negative freedom is a precondition for any more complex expressions of autonomy, my theory about politically legitimate intelligence operations should be insertable into political theories using more complex and detailed models of autonomy than I will develop here.

The model I use sees the consenter's autonomy expressed in specific arenas of thought, speech, and action in the form of rights. All abstract consenters are considered morally equal. I reject an atomistic (e.g., Hobbesian) model of autonomy that sees people as naturally autonomous outside some kind of settled political

---

[1] The ideas articulated in this section are developed in detail in my *Ethics of Interrogation*, chap. 2.

[2] Certain intelligence operations by illiberal states can be justified to the extent that they are deployed for the same reason they are deployed in liberal states: the protection of the state's inhabitants.

[3] Inhabitants are the relevant consenters rather than citizens of states, because hypothetical consent is modeled in reference to abstract conceptions of the human person rather than in reference to people of particular nationalities. For examples of hypothetical consent, see Kant, "On the Proverb," 61–92, 79; Rawls, *Theory of Justice*, 11; Waldron, "Theoretical Foundations of Liberalism," 127–50, 138; Waldron, "Special Ties and Natural Duties," 3–30, 25; Habermas, *Theory of Communicative Action*; Apel, *From a Transcendental Semiotic Point of View*.

community. Such a model conceives of government coercion as existing in tension with citizens' natural autonomy, a tension that is tolerated in exchange for the conveniences of living in a state.[4] Rather, the model I use sees autonomy as a nested concept, entailing as a necessary background an environment relatively free of rights violations and, so, a law-governed political entity (a formal state, in most empirical instances) with the coercive power to prevent and punish rights violations. Such a political entity is a necessary material precondition for a group of people to enjoy the full realization and expression of their rights over time consistent with equal rights expression. This is because an environment free of intentional or inadvertent rights infringements is a precondition for the realization and expression of one's autonomy in a given moment and over time. For example, one cannot build a house if one is being attacked or robbed of one's tools; and one would not even plan to build a house if one could not trust one's neighbors not to destroy or occupy it. Further, one will not develop the psychological faculties necessary for positive freedom (the capacity to deliberate on one's own and craft plans) if one is constantly in fear and want. So, provided certain constraints discussed below, coercive government actions, such as police activities, are in keeping with inhabitants' autonomy even when they restrict a person's autonomy in a particular instance. This follows, because the government actions are aimed in total at creating the environment relatively free of rights violations necessary for inhabitants to enjoy the full realization and expression of their rights consistent with universal and equal enjoyment.[5]

The underlying purpose of protecting inhabitants' autonomy creates the grounds for rejecting both certain government actions that are very harmful to autonomy and strategies meant to create an environment *perfectly* free of rights violations (because such strategies will likely cause intolerably high levels of rights infringements). The preferred moral framework I call the "security standard" endorses government tactics surviving a three0stage winnowing process. The standard (1) canvasses locally available tactics aimed at securing an environment relatively free of rights violations or the threat thereof; (2) isolates the most reliable, efficacious, proportional, and efficient tactics of those locally available; and (3) endorses the most rights-respecting among the tactics meeting the practical metrics of (2).

We can assume that any autonomous person would consent to domestic government actions aimed at securing a domestic environment relatively free of rights violations meeting the security standard. This consent will also justify actions by military and intelligence operators aimed at creating a domestic environment

---

[4] E.g., Nozick, *Anarchy, State, and Utopia*, chaps. 2–5. See T. M. Scanlon, *What We Owe to Each Other*, 19–25.

[5] To be clear, we might find a genuinely autonomous person outside a political community, marooned on a deserted island or living in a failed state like Somalia, but only if her formation occurred elsewhere. We do not expect children to grow up to be fully autonomous people in such environments.

relatively free of rights violations by defeating external threats to a state's security. Since all people in the world can be modeled as consenting to a regime of outward-facing security-seeking actions, a model consenter's consent to foreign operations by her security services also potentially justifies action by foreign agents targeting her. This dynamic can best be explained by discussing its domestic parallel. Hypothetical consent is permissive when it comes to the justification of police tactics meant to keep the model consenter safe. Considerations of how to secure the safety of a model consenter justifies a series of actions aimed at rights violators or potential rights violators. At the same time, a principle of reciprocity, justifying police behavior targeting the consenter if the consenter is suspected of perpetrating or planning rights violations, urges that police exercise restraint. So the consent that we imagine autonomous people extending to domestic security-seeking tactics takes into account that they might be the target of those tactics. The same reflexivity must apply to outward-facing security-seeking tactics since the security standard references an abstract autonomous person rather than a person of a particular nationality. By hypothetically consenting to outward-facing actions directed at foreign security threats, one would give leave to other governments to engage in outward-facing actions directed at foreign security threats, including oneself, if one is reasonably perceived to be a security threat. One cannot be modeled as consenting to illiberal governments, perhaps led by paranoid or sectarian leaders, monitoring foreign citizens who do not plausibly present a national security threat. To be clear, this equitable treatment of foreigners is arrived at by a different consent-based route than equitable treatment of one's neighbors. Whereas the actions of one's domestic law enforcement agencies can potentially be justified when such actions contribute to the necessary conditions for autonomy in one's own state, an adversary foreign state agent usually is not working to maintain conditions of autonomy in one's state, but rather the opposite. We can see that it would not make sense to model hypothetical consent to foreign agents' work if we also claim to justify domestic agents' actions opposing these foreign agents. Therefore, adversary foreign state agents' actions are potentially justified indirectly, as an entailment of consenting to one's own agents' outward-facing actions. By way of analogy, if one hires a lawyer to sue someone, one cannot begrudge the target of one's lawsuit hiring a lawyer to defend her rights and interests.

This reflexivity should encourage a conservative attitude toward intelligence collection from particular suspected foreign targets. The model consenter must use herself as a reference point, asking whether she can consent to her state agents using tactics abroad that, via the principle of reciprocity, she must also permit foreign agents to use against her. Using this approach, the rule of thumb should be that security agencies should use the same information-gathering tactics abroad that they use domestically. For example, if the security standard indicates that warrants issued by judges are necessary for a security service to intercept a particular domestic inhabitant's communications, the same treatment should apply to a foreigner targeted by the security service. Note, this standard marks a serious departure from current American practice, for example, where foreigners and US residents are subject to markedly different SIGINT practices and bodies of law.

That said, practical limitations on foreign agents acting abroad or the different nature of the target might suggest different tactics than their counterparts would use domestically, leading to greater or lesser infringements on the target's rights. For example, it might not be as feasible to have ground units watch a suspected militant in the Swat Valley or the Ugandan rainforest as it would be in a domestic suburb. This limitation might prompt the surveilling agency to use airborne surveillance platforms, which might be more privacy-infringing than ground-based options in that they can see over walls and into compounds ground units cannot. To say this more privacy-infringing tactic is consent-worthy under the security standard is to say the model consenter permits her adversary's security agencies to attempt the same in her country if it confronts the same practical limitations there.[6] I will return to this point later in the chapter.

A wide range of concrete practices could be justified if the security standard permits security services to conduct foreign operations employing the most reliable, efficient, rights-respecting, and so on tactics available to that service. The best locally available tactics justified by the security standard will vary depending on a given state's wealth, size, technological prowess, and ingenuity. If the standard then effectively permits all security actors to "do their best," the standard allows situations in which, for example, wealthy country A's intelligence services can conduct very discriminate, sophisticated, targeted, and automated intercepts of foreign intelligence targets' communications—so that very few innocent people have their privacy infringed or violated—while also permitting poor adversary country B's intelligence services to conduct relatively crude, indiscriminate intercepts that infringe on the privacy of far more innocent people. As an example of crude intelligence gathering, an American NGO employee, previously posted in Uzbekistan, told me that the Uzbek National Security Service (NSS) listens to and tapes all visiting foreigners' phone calls as a matter of course. Yet the NSS only has the capacity to record thirty minutes at a time on its antiquated analog equipment and so simply disconnects calls on the thirty-first minute.

Intelligence collection activities fail the security standard in particular instances if one state's adversary's best methods of intelligence collection are so crude as to be imagined to be intolerable to the inhabitants of the target state. In this case, intelligence officers would need to refrain from collecting from a certain state if their behavior would justify retaliation by the target state engaging in its crude collection methods. By way of analogy, military actions against a

---

[6] The adversary agency's permission does not mean agencies in the target state are not permitted to oppose their actions.

state fail the security standard even if otherwise just when the target state's only method of defense is use of WMDs.[7] That said, unlike military cases, it is difficult to think of an example of SIGINT that would be so rights-infringing as to be intolerable for any state to tolerate at the hands of its dangerous adversary if that was the price of garnering signals intelligence. Crude forms of SIGINT might not be consent-worthy if the reward for the risk was lower, such as if the target state did not pose a military threat to the collector state. The Uzbek case falls between these two clear extremes. Western states do have some security concerns in Uzbekistan potentially warranting intelligence operations directed at state and nonstate actors; the NSS apparently does not have the resources to monitor the communications of citizens of western states; and there are few western expatriates in Uzbekistan. All told, the security standard can probably justify western SIGINT operations against Uzbek targets. So watch out next time you are in Tashkent.[8]

12.2 The Right to Privacy

The signals intercept operations and accompanying analysis under discussion in this chapter are not as destructive as traditional military actions. Yet they are deeply troubling for their presumed infringements on people's privacy. Before discussing the tactics in detail, we need to clarify what is meant by infringements on, and violations of, privacy.

There are two definitions of mental privacy commonly used by philosophers: (1) a mental space of one's own, safe from external intrusion or disruption; and (2) a power to control the revelation of personal information. A certain mental

---

[7] This argument might create perverse incentives for small states in particular to invest in WMDs at the cost of improving their conventional forces. Then larger states would leave them alone for risk of incurring an indiscriminate military response. Yet perhaps this incentive is not so perverse if it reduces the likelihood of war.

[8] One might wonder if any states enjoy a unilateral right to collect against adversaries because of the illegitimate nature of the target government. Since the security standard is indexed to the protection of negative liberty, it justifies traditional policing and national security actions of even some illiberal and/or autocratic states. While the security standard does not justify repressive actions aimed at a government's nonviolent political or ideological opponents, it does justify the bread-and-butter responsibilities of a state aimed at protecting its inhabitants from street crime, piracy, terrorism, and foreign military attack. The security standard does not justify the coercive actions of states with governments that largely neglect ordinary inhabitants and use power largely to benefit ruling cliques. The coercive power of government is justified in order to create relatively crime-free environments for the benefit of the inhabitants of the state. As examples of states lacking the justification for coercive state actions, I would suggest: Amin-era Uganda, Mobutu-era Zaire, Duvalier-era Haiti, military-ruled Burma, present-day Equatorial Guinea, and North Korea.

space of one's own is thought to be a precondition for moral autonomy.[9] "Privacy is the condition of having secured personal space, personal space is space required to reason, and individuals have a fundamental moral right to reason as a means of securing autonomy."[10] We would not be able to plan for the future, develop a sense of self, or control our interactions with others if every thought was exposed prior to our decision to share it publicly.[11]

The power to control personal information helps protect the mental space where one should be free to reason and reflect. One would alter one's behavior, conversations, reading habits, and thoughts if one was concerned that one was under surveillance and, thus, was being forced to reveal ideas prior to their maturation.[12] The power to control personal information also puts one in control of one's intimate relationships, which are made intimate, in part, by the decision to divulge personal information to certain people.

Certain information is kept private because knowledge of it gives the knower power over the target. The information need not be what many cultures consider inherently private, such as information regarding the body, health, money, and sexuality, but also aggregated mundane information that in sum gives a portrait of the target's daily life.[13] Some private information can be damaging to the target in specific ways on account of the structure of society—leading her to lose her job, marriage, security clearance, health insurance, the trust of others, and so forth—and some information could be damaging if particular people wanted to harm her. For example, someone who wants to attack the target or rob her house would find her daily schedule of special interest. More broadly, the agent's collecting private information erodes the target's autonomy. There is now an asymmetry of knowledge and power between the target and the agent. He knows information about her normally only revealed to a friend, relative, or lover, but he is not any of those things. She did not choose to reveal this information to the agent, even though this is information she ordinarily only chooses to reveal to intimates. This knowledge can give the agent leverage over her in many interactions, as he

---

[9] Alfino and Mayer, "Reconstructing the Right to Privacy"; Stramel, *Same Sex*, 284, 285; McCloskey, "Privacy and the Right to Privacy"; Cohen, "Equality, Difference, Public Representation"; Nagel, *Equality and Partiality*, 142–43. Benn, "Privacy, Freedom, and Respect for Persons," 1–26, 3; Van Den Haag, "Definition: The Nature of Privacy," 149–168, 151. Alan F. Westin distinguishes four functions of privacy, one of which, "reserve," "protects the personality" by creating invisible walls between the person and the rest of the world. *Privacy and Freedom*, 32.

[10] Alfino and Mayer, "Reconstructing the Right to Privacy," 10.

[11] Bok, *Secrets*, 21–23. See also Westin, *Privacy and Freedom*, 34; Benn, "Privacy, Freedom, and Respect for Persons," 24–26; Simmel, "Privacy Is Not an Isolated Freedom," 71–87, 73.

[12] Van Den Haag and Benn make similar points: "Definition: The Nature of Privacy," 151, and "Privacy, Freedom, and Respect for Persons," 10, respectively.

[13] Nissenbaum, "Protecting Privacy in an Information Age," 559–96, 565; Solove, *Digital Person*, 146.

can use that knowledge she does not know he has to shape her perception of him, manipulate her, and make her irresistible offers. The asymmetry of knowledge is problematic even if it is not leveraged into some kind of invidious action. The agent has effectively coerced a level of intimacy from the target and taken from her the opportunity to choose how to present herself to him.[14] Thus, she is wronged even if she does not know about the privacy intrusion and even if the agent does not use the information to directly harm her.

In defining a privacy violation, I will focus on the second definition of mental privacy discussed above, the power to control personal information. I focus on this definition obviously because various kinds of communication intercepts remove the power to control personal information from the intelligence target. The first type of privacy (mental space) violation may also occur if the surveillance is discovered or assumed by the target since the knowledge that she is being watched will burden her thoughts.[15]

Having clarified the moral importance of privacy, we can now discuss the difference between harms, infringements, and violations related to privacy. Distinguishing these three things will be important to identify what is problematic about the cyberespionage techniques under discussion. One can suffer a harm in the arena covered by a right like privacy without being the victim of an infringement or violation. A woman suffers a harm associated with the involuntary disclosure of private information if she drops her purse in front of a coworker and some sensitive items spill out. The purse-owner feels the embarrassment, and the coworker has the knowledge that is normally associated with a rights infringement or violation. However, infringements definitionally involve an external imposition of harm or limitation through another person's action. Infringements that are not accidental, excused, or justified are rights violations, breaches of the agents' duties to respect others' rights.

I suggest a three-point definition of privacy infringement. First, privacy infringement involves the collection of a significant amount of information about the target that the target would not ordinarily reveal to a stranger. The context, the intent of the agent, and the choices of the target regarding what she considers private determines what is a significant amount. As discussion will clarify below, one datum might suffice if it is the sort of thing that most cultures consider private, such as information pertaining to the body, health, sexuality, and wealth. In other cases, aggregated facts—each innocuous on its own—together may reveal a portrait of the target she would not share, *as a whole*, with strangers or with certain strangers.[16] To be clear,

---

[14] Gross, "Privacy and Autonomy," 169–81, 172; Van Den Haag, "Definition: The Nature of Privacy," 152.

[15] Benn, "Privacy, Freedom, and Respect for Persons," 10.

[16] Benn discusses related matters, "Privacy, Freedom, and Respect for Persons," 4–6; Nissenbaum, "Protecting Privacy in an Information Age," 565, 589; Solove, *Digital Person*, 43.

while the aggregation of many mundane details might amount to an infringement, the collection or observation of a single detail would not.

Second, in order for the privacy infringement to be actual rather than potential, the information must be attached to a particular person. A person gains no power over anyone if he finds an unaddressed love letter on the street. He knows someone's heart is aflutter, but cannot use this information to gain any advantage over or insight into any particular person. Third, the sensitive information has to be eventually known by a person. This is in part a reflection of the normative fact that rights infringements are actions of people. Falling boulders, machines, or wild animals can harm people, but not infringe on their rights, since boulders, machines, and nonrational creatures are not capable of self-limitation of their actions based on a rational appreciation of a mutual web of rights and duties. A machine might scan and record someone's sensitive information, but the machine's storage of this information does not create an asymmetry of power between itself and the target, again, because the machine is not implicated in the web of rights apportioned equally by theory to each adult person. The elements of an infringement may be present when we consider the agenda of the human designer of the machine and the human analyst who may study the stored information.

12.3 Surveillance and Patrol

It is useful to introduce a distinction between two actions, patrol and surveillance, in order to understand the moral significance of keyword searches and data mining. It will be helpful to use examples of patrol and surveillance involving human observers as models for cyberoperations since we have stronger intuitions involving the former. In patrol, the agent monitors a particular area, alert for suspicious behavior or other types of danger. Patrol is not focused on a particular person. Examples of patrol might include a policeman walking a beat, an air marshal sitting on an international flight, or a naval task force sailing back and forth in a commercial shipping lane.

Patrol raises far fewer moral concerns than surveillance when conducted by a just liberal state. First, a state agent's patrolling is merely a more concerted expression of what everyone does every day: observing activities occurring in public view in one's immediate vicinity and reserving the right to respond if something untoward or dangerous appears to be happening. The state agent has a special obligation to do what the ordinary person has a permission and weaker duty to do in the event of some emergency. Second, the patroller's attention is not focused on any particular person, and so patrol does not trigger the moral concerns related to infringements and violations of a particular person's privacy. The patroller is not gaining power over a particular person; he is not taking a prurient interest in a particular person; nor is he doing anything to make a reasonable person feel threatened (on the contrary, the presence of law enforcement officers might well comfort a person in a just state). Third, the observed parties do not have a right not to be observed by a patrolling agent. The patrolling agent in a just state is not doing anything untoward. By exiting their homes, the observed parties tacitly consent to being observed by people on the street.[17] The patrolling agent is visible—often identified as a state agent—and so the observed party tacitly consents to be seen by him in particular. They cannot be modeled as tacitly consenting to be observed by an undercover state agent per se, but again, the patrolling agent is engaging in a permissible activity. It is reasonable to postulate an expectation of "privacy in public"[18]—a desire to not be closely observed in activities we perform in public but wish to keep private, such as shopping at a pharmacy or reading a letter on a train—but in most cases, the patrolling agent would not trespass this ambiguous border of privacy. While he might be more attentive than the layperson and this attention might press the boundary of privacy in public (e.g., a policeman might scrutinize the unusual bulge in someone's clothing or an airport guard might deliberately look at each face she sees for a second), it can usually be justified as a protective action in a just state. All these comments are restricted to patrols in a just liberal state. In a tyranny or other type of unjust state, political power is used to oppress inhabitants or a portion of the population for the benefit of the ruling clique or a privileged group. In this context, mere patrol serves to remind inhabitants of the scope of the government's power.

In contrast to patrol, surveillance raises a host of privacy concerns. Trailing someone, intercepting her communications, and watching her in and outside her home provide the agent with a profound degree of knowledge most cultures consider private. The agent gathers two kinds of information that strangers ordinarily do not know about one another and that people do not ordinarily reveal to strangers. First, the agent learns things that go on in the privacy of the target's home and in her communications—both occasions when she assumes her actions and words are private, only revealed to those she chooses. Second, the agent aggregates public actions to give a full picture of the target's daily life that no stranger (who might see her in a given moment) would know. While the individual elements of the target's daily public schedule are not necessarily sensitive (e.g., she picked up her dry cleaning, she bought coffee), their aggregation as "her daily schedule" is sensitive because it can give the agent significant power over the target.

Surveillance is a graver matter than many discrete privacy violations because it is definitionally expressive of a broader intention on the agent's part than intentions associated with discrete violations. It is difficult to think of benign reasons for ordinary citizens to engage in surveillance. Whereas a discrete privacy violation may result from the agent wanting to see the target naked, or find out a specific piece of

---

[17] Reiman, "Privacy, Intimacy, and Personhood," 2–44, 44.
[18] See Nissenbaum, "Protecting Privacy in an Information Age."

information about the target, the purpose of surveillance is to develop a portrait of the person, potentially inclusive of every facet of her life. This account of the target's life amounts to a major privacy violation, because the agent has gained the power that comes through unilateral knowledge of personal information with respect to (nearly) every facet of her life, not just one facet of her life (say, regarding her commercial habits).[19] In fact, whereas discussions of privacy are complicated by the fact that the boundaries of what is private are culturally constructed, surveillance would appear to be problematic in most cultures because it observes so many activities that might be considered private and because it attains knowledge of the target's life profile.

12.4 Tactics

We will now consider two types of intelligence-gathering operations potentially infringing on or violating their targets' privacy. I will speak about these operations at a certain level of generality, without ascribing them to specific agencies. This, because post-9/11 reports on the activities of various intelligence agencies are inconsistent, fragmentary, and frequently disavowed by people who are both in the position to know the truth about the operations and incentivized to lie about them.[20] Historically, initial reports of clandestine government operations often prove to be inaccurate. So we will consider two tactics as ideal types, with a presumed family resemblance to actual operation, past, present, or potential.

Keyword Searches—There are automated programs in existence that scour communication networks, collecting communications transmitted through fiber-optic cables or the electromagnetic spectrum. Supercomputers scan the intercepted data for "selectors": certain words, names, or phrases associated with potential intelligence targets. This form of data mining is different than traditional wiretaps, pen registers, and pen traps and traces in that these keyword searches are not directed at particular suspects and do not necessarily require the participation of the phone company to physically manipulate the routing of calls. Rather, keyword collections are more like vacuum cleaners that collect everything that is in the air and on the cables. Communications deemed of interest according to some automated algorithm are recorded and forwarded to human analysts who read them and decide whether they should be purged or forwarded for further intelligence analysis.

It will be helpful to separate consideration of keyword searches into automated search, analysis, and investigation phases. It seems important to segment the tactic

---

[19] See Solove, *Digital Person*, for concerns over the threat to privacy posed by the digital dossiers formed by the automated aggregation of mundane biographical details.
[20] http://www.washingtonpost.com/nsa-secrets; http://www.theguardian.com/world/the-nsa-files; Bamford, *Shadow Factory*; Inglis, "Remarks."

in this way—eventually referring to "keyword search and SIGINT-prompted investigation" to encompass the full action—since many people affected by the tactic may only be touched by the first or first two phases. Just the same, the program has to be assessed in its totality since the first two phases exist to collect data for exploitation in the third phase, and it is the third phase that is potentially the most controversial. Regarding the initial search now, I will consider whether this type of data mining amounts to a privacy infringement and, if so, whether this infringement can be justified.

The initial search does not meet the three criteria for a privacy rights infringement described in section 12.2. The initial capture of an email, blog post, cell phone conversation, or text with a flagged phrase in it captures what will often be a quite small amount of data, akin to a sentence one hears walking past someone who is talking on the phone. The whole communication is not yet read in a keyword search, but merely tagged and stored because of the suspect phrase. Second (and even if the intercept is a fairly comprehensive and self-contained communication), the communication is not attached to a particular person: it is merely associated with a phone number or ISP number. We can see that these first two elements do not necessarily amount to an infringement on privacy because the suspect communication may not even come from a human being; data of this sort could be an automated message sent by a computer. Third, no human has seen the communication yet; a computer scanned the communication and stored it. To the computer, of course, the suspect phrases are not even words, just electrons moving in a certain pattern. While this exposure might prompt someone in the target region to feel his rights have been infringed or violated since his private message was "opened" prior to its reception by the intended recipient, I think this is an emotional residue connected to the symptom, the harm, rather than the substance of a privacy right infringement or violation. High technology forces us to draw analogies based in the physical world; the closest analogue here, a spy opening one's letter, diverges in too many ways from the SIGINT tactic under discussion to be helpful. On a closer view, the components of a privacy infringement are absent with keyword searches. A human being who can understand what she is reading is not opening a complete, identifiable piece of correspondence.

The reader may object that my focus is too fine-grained here. There may be a genuine objection to the government *trying* to infringe on one's privacy by whatever means—the automated keyword search assembles some of the components of an infringement—and so we need to address below whether gathering the data making infringement possible is politically legitimate. A partial, preliminary response to this concern notes that to the extent that keyword searches are like patrols, the government is not trying to infringe on a particular person's privacy. Keyword searches are like patrols in that they are passive forms of collection, only leading to more invasive activities when something apparently untoward is observed. The difference between keyword searches and traditional patrols is that the latter are restricted to observing public activity, while the initial, automated keyword searches reach "into" messages the sender and receiver presumably intended to be private. As already discussed though, the "virtually invasive" nature of these collection methods alone do not necessarily amount to rights infringements. We will now consider the second stage of keyword search collection involving analysis of intercepted data by intelligence analysts.

Keyword search collection may provide enough information to amount to the first criterion of a rights infringement by the time a human being reads the intercepted communication. The analyst may listen to a conversation or read an email or text of sufficient length to give context to the suspect phrase and reveal sensitive information about the intelligence source. The automated program may also be programmed to collect a string of communications from the suspect source and so provide the analyst context in that way. In these events, the analyst still does not gain power over a specific person because the information is still likely associated only with a phone or ISP number (probably coded, at that) rather than a particular identified person. The analyst knows someone, somewhere, has been up to no good, but at this level, it is simply an account of actions without a connection to a specific person. From the analyst's perspective, the narrative would be indistinguishable from a dummy source—a copied passage from a spy novel or that unaddressed love letter found on the street—forwarded by his supervisor to test his analytical skills. Thus, keyword searches still do not involve rights infringements when the human analysis stage is included.

There will be a point with some collections, after a certain amount of aggregation, when the intercepted communications are disturbing enough to warrant further investigation into the source, utilizing all manner of investigative and intelligence collection techniques. At this point, other analysts and investigators likely become involved. Now all the criteria of a rights infringement are present: aggregated information is tied to a particular person and the information is read by human beings. To be clear, what constitutes an infringement is the investigation, which draws on, but goes beyond, the initial keyword search. We now need to consider possible justifications for this kind of rights infringement in order to determine whether SIGINT-prompted investigations are politically legitimate and so presumptively not rights violations.

Whether an action is a rights violation depends in part on the rights of the person affected by the action. The criminal or unprivileged irregular combatant[21] whose operational communications are intercepted does not have his moral rights violated because he lacks a right to contribute to criminal operations via those communications. Adversary military, privileged irregular combatant, or intelligence personnel have a right to discuss their operational plans with colleagues, since (according to

---

[21] An irregular combatant is irregular in affiliation (belonging to a nonstate group) and/or tactics (using guerilla rather than conventional military tactics). An unprivileged irregular is one who fails the criteria for moral and lawful belligerency: obeying a unified chain of command, carrying one's arms in the open, wearing identifying emblems, and obeying the laws and customs of war.

the traditional post-Westphalian just war tradition), these professionals do nothing legally or morally wrong in pursuing the national security goals of their states or nonstate entities. Yet since their adversaries have the same right to pursue the national security goals of their own states, those adversaries can engage in strategic behavior such as intercepting their enemy's communications.[22] The targeted service member or intelligence officer, therefore, is not wronged by having his operational communications intercepted. Further, assuming that the operationally significant information collected in the data-mining operation regards state secrets, foreign security personnel do not suffer personal privacy violations when their communications are intercepted any more than soldiers whose rifles are taken by the enemy suffer private property right violations.

Clearly, the cases of concern with keyword searches are the false positives, cases where the communications of innocent people are collected when their out of context remarks trigger automated collection. We have to focus on these false positives rather than legitimate intelligence "hits" if we are going to assess the political legitimacy of keyword searches, since even grossly inefficient and brutal tactics like arbitrary arrest and torture can occasionally stumble across a legitimate intelligence source. Rights-infringing investigations of suspects can be justified when they meet the security standard of being the practically best and least rights-infringing tactics locally available. A certain error rate is in principle permissible since security officials would not be doing their job if they only investigated known threats, to the exclusion of anticipating future threats. So intelligence agencies must likely, and may in principle, engage in some kind of collection from *suspected* intelligence sources in order to meet their mission of contributing to national security. The mandate to pursue suspected intelligence sources entails that some innocent people will be targeted.

So we need to consider if keyword searches and SIGINT-prompted investigations can meet the security standard of being the practically best and least offensive to targets' rights locally available. We will first consider how the tactic measures up in terms of deferring to targets' rights. Scholars lacking security clearance are somewhat hampered in considering this question because they are ignorant of other possible types of modern SIGINT. As already argued, the automatic collection and initial human analysis phases of the tactic do not amount to rights infringements. Keyword searches and SIGINT-prompted investigations do compare favorably in terms of rights deference compared with older methods of collection, like steaming open envelopes or tapping phone lines, in that these methods identify specific people and utilize human analysts in the first instance. Keyword searches are also more rights-respecting than human intelligence (HUMINT) collection designed to accomplish the same goal of identifying suspicious communication the state might

[22] See Skerker, *Ethics of Interrogation*, chap. 7.

not otherwise know to seek. HUMINT is morally fraught given that it often involves the corruption of the asset, the suborning of disloyalty, deception on the part of the recruiter, and great danger to both the asset and the recruiter. In the absence of clearly preferable alternatives and given the noninfringing nature of the first two phases of the tactic, I will tentatively say that keyword searches and associated investigations are sufficiently consent-worthy to meet the rights element of the security standard.

Having addressed the rights-respecting aspect of the security standard, it remains to be considered whether keyword searches meet the security standard's practical aspects of efficacy, reliability, efficiency, and proportionality. Regarding the program's efficacy, it has to be considered that imagery analysis—the analysis of imagery collected by satellites or reconnaissance aircraft—cannot substitute for signals intercepts since communications about military or terrorist operations are not necessarily accompanied by simultaneous physical actions. HUMINT can provide the same kind of behind-doors information about targets' communication, but is less efficacious than signals intelligence in several respects. HUMINT can be expected to be resource-limited compared to SIGINT since developing human sources is labor- and time-intensive and not consistently fruitful. It may be extremely difficult to cultivate human intelligence assets in all the locations one desires because certain government programs or installations employ small numbers of dedicated, highly vetted, highly monitored people intelligence officers will have great difficulty locating, much less "turning." Some states or groups are so isolated from the outside world that penetration by intelligence officers is all but impossible. By contrast, any of these selective, secretive, isolated groups, installations, or states are potentially vulnerable to signals intelligence collection.

Further, even for highly competent, well-funded agencies, the scope of HUMINT operations is limited by prior intelligence collection. Intelligence agencies only know to try to cultivate or collect from assets associated with adversary organizations or installations with which they are already familiar. The comparative benefit of wide-scale SIGINT is that it can alert agencies to threats they did not know were germinating.

The reliability of keyword searches is difficult to assess without disclosure of the types of searches conducted and the standard yield of useful intelligence they produce. Elsewhere, I argue that specific information about SIGINT capabilities and search terms should be classified lest whole avenues of intelligence collection be shut down by adversaries.[23] Therefore, I am forced to compare keyword searches conducted by intelligence agencies with those performed by civilians using Google and the like. Assuming that there are actually a tiny number of intelligence targets whispering into their satellite phones about nefarious things relative to the total human population, the comparison with civilian search engines (e.g., imagine searching for someone with a common name) suggests that intelligence searches would yield relevant intelligence along with a huge volume of false positives. Given the amount

---

[23] Skerker, "Moral Foundation for Government Secrecy."

of time and manpower that would be necessary to analyze every initial intercept containing a suspect word or phrase, one imagines that further automated filtering occurs prior to a human analyst seeing any collected information, including longitudinal collection of other intercepts from the same source and cross-referencing with geographic areas of interests. Thus, it is reasonable to assume that an exponentially smaller number of false positives reach the desks of analysts than are generated in the original collection. Finally, it would seem that this filtering coupled with human analysis is a reasonably reliable method of selecting targets for more focused investigation. While human analysts no doubt miss the significance of certain messages in which targets speak in unfamiliar codes, or think some innocuous conversation is laden with code words, it is hard to imagine a more reliable method of analyzing communications than to have trained analysts read them.[24]

There are also difficulties assessing the efficiency of keyword searches without access to the classified details of such programs. While it is not possible for someone without access to the relevant classified information to know if there are more efficient contemporary techniques available, it is possible to make some speculative comparisons between keyword searches and known historical alternatives. It must be more efficient for supercomputers to analyze data in milliseconds than have corps of analysts steaming open envelopes or listening to phone calls in real-time. It is also reasonable to assume that the keyword searches conducted by the best-funded intelligence agencies are of the most or nearly most efficient types currently available because efficiency (the rate at which collected intercepts can be analyzed and flagged for security-sensitive information) is something readily measurable by engineers employed by the agencies and by the inspectors who oversee the engineers' work. The gap between collected and analyzed information is a knowable quantity and, presumably, a matter of concern to intelligence analysts. Given presumed institutional interests in ever-increasing efficiency and incentive structures for engineers pegged to customer demands, it can be assumed that well-funded agencies would be able to get the most efficient collection methods.

It is no surprise that the values relevant to a proportionality calculation are also hazy, though not quite to the degree as to reliability and efficacy. In this case, proportionality has to be assessed in two stages. The potential good done by the program has to be considered by assessing the evils avoided, in other words, the scope of the

---

[24] To compare the reliability of this form of SIGINT with one form of HUMINT, it strikes me that there would be fewer false positives with this sort of collection than produced in interrogations. The analyst at NSA or GCHQ or a similar agency likely has an overwhelming number of noninvidious intercepts come across his or her desk every day. Given the relatively "dumb" nature and global span of collection, the analyst presumably expects to get far more false-positives seized by computers than actual intelligence sources. By contrast, interrogators might well be prey to confirmation bias, particularly with suspected High Value Targets whom the interrogator reasonably assumes was only interdicted after thousands of man-hours of analysis, investigation, and tactical planning.

threats posed by the state's adversaries. Then, the efficacy of the proposed program has to be assessed in order to know what percentage of the maximum potential good done can be theoretically accomplished. With respect to the first stage, applying the security standard in the case of intelligence collection is harder than applying it to preparations to meet a concrete threat, such as the threat of muggers in a particular neighborhood, because the "good done" element of the proportionality concern is undefined. It is difficult to know if one's security establishment is overdoing intelligence collection without knowing about the threats that are currently germinating. Yet this knowledge is only ascertainable through intelligence collection. That said, some crude estimates are possible: small, resource-poor countries with minimal international concerns (shipping, foreign bases, etc.) likely face fewer national security threats than large, wealthy, internationally involved countries. Countries that have been at peace for decades presumably face fewer threats than those engaging in antagonistic international actions or those hearing sustained, plausible threats from states or nonstate groups. Surinam or Bhutan, for example, probably face far fewer threats than say, Iran or Israel, and so are less able to justify a significant SIGINT collection capability. Still, due to the need to anticipate future threats, the security standard will justify collection efforts somewhat disproportionate to current, known threat levels.

Regarding the second stage of a proportionality assessment, selective disclosures by intelligence agencies as well as leaks present an ambiguous picture of efficacy of keyword searches at forestalling terrorist attacks or frustrating enemy military maneuvers. The public does not know about every counterterrorist or other type of military operation and does not know how many are predicated on key signals intercepts. And the public does not know how many intercepts relevant to security concerns were incorrectly analyzed or analyzed too late. No one knows the ratio of intercepted communications to the total number of security-sensitive communications sent between actors. Even if the "good done" portion of the proportionality calculation was clearly known, the calculation would still involve ambiguity because it involved a comparison of different values, say thousands of people's communications intercepted and read every year compared to a few thousand lives saved per year on account of frustrated terrorist plots or disrupted military maneuvers. Still, since keyword searches are minimally invasive and non-rights-infringing forms of patrol, I am inclined to think that in a state facing significant national security concerns, the proportionality calculation would favor the prospect of saving human lives even if the number of communications intercepted increased exponentially. I suspect the same holds true for SIGINT-prompted investigations amounting to justified rights infringements in just liberal states—when the process of filtering and analysis determining whether a SIGINT target is disclosed to human analysts meets the security standard itself. In other words, SIGINT-prompted investigations would fail the proportionality test if the filtering process was so lax that huge numbers of innocent people were subjected to detailed surveillance and investigation. Keyword search programs would also fail the proportionality element in low-threat environments if there was no way to ensure good behavior on the part of analysts—since analysts collecting or reading intercepts for puerile reasons would violate targets' rights.

In sum, keyword searches and SIGINT investigations are designed to gather information other imagery collection and HUMINT cannot; the programs appear to be more reliable than conceivable wide-scale collection alternatives; it is reasonable to assume that the relevant programs conducted by the few states with the resources to engage in them are fairly efficient; and the programs can be proportional when professionally run in states facing significant national security threats. I stressed above that many of my practical assessments are tentative given the secrecy surrounding the relevant programs and possible SIGINT alternatives. However, I have met my goal of establishing an abstract framework into which empirical details can be added. A particular program will fail to meet the security standard, for example, if it has a very high false positive rate or if its analysts are poorly trained; if the state has little realistic need for such a program;[25] and when practically better and more rights-respecting programs become available.

Metadata Analysis—There are many different types of metadata analysis, performed by entities ranging from commercial firms to political campaigns. Certain intelligence agencies have recently acknowledged gathering "telephony" metadata from telecom companies. This type of intelligence does not reveal the content of communications but the time, duration, and phone numbers or ISP numbers involved. One possible use would involve mapping the social network of a suspect drawing on stored tranches of metadata seized for the entire population in a region and gathered years prior to when anyone in the tranche was identified as suspicious. In what follows, I will first address concerns regarding collection and storage of metadata generally, and then focus on particular concerns with the government collecting and storing it.

A metadata map of all the communications in a particular region is not greatly sensitive unto itself. Modern communication technology means that communication patterns are invisible to the naked eye. It does not follow that individuals enjoy a legitimate expectation to avoid appearing to others as part of a telecommunication pattern. It will be helpful to draw an analogy with visible communication patterns to defend this claim.

Imagine a rookie policeman manning his post at the main intersection of a small town on the Canadian-US border in the early twentieth century. From his post, he observes the pattern of American and Canadian neighbors going from house to house to visit one another. He notes their pattern of communication including

---

[25] Programs that are weaker on the practical or rights-respecting fronts may be justifiable in states facing grave threats.

the origin and destination of communicators without knowing the content of their communication or the identity of the communicators (being new to the job and the town, he just notes general information like gender, age, height, hair color, etc.). For their part, there is no reasonable expectation on the part of the citizens not to be seen by others when they leave their houses in order to communicate with their neighbors. For the policeman's part, patrol by authorities is permissible, as discussed above, and a policeman's noting of a *pattern* in human traffic is merely an extension of patrol, in which discrete moments of the patrol day are linked together in the policeman's memory. The situation does not change if a machine employed by the police notes the pattern of communication. A machine's "memory," and perhaps its observation, will be vastly better than the patrolman's, but the citizens are still not specifically identified as individuals in such a pattern (machine memory will be discussed further below). Instead, they are just data points whose notable feature is taken from the pattern-noter's agenda. An observed person appears in the pattern as a "caller," "redhead," "pedestrian," "motorist," and the like. One does have a claim to control mundane pieces of information about oneself, but the pattern-relevant pieces of information are not being traced back to a unique person at this stage. So at this stage, so limited, the observation does not infringe on a person's privacy. The pedestrians have no right to demand that they not be objectified, in a sense, as a data point (e.g., as a pedestrian, voter, redhead, etc.) for someone else's observation or research. Again, it is worth emphasizing the distinction between patrol and surveillance. While there is no reasonable expectation against being noted as part of a pattern, there is a reasonable expectation against being surveilled (e.g., being followed all day), and in the course of this action, one's itinerary and associations being catalogued by a spy.[26] If being noted as part of a pattern is not rights infringing, we will address concerns related to the ways in which a machine-stored pattern could be later used below.

One cannot expect to go unnoticed when one visits another person's house in a populated area. So one arguing for the great sensitivity of telephony metadata needs to argue that patterns of communication become especially sensitive when the pattern is a product of technology permitting people to communicate without traveling from door to door. One needs to argue that the expectation of anonymity associated with such technology amounts to an expectation of privacy.[27] To be clear, one can be accustomed to anonymity, because of the nature of the technology, without that fact creating a normative expectation of privacy. There is an argument to be made that some people may reasonably expect privacy regarding the destination of at least their local calls when they choose to call instead of leave their homes to visit their

---

[26] The ACLU appears to conflate knowledge about associations garnered through patrol and surveillance in their June 11, 2013 complaint in the Southern District of New York District Court. *ACLU v. Clapper*. https://www.aclu.org/national-security/aclu-v-clapper-legal-documents.

[27] The distinction is George R. Lucas's.

interlocutor. One can certainly imagine instances when a caller does not want even the destination phone number revealed to certain parties. The phone number of a man's mistress (recognizable to the philanderer's wife) or the widely advertised number for a phone sex company, abortion clinic, or local welfare office could be sensitive in this manner. Yet first, obviously, these are exceptional cases; and second, they are moot when the person does not have a realistic option of physically visiting the recipient of their communications. If the person has no alternative but to write, call, or email, we cannot assume that the sender wants to keep the recipient's address secret merely by virtue of the fact that he uses technology making his communications invisible to the naked eye.

We still have to substantively determine if telephony metadata is so sensitive that we can expect its collection would amount to a privacy infringement. I will argue that is not sensitive to this degree. One should understand that telecom companies have access to communication metadata, and one presumably endorses their automated monitoring of Internet and phone traffic to prevent the overloading of servers or other technical glitches.[28] True, people rely on telephones and the Internet, and so may only grudgingly tolerate telecom companies' possession of metadata. Yet the toleration of telecom companies having this information also suggests that metadata is not seen as all that sensitive. We can imagine that people would not use phones or the Internet, or would demand the immediate disposal of metadata, if use of communication technology or storage of metadata was seen as a violation of privacy on par with the public exposure of one's sexual habits or health information.

Yet one might object that a telecom company's possession of metadata is very different from a government's possession of metadata. We may understand that phone companies know the numbers we have been calling—just like we know our doctors have our health records—but also assume that phone companies do not use that information for invidious purposes.[29] While tacit consent legitimates the telecom companies' possession and analysis of metadata for the purpose of facilitating communication, the objection continues, intelligence agencies have a completely different interest in analyzing metadata.

Granting that exposing information to one party does not imply consent to universal disclosure,[30] an intelligence agency's gaining access to an arena without the target's explicit consent is not inherently problematic. The security standard, for example, justifies the state gaining access to normally private material such as one's communications, home, and possessions through a warrant process. The challenge in justifying broad metadata collection and storage by the government is that the target is not a suspicious party or clearly identified adversary state agent. Instead, the

---

[28] The US Supreme Court argued that people did not have a reasonable expectation of privacy regarding telephone metadata for this reason in *Smith v. Maryland*.

[29] This argument exposes the shallowness of the Court's reasoning in *Smith v. Maryland*.

[30] Nissenbaum, "Protecting Privacy in an Information Age," 585; Solove, *Digital Person*, 43.

target set is a broad swath of the population, collected against with the thought that someone in that group might be contacted by a foreign security threat in the future. Contrary to due process, evidence collection occurs prior to the identification of a suspect.[31] I will reply to this objection along with the "right to be forgotten" below.

To this point, we have seen that merely collecting metadata about physical movement (presumably associated with communication) is not rights infringing and that we cannot assume a different assessment when it comes to collecting telephony metadata. There is evidence that people do not see their telephony metadata as especially sensitive. We have yet to address due process concerns with a government collecting it. Before addressing those concerns, it will be helpful to entertain a critique of metadata collection focusing on the long-term retention of metadata as the problem rather than the inherent sensitivity of the information. This critique goes along with the due process critique of collection since data that is collected but not stored can hardly be used as evidence in contravention to due process.

Such retention, one might charge, trespasses against a reasonable expectation of the ephemerality of one's communications and associations.[32] If one has a right to define oneself to strangers, then one's actions, communication, and associations should be allowed to dissolve into the past, as it were. One would take from others the chance to define and present themselves if one kept a record of others' every statement and association, even if one did not consult the data in real-time but merely stored it for possible future analysis. Not only would retrospective analysis challenge a person's real-time self-definition, but the threat of one's interlocutors checking the data would likely inhibit one's present interactions, associations, and self-definition.

Yet there cannot be a generalized expectation against the retention of past behaviors and communication—an expectation that one's behavior dissolves into the ether—because one cannot make demands, practically or morally, on another person's memory.[33] Noticing patterns is essential to learning. Noting and remembering patterns of a person's behavior or speech might be essential to realize that a person is manipulative or untrustworthy. Further, observing patterns of past behavior or speech can lead a person to be deferential to a friend's sensitivities or to anticipate her needs.

Is there then a legitimate, *special* expectation against computerized capture and retention of past behaviors and communication? Critics of metadata collection by

---

[31] One of the American Civil Liberties Union's concerns is that absent a warrant, the collection of metadata by the NSA amounts to an unreasonable search. See legal documents relevant to *ACLU v. Clapper* at https://www.aclu.org/national-security/aclu-v-clapper-legal-documents.

[32] The European Parliament passed a digital privacy law March 12, 2014 codifying a right to be forgotten. The law empowers citizens to force Internet companies to remove personal information from their servers.

[33] See Nissenbaum, "Protecting Privacy in an Information Age," 572.

intelligence agencies are not imposing demands on other people's memories but worried about digital storage. One arguing against metadata storage could point out we do not mind being passively observed by strangers in public because (a) there is an expectation that they will not ogle us and try to log every detail they see and (b) because we expect that they will not remember whatever they see for long. On one hand, people have a right to their memories; and on the other hand, those memories generally are not a threat to strangers. By contrast, automated data collection gets every pixel and syllable and can store that information for a near eternity.[34]

While this kind of record could be made for nefarious purposes, and could be abused even if it is made for good purposes, I do not think a near-permanent stored record poses an inherent normative problem. I will explain this point and then address the issue with potential abuse. Imagine an alien race interested in human beings for purely academic purposes. The aliens create machines that photograph and record humans' every moment. This information is studied merely to further the aliens' understanding of human beings and help alien academics make tenure. The aliens' creed forbids their interference with species on other planets. The aliens have sterling information assurance systems so that the collected information can never be leaked or used for purposes other than this kind of benign research. Neither privacy concern is in play here if the moral importance of controlling one's own personal information stems from the importance of choosing how to present oneself to others and protecting oneself from certain kinds of harm. The mere existence of a minute record of one's activities that will never be viewed by humans is not a problem. This leaves concerns about how the information will be used, and provided some justifiable use, how susceptible that benign process is to abuse. I will turn directly to these questions now and also address the due process question outlined above.

The security standard is a framework for identifying politically legitimate government coercive actions—government actions meant to protect inhabitants through consent-worthy means. The standard can take into account the risk of abuse of processes in its proportionality calculation. The collection and storage of domestic metadata for the purpose of retrospective analysis of networks associated with particular foreign suspects appears to be a good candidate to pass the security standard. I will focus first on the rights-respecting element of the standard. First, there is no generalized right not to be seen as part of a pattern, as argued above, nor a generalized right not to be remembered as part of a pattern. Second, the collection of metadata does not trespass on very sensitive areas. Collection does not reveal the specific content of communications; particular people are not identified; and, presumably, a computer algorithm (rather than a person) collects and organizes the metadata. Third, the means of acquisition of the metadata is not burdensome to the public. No one's daily behavior is disturbed through metadata collection.

[34] Ibid., 576.

Storage and retrospective searches of metadata also meet the practical elements of the security standard. As with targeted searches, retrospective searches of stored metadata would likely score well in terms of reliability and efficacy. Storing metadata increases the efficacy of metadata searches because collector algorithms now have more data to sift, promising more opportunities to find contacts of foreign security threats. That said, the practice of storing terabytes of metadata with the idea that a foreign security threat might have used the relevant telecom networks sometime in the last few years is terribly inefficient, like storing all the hay in the world in case someone realizes he lost a needle in a bale a year or two back.[35] Yet there may not be more efficient methods. The numbers called on a particular line can be efficiently collected in real time with a pen register, but this collection does not help to recover the calls made in the past. The most efficient method for retrospective investigation is hardly reliable or even possible in many cases: that is, asking the suspect whom he contacted regarding a covert operation. Regarding proportionality, the good of being able to see whom a foreign security threat contacted domestically seems very important and the harm of data collection is slight unto itself. However, a significant concern on this point is not simply the collection for the purpose of possible future analysis regarding foreign security threats, but the possible use by a government for ordinary domestic law enforcement or political oppression. We will return to the question of proportionality below, taking into account this risk.

One can envision metadata analysis becoming as common a response to arresting a criminal suspect as the examination of fingerprint records or arrest records is now. Again, the concern is that unlike the collection of fingerprints, metadata collection would violate due process by collecting evidence prior to reasonable suspicion of a crime occurring.[36] Further, while fingerprints can reveal whether one was present at other crime scenes, metadata coupled with subsequent database searches can do more in providing a fairly full biographical sketch of a person. This critique stands even though the information collected is not very sensitive and the mode of collection is noninvasive, because it violates what some consider the proper relationship of the liberal state and its inhabitants. In the words of US Supreme Court Justice Arthur Goldberg, the government should "leave the individual alone until good cause is shown for disturbing him."[37] In order to maintain the general respect for the individual, the state should maintain such a distance that it has to labor when it comes time to prove guilt in the criminal justice arena. It should not be in a constant state of laying the groundwork for every inhabitants' future prosecution. Regarding concrete effects, knowledge that one's metadata could be used in future criminal

---

[35] The analogy is one used by Inglis, "Remarks."

[36] It should be noted that some US states collect fingerprints of anyone who works with children, expressly for the purpose of tracking them if they are later accused of kidnapping or molestation. This practice would seem to offer fertile grounds of comparison with metadata storage.

[37] *Murphy v. Waterfront Commission.*

prosecution could have a chilling effect on one's innocent communications with people one suspects might be involved in criminal behavior and could certainly chill journalistic work or the sort of advocacy work that involves working with at-risk youth or ex-offenders. Some might harbor even graver fears that newly criminalized behavior could lead to retroactive prosecution via metadata analysis or that people could be persecuted for legal but politically unpopular views.[38] So a proportionality analysis has several significant risks to weigh against the possible counterterror or counterintelligence good done by metadata collection.

On these proportionality-relevant points, slippery slope arguments have to be tethered to a realistic assessment of risk, lest they be permitted to overturn every otherwise coherent argument and invalidate every policy prescription. I am sensitive to the argument that it is just plain unsettling that metadata might be stored by corporations or governments that can be used in combination with other archived information to develop detailed portraits of our lives. The concern is not necessarily over the inherent sensitivity of the information (all or most of it may be known by our relatives and friends) but the uncertainty of how it might be used by less trustworthy parties. In some states, the risks of metadata abuse may be high enough to make the possible harm both more likely and more costly than a terrorist attack or military or intelligence operation facilitated through contact with a domestic inhabitant. In other states, with strong histories of rule of law and cultures of responsible public service, the proportionality calculation may be resolved in the other direction. I do not think metadata collection and storage can be declared absolutely beyond the pale or absolutely essential to national security. Metadata collection offers its collectors a power promising certain benefits and bearing certain risks of abuse. Rather, I want to identify some of the possible risks and benefits associated with such a program and create a framework in the security standard for determining whether a metadata collection program should be implemented in a particular state given the balance of risks and benefits.

A practical solution addressing the due process and slippery slope arguments—and making us more confident of the proportionality of metadata collection and storage—would be for the government to refrain from collecting metadata and instead to require telecom companies to retain metadata for a set number of years.[39] The government could then obtain a warrant to collect the relevant metadata if it had a reasonable suspicion of a person. This maneuver appears to be in line with current due process procedures in some liberal states in which data more sensitive than metadata, like the content of calls and emails, can be obtained with a warrant. This solution no more admits of overreach by a government than any other

---

[38] Solove's main concern with digital dossiers is that they could be abused by an incompetent or oppressive state (*The Digital Person*).

[39] President Obama proposed a similar plan while this chapter was in press.

warrant process. Further, the proposed regime would only extend current practice, at least in the United States, since telecom companies are currently required to retain metadata for a period of time for regulatory purposes.

Conclusion

This chapter analyzed two types of SIGINT in reference to a moral framework based on respect for human autonomy. The security standard understands certain types of government coercion to support human autonomy by fostering an environment relatively free of rights violations. The standard assesses tactics by seeking a balance between achieving a positive security outcome and deferring to inhabitants' rights. The standard provides a framework for assessing whether the two types of SIGINT are justifiable in particular settings. The two forms of SIGINT may or may not be justifiable given the threats the state faces, the available technological alternatives, the professionalism of the analysts administering the program, and the justness of the government directing the program. Given the reflexivity built into the security standard, one can make an argument for the general, but not universal, permissibility of the two tactics. While a given tactic might pass the security standard on first analysis, the reflexive element of the standard permits adversaries to respond with a tactic that best approaches an in-kind response, even if the best they can muster is more rights-infringing than the tactics used by the more technologically advanced state. Thus, in particular concrete cases, an otherwise permissible intelligence-collection tactic may not be consent-worthy if the relevant adversaries' tactics are more rights-infringing than inhabitants in the first state could tolerate.

Bibliography

ACLU v. Clapper. S.D.N.Y. filed June 11, 2013. https://www.aclu.org/national-security/aclu-v-clapper-legal-documents.

Alfino, Mark, and G. Randolph Mayer. "Reconstructing the Right to Privacy." *Social Theory and Practice* 29 (January 2003): 1–10.

Apel, Karl-Otto. *From a Transcendental Semiotic Point of View*. New York: St. Martin's Press, 1999.

Bamford, James. *The Shadow Factory*. New York: Doubleday, 2008.

Benn, Stanley. "Privacy, Freedom, and Respect for Persons." *Nomos* 13 (1971): 1–26.

Bok, Sissela. *Secrets*. New York: Vintage, 1989.

Cohen, Jean L. "Equality, Difference, Public Representation." In *Democracy and Difference*, ed. Seyla Benhabib. Princeton: Princeton University Press, 1996.

Gross, Hyman. "Privacy and Autonomy." *Nomos* 13 (1971): 169–81.

Habermas, Jurgen. *A Theory of Communicative Action*. Boston: Beacon, 1985.

Inglis, John. "Remarks." Presented November 22, 2013 at the Center for Ethics and the Rule of Law Conference, University of Pennsylvania Law School, Philadelphia, PA.

Kant, Immanuel. "On the Proverb: That May be True in Theory." In *Perpetual Peace and Other Essays*, ed. Ted Humphrey, 61–92. Indianapolis: Hackett, 1983.

McCloskey, H. J. "Privacy and the Right to Privacy." *Philosophy* 55 (1980): 17–38.

Murphy v. Waterfront Commission. 378 US 52, 55 (1964).

Nagel, Thomas. *Equality and Partiality*. New York: Oxford University Press, 1991.
Nissenbaum, Helen. "Protecting Privacy in an Information Age: The Problem of Privacy in Public." *Law and Philosophy* 17 (1998): 559–96.
Nozick, Robert. *Anarchy, State, and Utopia*. New York: Basic Books, 1977.
Rawls, John. *A Theory of Justice*. Cambridge, MA: Belknap, 1971.
Reiman, Jeffrey. "Privacy, Intimacy, and Personhood." *Philosophy and Public Affairs* 6, no. 1 (1976): 26–44.
Scanlon, T. M. *What We Owe to Each Other*. Cambridge, MA: Belknap, 1998.
Simmel, Arnold. "Privacy Is Not an Isolated Freedom." *Nomos* 13 (1971): 71–87.
Skerker, Michael. *An Ethics of Interrogation*. Chicago: University of Chicago Press, 2010.
Skerker, Michael. "A Moral Foundation for Government Secrecy." Presented November 23, 2013 at the Center for Ethics and the Rule of Law Conference, University of Pennsylvania Law School, Philadelphia, PA.
Smith v. Maryland. 442 U.S. 735 (1979).
Solove, Daniel. *The Digital Person*. New York: New York University Press, 2006.
Stramel, James E. *Same Sex*. Lanham, MD: Rowman & Littlefield, 1999.
Waldron, Jeremy. "Theoretical Foundations of Liberalism." *Philosophical Quarterly* 37, no. 147 (1987): 127–50.
Waldron, Jeremy. "Special Ties and Natural Duties." *Philosophy and Public Affairs* 22, no. 1 (winter 1993): 3–30.
Westin, Alan F. *Privacy and Freedom*. New York: Atheneum, 1967.
Van Den Haag, Ernest. "Definition: The Nature of Privacy." *Nomos* 13 (1971): 149–68.

Name Index

Abney, Keith, 94n26, 109n68
Adam, Tanja C., 169n30
Adee, Sally, 79n24
Albert, Stuart, 118n13
Albright, David, 182n12
Alfino, Mark, 257nn9–10
Allen, Anita L., 152, 152n50
Allende, Salvador, 237
Allhoff, Fritz, 5n18, 9, 37n10, 91n9, 193n56
Andres, Richard B., 183n15
Andress, Jason, 177n3
Anselm, Saint, 89n1, 92n19
Antoniou, Chris, 119n17
Antonius, Daniel, 162n9
Anwer, Jared, 231n11
Apel, Karl-Otto, 252n3
Appelbaum, Jacob, 211
Aquinas, Thomas, 57, 148n35
Arendt, Hannah, 172, 172n39
Ariely, Dan, 184n19
Ariely, Gil, 166n24
Aristotle, 24, 122, 122n27, 218, 218n58
Arquilla, John, viiin4, xii, 6, 6n19, 14n2, 28n35, 56n2, 189n42, 202n6
Augustine, Saint, 203, 207

Baase, Sara, 131n53
Bachrach, Judy, 231n11
Bamford, James, 261n20
Barboza, David, 190n43
Barrett, Edward, 19, 19n17, 25n32, 29, 109n68
Baxter, Richard, 203n8, 209n30
Beard, Matthew, 7, 8, 139, 140n6
Beaton, Alan, 166n20
Beauchamp, Tom L., 63n21
Bekey, George, 109n68
Bellamy, Alex, 203n9
Bellinger, John B., III, 48n80
Benbaji, Yitzhak, 100n46

Benn, Stanley, 257n9, 257nn11–12, 258nn15–16
Ben-Porath, Yossef, 169n29
Berson, Thomas, 3n4, 4n8
Betz, David, 125n33
bin Laden, Osama, 230
Bitterman, Haim, 171n36
Blackler, John, 240n37, 244n46
Bleich, Avi, 164n17
Bleich, Avraham, 164n17
Boemeke, Manfred, 118n11
Bok, Sissela, 257n11
Bonanno, George A., 171n35–36
Boonin, David, 98n40
Booth, Kenneth, 216n49
Boothby, William, 206, 208–210, 208n28, 209n29, 210n32
Boucher, Richard, 38n18
Bradley, Ben, 98n40
Bradley, Gabriel, 142–143, 143n11
Brannan, Paul, 182n12
Braun, Megan, 57n7, 65n27, 91n9
Brenner, Joel, 19–20, 20n20
Brenner, Susan, 127n37
Bridoux, Jeff, 123n29
Broad, Charlie Dunbar, 92n19
Broad, William J., 183n14
Brown, Barbara B., 166n20
Brunstetter, Daniel, 57n7, 65n27, 91n9
Bumiller, Elisabeth, 37n10
Burne, Alfred H., 81n32
Burstein, Paul, 171n37
Bush, George W., 51n93
Butler, Amy, 79n23

Campbell, John P., 184n20
Canetti, Daphna, 7–8, 157, 168n27, 169n32, 170n33, 232n15
Canetti-Nisim, Daphna, 162n10
Carnelley, Katherine B., 171n34

Name Index

Carr, Jeffrey, 26n33, 126n35
Carrubba, Clifford J., 222n74
Chatterjee, Deen K., 26n33
Chen, Thomas M., 97n31
Chertoff, M., 61n16
Chesterman, Simon, 129, 129n47
Chien, Eric, 97n35
Childress, James F., 63n21
Choi, Incheol, 194n59
Christopher, Paul, 57n5, 64n24
Chu, Beong-Wan, 166n22
Cimbala, Stephen, 118m13
Clark, David, 3n4, 4n8, 180n7
Clarke, Richard A., 19–20, 20n20, 26–27, 115n3, 211, 211n34
Clausewitz, Carl von, 2, 2n3, 3, 76–77, 76n7, 77n9, 80, 84n37, 85, 86, 86n41, 86n43, 91–92, 91n10, 91n13, 92n17, 92n21, 177n2
Clinton, Bill, 1
Clover, Charles, 14n3
Coates, A. J., 238n33
Cohen, Jean L., 257n9
Conner, Mark, 169n30
Cook, James, 58n8, 60
Cook, Martin, 128n43
Cornish, Paul, 177n3
Corrin, Amber, 61n15

Dam, Kenneth, 6n19, 56n2
Danks, David, 8, 177, 178n4, 187n30–32, 188n35, 236n27
Danks, Joseph H., 8, 177, 178n4, 187n30–31, 188n35, 236n27
Darley, William M., 91n14, 92nn16–18, 92n21
Darwish, Ragy, 163n13
Davis, David R., 218n59
de George, Richard, 146, 146n31
Delbrück, Hans, 80, 80n26
Delgado, Mauricio R, 169n29
Denning, Dorothy E., 14n2, 103n54, 109n68, 151, 151n46, 192n52
Diamond, Gary M., 162n11
Dimassi, Hani, 164n15, 164n18
Dinniss, Heather Harrison, 40n36, 116, 116n4
Dinstein, Yoram, 46n64, 85n40, 201n1, 203nn11–12
Dipert, Randall R., ix, 5nn16–17, 7, 15, 15n6, 18, 26, 56, 56n3, 60, 61n16, 62n19, 64nn23–24, 70n38, 111n71, 128n43, 155n63, 180n6, 232, 232n16, 243n43
di Vitoria, Francisco, 148n35
Dobbins, James, 122n25
Doswald-Beck, Louise, 46n66, 47nn73–74, 47n79, 48n83, 49n87
Dunlap, Charles J., 15n5, 147, 147n34, 149n37, 153–154, 154n56

Eberle, Christopher, 19, 19n17, 25n32, 29, 231n12
Enemark, Christian, 155, 155n64
Epel, Elissa S., 169n30
Espiner, Tim, 14n3

Falliere, Nicolas, 97n35
Farajalla, Nadim, 163n13
Farhood, Laila, 164n15, 164n18
Farwell, James P., 97n31
Fattouh, Bassam, 163n13
Fearon, James, 219n63
Feaver, Peter D., 91n9, 201n5
Feinberg, Joel, 98n40
Feit, Neil, 99n45
Feldman, Fred, 98n41, 102n53
Fidler, David P., 191n49
Finn, Peter, 230n8
Floridi, Luciano, 6n20, 15–16, 16n7, 65, 124n31
Follath, Erich, 230n9
Ford, John C., 154n60
Ford, S. Brandt, 57n7, 91n9, 228n1
Ford, Shannon, 9, 13n1
Fortuna, Virginia Page, 219n63
Freidman, Allan, 210n32
French, Shannon E., 140, 140n3, 141–142, 141n7, 142nn8–10, 149, 149nn38–39
Friedland, Nehemia, 169n29
Friedman, Allan, 28n35, 115n1, 220n66, 221n71, 229n2, 235n23
Fulghum, David A., 79n23
Fuller, J. F. C., vii, viii

Garfield, R., 172n38
Gelkopf, Marc, 162n6, 164n17
Gellman, Barton, 211n37
Giles, Keir, 40n32
Gini, Gianluca, 166n23
Goldberg, Arthur, 273
Goldman, Jan, 65n28
Gordon, Ian, 235n26
Gottstein, Ulrich, 172
Gouin, R., 166n22
Goyal, Ravish, 107n64
Graham, David E., 28, 28n36
Graham, Jennifer E., 169n32
Green, Stuart P., 239n34
Greig, Michael J., 220n65, 223n76
Grimes, Roger A., 154nn57–58
Gross, Hyman, 258n14
Gross, Michael Joseph, 76n3, 127nn40–41, 128n42
Gross, Michael L., 7–8, 157, 232n15
Grotius, Hugo, 57, 90n7, 215n47
Gu Chunhui, 190
Guelff, Richard, 117n8
Guerrero, Alexander A., 188n37

Habermas, Jurgen, 252n3
Hampson, Fen Osler, 118n13
Hardin, R., 217n55
Harding, Luke, 229n5, 233n19
Hare, R.M., 23n28
Harman, Elizabeth, 98n42
Harris, Paul B., 166n20
Harris, Victor A., 194n57
Hart, B. H. Liddell, 201n4
Hastings, Max, viin2
Hayden, M., 61n16
Haynes, William J., II, 48n80
Heller, Kevin Jon, 181n8
Henckaerts, Jean-Marie, 46n66, 47nn73–74, 48n79, 48n83, 49n87
Henschke, Adam, 5n18, 9, 13n1, 67n30, 234n20
Herald, Chandos, 81, 81n34
Herring, George, viin3
Herzog, Stephen, 182n10, 183n16, 184n21
Hobbes, Thomas, 90n7
Hobfoll, Stevan E., 170n33
Hoff, Dianne L., 166n23
Holden, Dan, 14n3
Holder, Eric, 191
Horowitz, Michael, 220nn67–68, 221, 221nn69–70, 222
Howie, Luke, 163n12
Huang Zhenyu, 190
Hume, David, 22, 22–23n28
Hunker, Jeffrey, 180n7
Hussein, Saddam, 118, 238
Hutchinson, Bob, 180n7

Inglis, John, 261n20, 273n35

Janoff-Bulman, Ronnie, 171n34
Jenkins, Ryan, viii, 5n18, 7, 18n16, 26n33, 68n32, 89, 90n8, 91n9, 97n31, 149n41, 153, 153n55
Jones, Edward E., 194n57
Jonsen, Albert R., 24n30
Jost, John T., 171n35

Kagan, Shelly, 102n52
Kahneman, Daniel, 184n19
Kanke, Robert K., 20n20
Kant, Immanuel, 23n29, 63, 69, 216, 216n48, 252n3
Karatzogianni, Athina, 125n32
Kaska, Kadri, 78n15
Keegan, John, xin9
Keinan, Giora, 169n29, 169n31
Klein, Gary, 177n2
Kleinig, John, 246n50
Klocksiem, Justin, 98n40
Knake, Robert K., 211, 211n34
Knox, Frank, vii
Koh, Harold Hongju, 42n43, 44, 44n54

Kolb, Joachim, 163n13
Kushner, David, 182n11, 192n53

Landau, Susan, 180n7
Landauer, Rolf, 90n8
Langner, Ralph, 97n32, 97n36, 182n11
Latimer, Jon, 201n4
Lazar, Seth, 102n53
Lee, J., 167n25
Leopold, Jason, 234n21
Leverick, Fiona, 110n69
Levi, M., 217n55
Li, Qing, 166n22
Libicki, Martin, 6n19, 14n2, 64n25, 127n38
Lieberthal, Kenneth, 6n19
Lin, Herbert S., 3n4, 4n8, 6n19, 15n5, 56n2, 104n57, 150, 150n43, 245n47
Lin, Patrick, 5n18, 9, 37n10, 109n68, 193n56
Lindgren, G., 164n14
Lindsay, Jon R., 211n36
LoCicero, Alice, 162n7
Loewenstein, George, 183n18, 188n34
Lowther, Adam B., 16n8, 22, 22n25
Lubell, M., 217n55
Lucas, Edward, 234n20
Lucas, George R., Jr., 5nn16–17, 6, 13, 21n23, 22n26, 28n35, 59n10, 64, 64n26, 87n45, 97n33, 109n68, 128n42, 152, 152n49, 235n25, 269n27
Luck, Edward, 118n13
Lynn, W. J., 126n35

MacIntyre, Alasdair, 23, 23n29
MacMillan, Margaret, 118m11
Madden, Mike, 205–206, 205n19, 206nn20–22, 213
Maguire, Mike, 166n20
Maness, Ryan, 211n34
Marchant, Gary E., 22n26
Margulies, Jonathan, 180n7
Martel, William, 118n12
Masri, Rania, 163n13
May, Larry, 148n36, 153, 153n52
Mayer, G. Randolph, 257nn9–10
McCloskey, H. J., 257n9
McConnell, M., 61n16
McDougal, Myres S., 47n69
McMahan, Jeff, 94nn24–25, 100n46, 110n69
Merkel, Angela, 219, 236–237
Messmer, Ellen, 61n15
Michel, Thorsten, 216n49, 217n51
Miller, J. G., 194n58
Miller, Seumas, 4n14, 8, 228, 230n7, 235n26, 240n37, 243n43, 244n46, 246n49, 249n36
Milson, Andrew J., 166n22
Mishler, Alan, 177n1
Mitchell, Sidney N., 166n23
Monaghan, Andrew, 40n32

Moore, Gordon E., 23n28, 25n31
Mullins, Michael, 235n24
Murchu, Liam O., 97n35
Murray, Williamson, 117n9

Nagel, Thomas, 257n9
Nakashima, Ellen, 183n16
Nardin, Terry, 217n56, 218, 218n57
Nazario, Jose, 76n6
Newman, Emily, 169n30
Newton, Kenneth, 216, 216nn49–50, 217nn52–53
Nisbett, Richard E., 183n17, 194n59
Nissenbaum, Helen, 257n13, 258n16, 260n18, 270n30, 271n33, 272n34
Norcross, Alastair, 99n45
Norenzayan, Ara, 194n59
Nozick, Robert, 253n4
Nuwayhid, Iman, 164n18

Obama, Barack, 214n43, 219, 274n39
O'Connor, Daryl B., 169n30
O'Donoghue, Ted, 183n18, 188n34
O'Driscoll, Cian, 119n15
Olson, Parmy, 167n26
Oman, Charles, 80, 80n27
O'Meara, Richard M., 20n22
O'Neal, John R., 218n59
Oppenheim, Lassa, 203, 203n10, 207
Orend, Brian, 2, 2n1, 7, 60, 64p24, 90n7, 94nn25–26, 99n46, 115n2, 117n8, 119n16, 120nn19–20, 121nn21–24, 122nn25–26, 123n28, 124n30, 128n45, 131n52, 132n54, 148nn35–36
Osiel, Mark, 142, 142n9, 242n39
Owens, William, 6n19, 56n2

Pace, Thaddeus W., 169n32
Page, Benjamin I., 171n37
Paletz, Susannah, 177n1
Pape, Robert, ixn5, 220n67, 222
Perlroth, Nicole, xn7, 50n88, 190n43
Perry, David L., 152n51, 154–155, 155n62, 236nn28–29, 237, 237nn31–32
Peterson, Andrea, 211n37
Petit, Philip, 216n49
Pfaff, Tony, 152, 152n48
Piazza, Pier V., 169n30
Pillar, Paul, 118n13
Plaw, Avery, 106n60
Pliny, 215–216
Popper, Nathaniel, 187n33
Porcelli, Anthony J., 169n29
Posen, Barry, 219n63
Powell, Robert, 222n74
Pozzoli, Tiziana, 166n23
Preston, Stephanie D., 169n29
Purves, Duncan, 95n29, 98n40
Putnam, Robert, 217n55

Quester, George, xi, xin10
Quinlan, Michael, 147n33, 148n35

Rabin, Matthew, 183n18, 188n34
Rabkin, Ariel, 59n10
Rabkin, Jeremy A., 59n10
Rathbun, Brian., 216n49, 217n51
Rawls, John, 101n49, 121n24, 252n3
Reagan, Ronald, xi
Regan, Patrick M., 220n65
Reichberg, G., 57n5
Reiff, David, 118n10
Reiman, Jeffrey, 260n17
Reisman, W. M., 119n17
Reiter, Dan, 220nn67–68, 221, 221nn69–70, 222, 222n74
Rid, Thomas, 3, 3n6, 4n9, 7, 19, 19n18, 25, 28, 59n11, 75–80, 76n1, 76nn4–6, 77n8, 77nn10–12, 78n14, 78nn16–19, 79nn21–22, 79n25, 83, 84n39, 85, 86–87, 104n56, 127n41, 149n40, 151, 151nn44–45, 153, 153nn53–54, 154n59, 192n53, 205, 205n18, 229n2, 230nn8–9
Roberts, Adam, 117n8
Roberts, Lynne D., 166n21
Rodin, David, 100n46, 223–224, 223n78, 224n79
Roff, Heather M., 8, 68n33, 68n35, 91n9, 155n61, 185n23, 189n38, 201
Rogers, M. K., 192n52
Rohan, Brian, 4n10
Rohozinski, Rafal, 97n31
Ronfeldt, David, viii, viiin4, xi, 14n2, 189n42
Roscini, Marco, 40n36
Rosen, Gideon, 188n36
Rosenzweig, Paul, 125n32, 177n3
Ross, Lee, 194n57
Ross, W. D., 57n6, 102n52
Roth, Andrew, 189n41
Rousseff, Dilma, 50n91
Rowe, Neil C., 14n2, 37n10, 56n2, 60n12, 68, 68n34, 69n36, 97n34, 106n61, 110n70, 180n6, 189n38, 193n56, 206, 207n24, 235n25
Rubel, W. R., 21n23
Russett, Bruce, 218n59

Saddam Hussein, 118, 238
Sanger, David E., 4n13, 20n21, 37n10, 61–62, 183nn13–14, 183n16, 190n43, 190n46, 192n51, 223n77, 229n3
Sawyer, J. P., 222n74
Scanlon, T. M., 253n4
Schmidt, Michael S., 4n13, 190n46
Schmitt, Michael N., x, 3n7, 5nn15–16, 6–7, 15n5, 16n10, 28, 28n36, 34, 36nn8–9, 41nn37–41, 42n43, 42nn45–46, 43n51, 43nn52–53, 45n58, 56n4, 60, 60n13, 83, 105n58,

116, 116n5, 125n34, 126n36, 128n46, 129nn48–51, 140nn4–5, 144nn14–20, 145nn21–26, 146nn27–30, 147n32, 151n47, 160n1, 161n2, 178n5, 189n39, 207, 231n10, 236n29, 242n40
Scholz, J. T., 217n55
Seligman, A. B., 217n55
Shackelford, Scott J., 183n15
Shakarian, Paulo, 128n44
Shanker, Thom, 61–62
Shapiro, Robert Y., 171n37
Shariff, S., 166n22
Sharp, Tracy, 165–166, 166n19
Sherman, William T., 64n24
Shiffrin, Seanna, 98n42
Shue, Henry, 121n24
Silver, Roxane Cohen, 162n8
Simmel, Arnold, 257n11
Sinclair, Justin, 162n5, 162n9
Sinclair, Samuel J., 162n7
Singer, Peter W., 6n19, 28, 28n35, 115n1, 210n32, 220n66, 221n71, 229n2, 235n23
Skerker, Michael, 4n14, 8, 242n39, 251, 252n1, 264n22, 265n23
Slantchev, Branislav, 222n74, 223n76
Smail, R. C., 80–81, 80n28
Smith, Alastair, 219n63, 222n74
Smith, Peter K., 166n22
Smith-Spark, Lauren, 219n61
Snowden, Edward, 20n21, 30n37, 152, 211, 229, 233–234, 251
Snyder, Glenn H., 220n67
Solis, Gary, 117n8
Solomon, Zahava, 164n17
Solove, Daniel, 257n13, 258n16, 261n19, 270n30, 274n38
Soltani, Ashkan, 211n37
Sourander, Andre, 166n23
Sparrow, Robert, 13n1, 109n68, 185n22, 186n29
Stam, Allan C., III, 219n63, 222n74
Stanley, E. A., 222n74
Stark, Holger, 230n9
Steinhoff, Uwe, 90nn6–7, 91n10, 154n60
Stevens, Tim, 125n33
Stramel, James E., 257n9
Strauss, Nicole L., 164n15, 164n18
Strawser, Bradley J., 9, 16, 87n44, 103n54, 106n60, 109n68
Stroll, Avrum, 57n6
Sun Kailiang, 190
Sun Tzu, 177n2, 201, 201nn2–3
Sutton, Willie, 26
Syse, H., 57n5

Taddeo, Mariarosaria, 6n20, 15–16, 16n7, 16n11, 65, 124n31
Tavernise, Sabrina, 189n41

Thomas, Hugh, xn8
Thomson, Judith, 98n40
Tiel, Jefferey R., 152, 152n48
Tikk, Eneken, 78n15
Tondini, Matteo, 123n29
Toulmin, Stephen E., 24n30
Treverton, Gregory, ixn6
Turing, Alan, 1

Valeriano, Brandon, 211n34
Van Den Haag, Ernest, 257n9, 257n12, 258n14
VanPelt, Dusty, 9
Verbruggen, J. F., 80–81, 80n28, 81n29
Vihul, Liis, x, 5n16, 6–7, 34, 45n58, 78n15

Waismel-Manor, Israel, 7–8, 157, 232n15
Waldron, Jeremy, 172–173, 172n40, 173n41, 252n3
Wall, Andru E., 189n40
Wall, Robert, 79n23
Wallach, Wendell, 13n1
Walrond, Christina, 182n12
Walter, Barbara, 219n63
Waltz, Kenneth, 64n24
Walzer, Michael, 63, 63n22, 64n24, 90n7, 91n9, 94n25, 99n46, 108n66, 116n6, 119n14, 119n16, 148n35, 239n36
Wang Dong, 190
Waterman, Shaun, 83n36
Waxman, Matthew C., 180n7
Wen Xinyu, 190
Werner, Suzanne, 219n63, 222, 222nn74–75
Westin, Alan F., 257n9, 257n11
Wheeler, Nicholas J., 216n49
Whetham, David, 3n7, 7, 19n19, 75, 81nn30–31, 81n33, 86n43
Whitehead, Alfred North, 22, 22n28
Wierenga, Edward, 92n19
Wilson, Timothy DeCamp, 183n17
Wingfield, Thomas, 15n5
Winterfeld, Steve, 177n3
Wolford, Scott, 222n74
Woods, Kevin, 117n9
Woolley, Anita Williams, 184n20
Wu, Xu, 127n40

Yadron, Danny, 61n15
Yannakogeorgos, Panayotis A., 16n8, 22, 22n25, 58n9, 90n8
Yehuda, Rachel, 161n3, 162nn4–5
Yuen, Amy, 219n63, 222, 222nn74–75

Zartman, I. W., 223n76
Zemishlany, Zvi, 164n16
Zenko, Micah, 106n60
Zetter, Kim, 221n72
Zsambok, Caroline E., 177n2

Subject Index

ABH (Actual Bodily Harm), 84
Absolute vs. real war, 86
Access, issue of, in attribution problem, 236
Accountability, 4, 13–15, 19n17, 21, 126, 131, 133, 246. *See also* Attribution
ACLU (American Civil Liberties Union), 269n26, 271n31
Actors and agents: cognitive limitations of, 179, 183–185; in cyberspace, problem of, 15; motives for, 192; nature of, 178; patrolling agents, 259–260; unprincipled belligerents, 116. *See also* Cyberwarriors
Acts of violence, attacks as, 43
Acts of war: collateral damage as, 231; covert political actions as, 236; levels of harm in, 231–232; Operation Orchard as, 230; sabotage and cybersabotage as, 25; Stuxnet worm as, 83; Tallinn cyberattacks as not, 230; use of phrase, 37. *See also* Cyberwar and cyberwarfare; War and warfare
Actual Bodily Harm (ABH), 84
Ad bellum requirements, for just war theory, 108–111
Additional Protocols. *See* Geneva Conventions, Additional Protocols
Advanced persistent threats (APTs), 211
Advanced Research Projects Agency (ARPA, later DARPA), 1
Afghanistan: as model of rehabilitative ending of war, 122; UN Security Council authorization for U.S. intervention in, 28
Aftermath of cyberattacks, 115–135; additional regulating principles of, 130–132; conclusions on, 132–133; justice after war, 116–119; overview of, 115–116; regulating principles of, 120–122; *Tallinn Manual* principles, implications of, 124–130; use of models for, 122–124
After Virtue (MacIntyre), 23n29

Agents. *See* Actors and agents
Aggression, 94n25, 108. *See also* Violence
Aggressors, 5, 120–124, 131, 132
Air Force Research Institute (U.S.), 16
Air gaps, 212
Air power, viii, 235; Hague Rules on air warfare, 44n57, 204n17. *See also* Bombs/bombing
Alcoa, cyberattacks against, 190
Allen, Anita L., 152
Allende, Salvador, 237
Allied forces (WWII), unethical practices of, vii
Allies (general), nature of relationships of, 218–219
Ambassadorial immunity, 22n24
American Civil Liberties Union (ACLU), 269n26, 271n31
Analysis phase (of keyword searches), 263
Anonymity, ix–x, 269–270. *See also* Attribution
Anonymous (hacker network), 167–168
Anselm, Saint, 89n1, 92n19
Anticipatory anxiety, 162
Antipersonnel landmines, 44
Appelbaum, Jacob, 211
Apple (company), Chinese attacks on, 127
Aquinas, Thomas, 57, 148n35
Aquinas Lecture (MacIntyre), 23n29
Arabian-American Oil Company, x
Archai (first principles), 23–24
Arendt, Hannah, 172
Aretaic elements (in codes of honor), 143, 153
Aristotle, 23–24, 218
Armed attacks, 36–37, 42, 62, 83
Armed conflicts, 35, 36. *See also* War and warfare
Armed interventions short of war, 238
Arms. *See* Cyberweapons; *names of types*; Nuclear weapons
ARPA (Advanced Research Projects Agency, later DARPA), 1
Arquilla, John, xii, 28n35
Artifacts, Art Works and Agents (Dipert), 62n17

283

Subject Index

Assassins and cyberassassinations, 141, 150, 154–155
Asylum Case (*Colom. v. Peru*, ICJ), 49
Asymmetry of knowledge, 257–258
Attribution: of anti-Tallinn cyberattacks, 4; of assassinations, 155; Bayesian model of, 181n9; complexity of law of, 45; counterattack vulnerability and, 98n39; evidence and forensics, 235, 236; fundamental attribution error, 194; key elements of, 194; problem of, 5, 15, 19n17, 59, 106, 128; problem of, covert political action and, 235–238; problem of, humans and, 179–185; problem of, in war, 85, 87; risks of faulty, x; *Tallinn Manual* on, 126; in wars, 77
Augustine, Saint, 203
Australia: information sharing by, 247; trust of U.S. by, 248n52
Automated keyword searches, 261–268
Automaticity, 185–188
Autonomy: autonomous robots, war between, 89n3; coercive state actions and, 252–254; collection of private information as erosion of, 257–258; of individuals, 232nn15, 17

Back-door booby traps, 25
Barrett, Ed, 19
Battles, as core of military campaigns, 80–81
Bayesian model, 181n9
Beard, Matthew, 139
Best practices (Montreaux Document), 20–21
Bhutan, 267
Binary bullets. *See* Cyberwar and cyberwarfare
Binding legal norms, locus of, 34
Bin Laden, Osama, 230
Blackmail, 152–153
Blitzkriegs, xi
Board games, deception in, 69
Bombs/bombing: aerial bombardment, ix, 220–221; debate over air power strategic bombing, ixn5; GPS-guided, 106; Israeli-conducted bombing raids, 4, 27, 79, 220n66, 229–230; logic bombs, 211, 211n34, 233; World War II, vii
Boothby, William, 208–210
Borders in cyberspace, 67–68
Bosnia, UN peacekeeping operations in, 238
Bosnian civil war (1992–1995), 118
Botnets, 58
Bradley, Gabriel, 142–143
Brazil, 50
Brenner, Joel, 19
Britain. *See* United Kingdom
Bronze Soldier of Tallinn, 77–78

Budapest (International) Convention on Cyber Crime (CCC, 2001), 14–15, 28
Bugs (software problems), 154, 193, 212n42
Burden of proof, 108
Bush, George W., 51n93

Camouflage, 203
Canada, information sharing by, 247
Canetti, Daphna, 157
Capital, social vs. political, 216n49
Card games, deception in, 69
Carr, Jeffrey, 26n33
Casus belli, 126, 237
Causation, 85, 205, 215
CCC (Convention on Cyber Crime, 2001), 14–15, 28, 39, 45
Cell phones, outcomes of cyberattacks on, 159
Central Intelligence Agency (CIA), 237, 238
Ceteris paribus clauses, issues of, 103–107
Chain reaction challenge, 187
Chandos Herald, 81
Charter (UN). *See* UN Charter
ChatMor (fict. company), 191
Chesterman, Simon, 129
Chevauchées and cyber *chevauchées*, 75–88; cyberwarfare, as war, 85–87; cyberwarfare, impossibility of, 76–80; cyberwarfare, nature of, 75–76; hypothetical example of, 82–85; medieval warfare and, 75, 80–82
Chile, CIA activities in, 237
China: application of international law to cyberspace, response to, 52; cyberespionage accusations against, 4, 185–186, 190–191, 229, 234–235; on cyberspace, nature of, 20n21; espionage, preference for, 127
China Daily's editorial against U.S. cyber espionage, 20n21
CIA (Central Intelligence Agency), 237, 238
Citizens, states' relationship to, 59
Civil disturbances, 36
Civilian computers, military computers vs., 104–105n57
Civilian objects, question of data as, 43
Civilian operatives, 149
Civilian status. *See* Noncombatant immunity
Civil wars, 219, 223n76
Clarke, Richard A., 19, 26–27, 211
Clausewitz, Carl von: on ideal wars, 91–92; influence of, 80; on threats of action, 84n37; on violence, limits to, 92n21; on war, 2, 76–77, 86, 91
Cleanliness (cybersabotage virtue), 153, 154
Clinton, Bill, 1
Clustering of nation-states, 247–248
Cluster munitions, 44
CNE (computer network exploitation), 189, 236n29

Subject Index 285

The Code of the Warrior (French), 141–142
Codes of conduct (codes of ethics, codes of honor), for cyberwarriors, 139–156; conventional warrior codes of honor, inapplicability of, 147–150; cyberwarriors, central roles for, 150–155; formal codes of conduct, 143–147; need for, 141–143; overview of, 139–141
Coercion, 152
Coercive state action, 251–256
Cognitive limitations, 179, 183–185, 187–188
Cold War: covert political operations during, 127, 237; nuclear standoff of, x, 1
Collateral damage, 145, 154, 231, 245
Collectivistic cultures, 194
Combatants: civilian contractors as, 149n37; irregular, 263; spies vs., 190. *See also* Cyberwarriors; Noncombatant immunity
Commitment problem (in international relations), 220n64
Commons, cyberspace considered to be, 20n21
Communist-controlled labor unions, 237
Compensation payments, postwar, 121, 131–132, 133
Completion, as justification for laws on war termination, 117
Computer network exploitation (CNE), 189, 236n29
Computers: honeypots, 69n36, 211; military vs. civilian, 104–105n57; personal, outcomes of cyberattacks on, 159; security of, civilians and, 107. *See also* Malware
Conflicts. *See* Cyberwar and cyberwarfare; War and warfare
Consent-worthiness of policies, 252
Conventional warriors. *See* Warriors
Convention on Cyber Crime (CCC, 2001), 14–15, 28, 39, 45
Conventions. *See* Treaties
Cook, James, 58n8, 60
Cookie hacks, 211
Cooperative Cyber Defence Centre of Excellence (NATO), 41
Costs: of cybersecurity, 61; of cyberwar, xi; of targeted cyberweapons, 104n56
Council on Foreign Relations, 106n60
Covert political actions and cyberactions, 228–250; attribution, problem of, 235–238; espionage vs., 240; morality of, 239–248; overview of, 228–229; tactics deployed in, 236; term definitions, 229–235. *See also* Cyberespionage
Creativity of cyberwarriors, 145–146, 154
Criminal justice. *See* Justice
Criminal law convictions, 181
The Critique of Judgment (Kant), 23n29
Crown Prosecution Services (UK), on Assault Occasioning Actual Bodily Harm, 84
Crude intelligence gathering, 255

Cultures, collectivistic vs. individualistic, 194
Customary International Humanitarian Law study (ICRC), 48
Customary international law (*ius gentium*), 21–22, 45–51, 218
Customary space law, 47
Cyberactivities: as acts of war, 37; coercive state actions, politically legitimate forms of, 251–256; cybereffect, description of, 214n43; humans' roles in, 178–179; justification for, 110; politically legitimate, 252; problem of visibility of, 49; war crimes, deaths from, 214. *See also* Acts of war; Covert political actions and cyberactions; Cyberwarfare; Terrorism and cyberterrorism
"'Cyberation' and Just War Doctrine: A Response to Randall Dipert" (Cook), 58n8
Cyberattacks: as acts of war, 28–29; Additional Protocol I on, 43; armed attacks, 36–37, 42, 62, 83; attack signatures, 186; as casus belli, 126; definition of, 150–151, 178; emerging norms on, 29–30; list of recent, 229; nature of, 60; nonlawyers' use of term, 35; not acts of force, 77; reversibility of, 90n5; speed of, 185, 194; types and outcomes of, 159. *See also* Aftermath of cyberattacks; Cybertheft; Estonia and Tallinn cyberattacks
Cyberbullying, 166
Cyber Command (U.S.), 60
Cybercrimes, 25–26, 126, 133
Cyberdefenses, standard, 186
Cybereffect, description of, 214n43
Cyberespionage: covert political actions and cyberactions vs., 240; cyberconflict as espionage, 25; differences from cybertheft, 233; as harm type, 126; industrial cyberespionage, 234; Internet information and, 65; morality of, 57–58, 246–248; principles guiding, 145–146, 188–189, 251; as role of cyberwarriors, 150–153; *Tallinn Manual* on, 189, 236n29. *See also* Cybersurveillance; Humans in cyberoperations, espionage, and conflict; Moral concerns of cyberespionage
Cyberfire, immunity from. *See* Psychological and physiological effects
Cybersabotage: as harm type, 127–128; Operation Orchard, 4, 27, 79, 220n66, 229–230; property destruction as, 205, 213–214; as role of cyberwarriors, 150–151, 153–155
Cybersocial contract, 243, 244, 245–246, 247
Cyberspace (cyberdomain), 2–3, 15, 18
Cybersurveillance: domestic vs. foreign, 255; emerging norms for, 30n37; patrol vs., 259–261, 269. *See also* Cyberespionage
Cybertheft, 121, 131, 133, 233

Cyberwar and cyberwarfare: aftermath of cyberattacks, 115–135; code of cyberwarriors, 139–156; concept of, viii–ix, 3, 3n5, 90–91; costs of, xi; cyber *chevauchées*, 75–88; emerging norms for, 13–33; ending of, 117–118, 219–224; escalation to physical conflict, 4; ethical issues of, ix, xi, 56–72, 251–276; historical background of, 14–16; humans in cyberoperations, espionage, and conflict, 177–197; as ideal war, 89–114; international legal norms for, 34–55; locating end of, problems of, 119; maximally discriminate, 96–98; maximally justified, 108, 111; maximally proportionate, 93, 95; nature vs. character of, 86; outcomes of, 119; overview, 1–10; paradox of, xi–xii; privacy rights during, 246–247; psychological and physiological effects, 157–176; Rid's criteria for, 76–77, 83; ruse and deception in, 201–227; as social phenomenon, 75–76; strategic, viii–ix; success of, 99n44, 108n65; types of, 151; victory in, 119–120n18; worse than conventional war, 105n59; zero-sum, 222. *See also* Cyberattacks; Ideal war, cyberwarfare as; International law and legal norms; Norms for cyberwarfare, emerging; Perfidy, ruse, and deception in cyberwarfare; *Tallinn Manual*

Cyber Warfare and the Laws of War (Dinniss), 116
"Cyberwarriors: Activists and Terrorists Turn to Cyberspace" (Denning), 151
"Cyber War Will Not Take Place" (Rid), 76
Cyberweapons: arms control, x–xi; cyber perfidy and, 207–215; difficulty of controlling, 221; enthralling nature of, vii; GPS-guided bombs, 106; heat-seeking missiles, 186n27; kinetic armaments, timing of use of, 185n24; as maximally proportionate, 98; responsible development of, 145; reverse engineered cyberweapons, 105, 107; smart arms race, 146; *Tallinn Manual* on, 62n18; treaties on, U.S. nonparticipation in, 44n56
Cyboteurs. *See* Cybersabotage

Dahout model (of punishment bombing), 221
Damage. *See* Harm
Danks, David, 177
Danks, Joseph H., 177
Darley, William M., 92n17
DARPA (formerly Advanced Research Projects Agency, ARPA), 1
Data. *See* Keyword searches; Metadata and metadata analysis
Data mining. *See* Moral concerns of cyberespionage
DDoS. *See* Distributed denial of service (DDoS) attacks

Death: from cyberoperations, as war crimes, 214; killing in war, 60, 99–100n46, 149; in Russo-Georgian War, xi; from strategic cyberattacks, ix
Deception, 68–69, 208, 215–224. *See also* Perfidy, ruse, and deception in cyberwarfare
Decyberization, 131, 133
Deep Web, 66n29
Defense Department, on cybereffects, 214n43
Defense Intelligence Agency, on Snowden affair, 234
Defensive actions, moral legitimacy of, 179
De George, Richard, 146
Delbrück, Hans, 80
Denial of service attacks, 210n33. *See also* Distributed denial of service (DDoS) attacks
Denial operations (aerial bombardment type), 220, 221
Denning, Dorothy, 151
Deontology, 102n52
Der Spiegel on NSA spying, 212n40
Desert: of harm vs. liability to harm, 94–95; individual's, 94n26
Desert adjusted hedonism, 102n53
Deterrence, x, 58, 64, 70, 244, 248
Diaya-al-Sahir, Syria, Israeli attack on, 27
Digital privacy law (European Parliament, 2014), 271n32
Dinniss, Heather Harrison, 116
Dipert, Randall R., ix, 15, 56, 232
Diplomatic immunity, 22n24
Dirty hands actions, 228, 239, 240, 242, 244. *See also* Covert political actions and cyberactions
Discernment, cyberespionage and, 152
Discrimination (target selection capability): of covert political actions, 241, 242, 244–245, 248; cyberweapon programming and, 104; in postwar punishment, 126, 133; as principle for code of conduct for cyberwarriors, 144–145; proportionality vs., 101–102. *See also* Noncombatant immunity; Proportionality
Disinformation, 65, 126–127
Dispositionism, 194
Distress signals, false, 204
Distributed denial of service (DDoS) attacks, 210n33. *See also* Estonia and Tallinn cyberattacks
Domestic law, treaties and, 37
Doomsday devices, 186n27
Double-effect, doctrine of, 154n60
Drones (unmanned aerial vehicles, UAVs), 62, 106, 109n68, 180–181
Dual usage infrastructure, x, 104, 153, 244–245
Dublin Treaty (1990), 44n56
Due process, 271, 273–275
Dummy constructions, 203
Dunlap, Charles J., 147, 153–154

East Timor, UN peacekeeping
    operations in, 238
Eberle, Christopher, 19, 25n32
Economic issues: economic coercion, UN Charter
    on, 63n20; economic damage, need for moral
    theories on, 57; economic espionage, lack of
    international law on, 191; economic warfare,
    171–173, 238; economic wellbeing, political
    liberty vs., 101n49
Efficacy and efficiency: of keyword searches,
    265–267; of metadata analysis, 273
Emblems, protected, false use
    of, 146, 204–205, 208, 213, 215
Emergent norms, 20–23, 31. *See also* Norms
    for cyberwarfare, emerging
Ending of wars, 117–118, 219–224. *See also*
    Aftermath of cyberattacks
Enemark, Christian, 155
Enemies: Grotius on right of killing of, 215;
    punishment of, 94; trust between, 201–202,
    203. *See also* Attribution; Retribution and
    retaliation
Enemy indicators, false use of, 146
Enlightened egoism, 63
Equilibrium state, 243, 244
Equivalent harm, principle of, 30
Escalation of cyberwarfare to physical conflict, 4
Espionage, applicability of international law to,
    19n16. *See also* Cyberespionage
Estonia and Tallinn cyberattacks: clean
    cyberenvironment, beliefs on, 48;
    compensation for, 132; description of, 4,
    26–27, 229; not act of war, 77–78, 230;
    as possible cyberwar, 115; problem of
    attribution of, 128, 181–182, 184, 189
Ethical issues, 56–72; automaticity and, 186–187;
    conclusions on, 69–70; cyberwarfare,
    ethical distinctness of, 59–62; cyberwarfare,
    implications for ethics, 171–173; of Deep
    Web, 66n29; humans' importance for ethical
    decisions, 193–195; just war theory, 64;
    killing in war traditional vs. revisionist views
    on, 99–100n46; legal issues vs., 63–64; moral
    principles, application to cyberwarfare,
    16–20, 65–68; overview, 56–59; perfidy and
    deception, 68–69. *See also* Codes of conduct;
    Humans in cyberoperations, espionage, and
    conflict; Moral concerns of cyberespionage;
    Morality
European Parliament, digital privacy law, 271n32
Evil, war as, 105, 108
Excessiveness, avoidance of, 145
Exploitation and cyberexploitation, 150, 152–153
External vs. internal justice, 59n10
Extradition, 58, 67n31
"Eye for an eye and tooth for a tooth" (reciprocity)
    principle, 242–246, 247, 254

False codes, 203
False flags and false-flag agents, 152, 204
False positives in keyword
    searches, 264, 265–266
Fear, as effect of terrorism, 162–163
Fides etiam hosti servanda (trust must be kept
    with the enemy), 203
Financial networks, outcomes of cyberattacks
    on, 159
Fingerprints, collection of, 273n36
First Gulf War (1990), 172
First principles (*Archai*), 23–24
First resort, strategic cyberattacks as, ix. *See also*
    Last resort criterion
FISA (Foreign Intelligence Surveillance Act)
    Amendments Act (2008), 246
Five Eyes, 247
Fixed laws, movement to emergent norms, 20–23
Flag of truce, false use of, 146
Flame (computer virus), 98n37, 107n64
Floridi, Luciano, 15–16, 65
Force: cyberoperations as physical force, 90n8;
    interpretation of use of, in treaties,
    41–42; as last resort, ix, 60, 65; use of,
    30, 36–37, 42; war as act of, 76. *See also*
    Collateral damage
Foreign Affairs Ministry (PRC), 191
Foreign Intelligence Surveillance Act (FISA)
    Amendments Act (2008), 246
Foreign operations, consent to, 254
Formal codes of ethics. *See* Codes of conduct
Form of War, 91n15
Forum on International Cybersecurity, Sixth
    Annual European, 18n15
Four Horsemen of the Apocalypse, 89
France: labor strikes in, 237; response to PRISM
    surveillance program, 50
Fraud, 233
Freedoms, negative and positive, 252, 253
French, Shannon E., 140, 141–142, 149
French National *Gendarmarie*, 18n15
F-35 Joint Strike Fighter plane, 235
Fuller, J. F. C., vii, viii
Fundamental attribution error, 194
Future self-projection bias challenge, 188, 194
Future wars, prevention of, as justification for laws
    on war termination, 118
Fuzzers (software programs), 154

Games, deception in, 69
GCHQ (Government Communications
    Headquarters, UK), 211, 251
Geneva Conventions (1949): cyberconflicts,
    applicability to, 40, 63; on mental
    suffering, 171; on perfidy, 204; subjects of,
    35. *See also* Geneva Conventions, Additional
    Protocols

Geneva Conventions, Additional Protocols (1977): on attacks, 43; on civilian objects, 43; cyberconflicts, applicability to, 40; on perfidy, 204, 206n21; on protected emblems, use of, 204n17; on ruses, 203n11; subjects of, 35

Genocide, 108

Georgia, cyberattacks on Ossetia, viii, xi, 4, 27

Germany: post–WWII success of, 123, 124; response to PRISM surveillance program, 50; U.S. spying on, 219, 236–237

GGE (UN Group of Governmental Experts), 40

God, existence of, 89n1

Goldberg, Arthur, 273

Google, Chinese attacks on, 127

Gottstein, Ulrich, 172

Government Communications Headquarters (GCHQ, UK), 211, 251

Graham, David, 28

Greater good, 239

Gross, Michael L., 157

Grotius, Hugo, 57, 90n7

Group of Governmental Experts (GGE, UN), 40

Groups, rationality of, 184

Guatemala, CIA actions in, 238

Gu Chunhui, 190

Gulf War (1990), 172

Hackers. *See* Actors and agents; Warriors and cyberwarriors

Hacking: analogies for, 66–67; day-to-day variety of, 154. *See also* Cyberespionage

Hacktivism, 151, 166

Hague Conferences (1899 and 1907), 46n62

Hague Convention IV (1907) and Regulations (Rules), 44n57, 46, 203n11, 204n17

Haiti, 132

Hamas, cyberthreats from, 166

Hare, R. M., 23n28

Harm (damage): Actual Bodily Harm, 84; counterfactual comparative account of, 98n40; cyberharm, 232; description of, 98–99; direct, as terrorism, 161; disproportionate, noncombatant deaths as, 158; incidental harm, importance of, in cyberwarfare, 58; indiscriminate, 104; infringements vs., 258–259; from involuntary disclosure of private information, 258; lesser forms of, need for moral theories on, 57–58; liability-adjusted, 102; liability to vs. desert of, 94–95; morally weighted, 102n53; objective state account of, 98n42; perfectly discriminate, 110; perfidy and, 213; property destruction, 163, 205, 213–214; proportionality of, 93–94; quantity and quality of, from cyber vs. kinetic weapons, 59; reduction of, through cyberwarfare, 153; success of cyberwarfare vs., 108n65; totally reversible, 98; types of, 126–127, 232; in war, 60, 178–179, 231–232; by warriors and cyberwarriors, 149. *See also* Collateral damage

Heartbleed bug, 212n42

Heat-seeking missiles, 186n27

Hell, war as, 89

Hermeneutical gymnastics, 18, 24

Historical background of cyberwarfare, 14–16

Hobbes, Thomas, 90n7

Holder, Eric, 191

Honeypots (computer traps), 69n36, 211

Honor of soldiers and cyberwarriors, 143, 161. *See also* Codes of conduct

Horowitz, Michael, 221, 222

Huang Zhenyu, 190

Human rights, 35, 121, 153, 232n15, 246, 248

Humans in cyberoperations, espionage, and conflict, 177–197; automaticity, problem of, 185–188; conclusions on, 193–195; the human, importance of, 177–180; motives, importance of, 180–185; rules of law, cyberwarfare and, 188–193

Hume, David, 22–23n28

HUMINT (human intelligence), 264–268

Hundred Years' War, 75, 81

Hussein, Saddam, 118, 238

Iceland, 2

ICT. *See* Information and communication technology

Ideal war, cyberwarfare as, 89–114; *ceteris paribus* clauses, issues of, 103–107; cyberwarfare as realization of ideal war, 96–103; ideal war, possibility of, 90–95; just war theory, *ad bellum* requirements for, 108–111; overview of, 89–90

Identity theft, 165–166

Ignorance, as excuse, 188

IHL (international humanitarian law), 5, 19, 35–36, 43

Imagery analysis, 265

Immunity, diplomatic, 22n24

Immunity from cyberfire. *See* Psychological and physiological effects

Imperfect sciences, 23n29

In bello moral continua, 90–95

Incidental harm, 58

Incommensurability of proportionality vs. discrimination, 101–102

Inconvenience, 161

Indicators for espionage, 146

Indiscriminate harm, 104

Individualistic vs. collectivistic cultures, 194

Individuals: individual autonomy, violations of, 232nn15, 17; individual desert, 94n26; outcomes of cyberattacks on, 159; terrorism against, 167. *See also* Cyberwarriors; Noncombatant immunity

Industrial control, vulnerability of, 2
Industrial cyberespionage, 234
Industrial web model (of punishment bombing), 221–222
Inference problems, attribution as, 181
Infiltrations (cyberoperations), 211
Influence operations, cyberattacks and, 76n6
Information: five pillars of, 92n17; partial vs. universal disclosure of, 270; personal information, control of, 256–257
Information and communication technology (ICT), x, 40, 229–235, 241, 244–245
Information warfare. *See* Cyberwarfare
Infoworld (journal), cyberwarrior interview on, 154
Infrastructure, effects of cyberattacks on, 164–165
Infringements, harm vs., 258–259
Innocent bystanders, 102n53, 264. *See also* Noncombatant immunity
Innovation, 154
Inside Cyber Warfare (Carr), 26n33
Institutional harm, 232
Instrastate conflicts, settlements of, 223n76
Intellectual property (IP), 19n17, 131; theft of, 60, 126, 190, 233–234. *See also* Cybertheft
Intelligence capabilities, 104–105, 255–256
Intelligence operatives. *See* Spies
Internal vs. external justice, 59n10
International agreements. *See* Treaties
International armed conflict, description of, 36
International Committee of the Red Cross (ICRC), 17, 48
International Court of Justice (ICJ), 34, 42, 45, 46–47, 49
International Criminal Court Statute, 38
International Criminal Tribunal for the Former Yugoslavia, 44
International Group of Experts, 41–42, 43, 83
International Information Security Agreement, 39–40
International law and legal norms, 34–55; applicability to cyberspace, 8–19; conclusions on, 52; coverage of cyberwarfare, 125; cyberconflict and treaties, 39–45; humanitarian law, 5, 19, 35–36, 43; lack of, against economic espionage, 191; overview, 34–35; on perfidy, 68; relevant vocabulary, 35–37; on sabotage, 25; treaty law, 37–39; on war termination, 116–117; as written norms, 31. *See also* Customary international law
International relations (IR), norms in, 21–22
International society, nature of, 218
International Telecommunication Regulations (ITU), 38, 40
Internet: crime in, 25–26; possible evils of, 105; protection of information from, 65; *Tallinn Manual's* understanding of, 165; ubiquity of access to, 1–2

Interpretative declarations (to treaties), 39
Intimate relationships, 257
Intrusions (cyberoperations), 210
Intuitions, use in ethics, 64
Investigation, postwar, 126
Investigative phase (of keyword searches), 263–265
IP (intellectual property), 233
IPv6, 106n61
Iran: level of possible threats to, 267; Stuxnet worm, response to, x, 64, 192–193. *See also* Nuclear enrichment program; Stuxnet
Iran-Iraq War (1979–1989), vii–viii, 117
Iraq: effects of sanctions on, 172; U.S. sanctions against, 238
Iraq War (2003): control of cyberweapons during, 221; as model of rehabilitative ending of war, 122
IR (international relations), norms in, 21–22
Irregular combatants, 263
ISightPartners, 192–193
Israel: bombing raid on Syria, 4, 27, 79, 220n66, 229; legality of covert political activity within, 239n34; level of possible threats to, 267; rocket attacks on, public responses to, 162, 164; Stuxnet worm, possible responsibility for, 79, 128, 192
Italy: labor strikes in, 237; Spanish Civil War, involvement in, x
ITU (International Telecommunication Union), 38, 40
Ius. See entries beginning "Jus"
Ius gentium. See Customary international law

Japan, post–WWII success of, 123
Jenkins, Ryan, 18n16, 26n33, 89, 153
Johvi, Estonia, riots in, 78
Joint security treaties, 45
Journal of Military Ethics, special issue on cyberwarfare, 16
Journal of Strategic Studies, Rid article in, 76
Jurisdiction, 144
Jurisdictional equivocation, 19
Jus ad bellum (justice of war), 36, 57, 117, 119, 185, 223
Jus ad vim (use of force short of war), 91
Jus gentium (law of nations), 21–22, 218
Jus in bello (justice in war): *jus post bellum* phase vs., 119, 120n18; *jus terminatio* and, 224; just war theory and, 57; laws of, 117; rational actors, reliance on, 185. *See also* International law
Jus post bellum (justice after war): application of *Tallinn Manual* principles to, 124–130; description of, 116–118; *jus terminatio*, 223–224; regulating principles of, 120–122, 130–132; reversibility of cyberattacks and, 90n4; starting assumptions behind, 119; use of models for, 122–124

Subject Index

Just and Unjust Wars (Walzer), 63, 99n46
Just cause criterion, in just war theory, 62, 65
Jus terminatio (just termination), 223–224. See also *Jus post bellum*
Justice, 59, 63; criminal justice comparisons, 94n26, 242, 244, 273; internal vs. external, 59n10. See also *entries beginning "Jus"*
Justice Department, actions against China, 4
Just liberal states, 259–260, 267
Just soldiers, 103
Just war theory: *ad bellum* requirements for, 108–111; applicability of, 5, 56–58, 65–68, 228–229, 245; context for origins of, 69; core principles of, 65; description of, 64; dirty hands actions vs., 240; justification for just war, 148; problematic status of, 63; on proportionality, 93; similarity to *Tallinn Manual* principles, 147. See also *Jus in bello*; Laws of armed conflict

Kant, Immanuel, and Kantianism, 23n29, 63, 69
Kaspersky, 61
Keystroke logging, 211n34
Keyword searches, 251, 261–268
Kill commands (in cyberweapons), 107
Killer robots, 31
Killing in war, 99–100n46, 149. See also Death
Killing in War (McMahan), 100n46
Knake, Robert K., 211
Knowledge: asymmetry of, 257–258; in modern warfare, vii
Knox, Frank, vii
Koh, Harold, 44
Kohtla-Jarve, Estonia, riots in, 78
Korean War, vii

Laboratory assessments of psychological effects of terrorism and cyberterrorism, 167–171
Labor unions, communist-controlled, 237
Landmines, 185n24
Langner, Ralph, 97
Last resort criterion, ix, 60, 65, 109, 109n68, 172
Law: on attribution, complexity of, 45; best practices vs., 21; codes of conduct vs., 142–143; criminal law convictions, 181; customary law, 218; customary space law, 47; evolution of, 24; function of, 117; implications of cyberwarfare for, 171–173; of international armed conflict, 201; legal issues vs. ethical issues, 63–64; occupation law, 129n47; rules of law, cyberwarfare and, 188–193; treaty law, 37–39. See also Customary international law; International law and legal norms; *entries beginning "Jus"*
Law enforcement moral framework, 228–229, 240–241, 244

Laws of armed conflict (LOAC), 5, 139–140, 144, 203. See also Just war theory
Laws of international armed conflict (LOIAC), 201
Lazar, Seth, 102n53
Learning, patterns and, 271
Lebanon, Second Lebanon War (2006), 163–164
Legality: of covert political actions, 239–241; of cyberactivities, states' positions on, 50–51
Legal obligation (*opinio juris*), 46, 47–48
Lethal autonomous systems, 30n37
Letters of Marque, 59n10
Lexical priority, 101, 102n51
Liability-adjusted harm, 102
Liability to harm vs. desert of harm, 94–95
Liberal states, intelligence operations by, 252
Lighting of ships (deceptive), 204, 206
Lin, Harbert, 150
LOAC. See Laws of armed conflict
Logic bombs (malware), 127, 211n34, 233
LOIAC (laws of international armed conflict), 201
Losers (of wars), possible concerns of, 117–118
Low boil cyberwarfare, 99n44, 107n62
Lowther, Adam, 22
Lucas, George R., Jr., 13, 28n35, 64, 152

Macafee, 61
MacIntyre, Alasdair, 23
Madden, Mike, 205–206, 213
Mala in se, applied to cyberware, 109n67
Malware: Flame (computer virus), 98n37, 107n64; logic bombs, 127, 211n34, 233; viruses and worms, 211n34; weaponization of, 62. See also Stuxnet
Management Directive #424 (NSA), 152
Mandiant (cybersecurity company), anti-Chinese accusations, 4, 185–186, 190, 229, 234–235
Maness, Ryan, 211n34
Marines' Code, 142
Mass, maximum of, 92
Mass casualty terrorism, 167
Maxima: maximally discriminate war, 96–98; maximally just causes for war, 108, 111; maximally proportionate war, 93, 95; of properties, 92–93
May, Larry, 153
McMahan, Jeff, 100n46
Means of warfare, *mala in se*, 109n67
Medical infrastructures, outcomes of cyberattacks on, 159
Medieval warfare, *chevauchées* and, 80–82, 85
Mental privacy, 256–258
Mental suffering: from cyberterrorism, 160–161, 164–167. See also Psychological and physiological effects
Mercenaries, 21
Merkel, Angela, 219, 236–237

Metadata and metadata analysis: collection of, 246, 251; description of, 229n6, 251; longterm retention of metadata, 271–273; moral concerns of, 268–275; right to privacy and, 234; Supreme Court on privacy of telephone metadata, 270n28

Microsoft, Chinese attacks on, 127

Middle Ages, *chevauchées* and warfare in, 80–82, 85

Militarism, predictors for, 171

Military: ethics of, applicability to cyberspace, 69; military alliances, nature of relationships of, 218–219; military computers, civilian computers vs., 104–105n57; *opinio juris* derived from manuals, 48; practices of, relationship to technology, 1; robotics, threshold problem and, 109n68. *See also* Cyberweapons; War and warfare

Miller, Seumas, 228

Minimally just political communities, 121

Minimal violence, 93

Misinformation, 203

Models: Bayesian model of attribution, 181n9; of postwar settlements, 120–123; of punishment bombing, 221–222

Modern conventional warfare, possible evils of, 105

Montreaux Document (2008), 17–18, 20–21

Moore, Gordon E., 23n28, 25

Moral concerns of cyberespionage, 251–276; coercive state action, politically legitimate forms of, 251–256; conclusions on, 275; keyword searches, morality of, 261–268; metadata analysis, morality of, 268–275; overview, 251; privacy, right to, 256–259; surveillance and patrol, 259–261

Morality and moral issues: consistency of morality, 215; of covert political cyberactions, 239–248; espionage, moral foundations for, 252; moral autonomy, preconditions for, 256–257; moral dilemmas of incommensurable goods, 102n51; moral equality, Walzer on, 100n46; moral injuries, 84; morally perfect wars, 91; moral philosophy, impact on cyberwarfare, 16–20; moral principles, application to cyberwarfare, 65–68; in war, source of traditional theories of, 60. *See also* Codes of conduct; Ethical issues; Ideal war, cyberwarfare as

The Morality of War (Orend), 99n46

Motives: human, complexity of, 192; importance of, 180–185

Multilateral treaties, 39

Münster (Westphalian) Treaty (1648), 62

Murderers, 94n26, 149

Mutually Assured (nuclear) Destruction, 64

Naked soldier case (of ethics in war), 99–100n46

Nardin, Terry, 218

National Academies, definition of cyberspace, 2–3

National Security Agency. *See* NSA

National Security Service (NSS, Uzbekistan), 255

NATO (North Atlantic Treaty Organization), 16n11, 27, 28–29, 41, 218–219

Naval warfare, 204

Necessity, covert political actions and, 241

Needle exchanges, 109n68

Negative freedoms, 252

Netherlands: on application of international law to cyberspace, 52; on armed attacks, 42; on defensive use of force, 42

Neutrality, false claims of, 146

Newness, uncertainty associated with, 23–24

NEWSCASTER (social networking campaign), 192–193

NewsOnAir.com, 193

Newton, Kenneth, 216

New Zealand, information sharing by, 247

9/11 Twin Towers attacks, 230

Noncombatant immunity: cyberwar and cyberterrorism and, 158–161; description of, 157; economic warfare and, 173; maintenance of, in cyberwar, ix; morality of war and, 57; in occupations, 129; as principle for cyberwarfare, 133, 144

Noninternational armed conflict, 36

Nontool instruments, 62n17

Norms for cyberwarfare, emerging, 13–33; and ethics and moral philosophy, 16–20; historical background, 14–16; movement from fixed laws to, 20–23; overview, 13; and uncertainty, 23–31. *See also* International laws and legal norms

North Atlantic Treaty Organization (NATO), 16n11, 27, 28–29, 41, 218–219

North Korea, technical assistance to Syria, 27

North Sea Continental Shelf (ICJ case), 46–47

Norway, 2

NSA (National Security Agency): ACLU on metadata collection by, 271n31; automatic attacks by, 212; cookie hacks, use of, 211; damage to, from Snowden's actions, 233; Management Directive #424, 152; metadata collection from Verizon, 229, 246; PRISM surveillance program, 50, 229, 233–234, 246; QUANTUMINSERT, use of, 211–212; Snowden charges against, 251

NSS (National Security Service, Uzbekistan), 255

Nuclear enrichment program (Iran), 29, 78, 79, 96, 128, 180, 182, 229

Nuclear weapons, vii, xi, 24, 27, 58, 64, 105n59, 232, 240

Obama, Barack, 219, 274n39

Obeying Orders: Atrocity, Military Discipline, and the Law of War (Osiel), 142

Subject Index

Obligatoriness: of cyberwarfare, 99–100, 107; of war, 95
Occupation and occupation law, 128–130
Olympic Games (surveillance and sabotage operation), 27
Oman, Charles, 80
Ontologies on information-theoretic entities, 62n19
Operation Orchard (Israeli bombing raid on Syria), 4, 27, 79, 220n66, 229–230
Operations. *See* Cyberactivities
Operatives and cyberoperatives, 143n12
Opinio juris (legal obligation), 46, 47–48
Oppenheim, Lassa, 203
Orend, Brian, 2, 90n7, 94n25, 99n46, 115
Osiel, Mark, 142
Ossetia, Georgia, cyberattacks on, viii, xi, 4, 27, 29
"Other things being equal" clauses, issues of, 103–107
Ottawa Convention (1997), 44n56
Overlapping consensus (postwar settlement model), 120–121
Owned (private) space, 66–67

Packet sniffers, 211n34
Pape, Robert, 222
The Paquete Habana (Supreme Court case), 46
Passwords, 61, 107n63, 193n55, 204
Patrol: keyword searches vs., 262–263; surveillance vs., 259–261, 269
Patterns: learning and, 271; in metadata analysis, 269
PBS, *Wired Science* episode on Estonian cyberattack, 14n3
Peacetime, deception during, 216
Peace treaties, 118
People's Liberation Army (PLA), accusations against, 4, 186, 189
Perceptions: deception and, 208; importance of, 191
Perfidy, ruse, and deception in cyberwarfare, 189, 201–227; conclusions on, 224; cyber perfidy, cyberweapons and, 207–215; deception and trust, strategies of, 215–224; LOIAC on, 201; nature of, 204–207; overview of, 201–203; prohibition of, 202; restrained, 145–146; Rowe on, 68–69, 207n24; ruse, nature of, 203–204; in warfare, results of, 223
Perry, David L., 154–155, 236n29, 237
Persian Gulf War (1991), 122
Persistent objection doctrine, 48–49
Personal computers, outcomes of cyberattacks on, 159
Personal information, control of, 256–257
Personal space, 257
Pfaff, Tony, 152

Philosophers: deterrence, rejection of, 64; non-participation in *Tallinn Manual* process, 18–19
Philosophy. *See* Morality
Phishing, 193n55
Physical intrusion, lack of, in cyberwarfare, 61–62
Physiological effects. *See* Psychological and physiological effects
Pivots, 58
PLA (People's Liberation Army), accusations against, 4, 186, 189
Pliny, 215–216
PMCs (private military contractors), 17, 20–21, 148–149
Police, 244, 253, 254, 268–269
Politics: political capital, social capital vs., 216n49; political coercion, 63n20; political cyberoffenses, 78; political decision-making, 170–171; political liberty, economic wellbeing vs., 101n49; political trust, social trust vs., 216n49, 217n52. *See also* Covert political actions and cyberactions
Pop-up ads, 203
Positive freedom, 253
Possession, rights of, 81
Postcyber: meaning of, 115. *See also* Aftermath of cyberattacks
Postcyber Seven, 132–133
Posterior Analytics (Aristotle), 23–24
Post-traumatic stress disorder (PTSD), 161–162
Postwar period, 120–123, 129–130, 133. *See also* Aftermath of cyberattacks; *Jus post bellum* (justice after war)
Practical association, 218
Precommitment problem, 220
Predictions of cyberactions, 187–188
Preemption (prevention), 30, 58, 60, 118
Principle of Unnecessary Risk, 103n54
PRISM surveillance program, 50, 229, 233–234
Prisoners-of-war (POWs), 121
Privacy, right to: anonymity vs., 269–270; cyberespionage and, 152–153; ethical issues related to, 59; as human right, 246; intelligence-gathering and, 247–248; metadata and, 234; patrol vs. surveillance as violations of, 256–260; privacy infringement, definition of, 258–259; of private citizens in authoritarian states, 248; telephone metadata, privacy of, 270n28. *See also* Metadata and metadata analysis
Private enterprises, 191
Private military contractors (PMCs), 17, 20–21, 148–149
Private (owned) space, 66–67
Private property, 130
Private security companies (PSCs), 17

Probability of success criterion, in just war theory, 65, 110–111
Problem of attribution. *See* Attribution
Propaganda, 203
Properties, maxima of, 92–93
Property destruction. *See* Harm
Proportionality, principle of: of covert political actions, 241; of cyberattacks, 160; discrimination vs., 101–102; human agents and, 179; intrinsic maximum for, 93; as just war theory criterion, 65, 111; of keyword searches, 266–267, 268; of metadata analysis, 273–274; nature of, 93–94n23; of Stuxnet attack, 29; of war, cyberwarfare's impact on, 109. *See also* Discrimination
Prospective reciprocity principle, 243–248
Protected emblems, false use of, 146, 204–205, 208, 213, 215
Protected persons, in occupied territory, *Tallinn Manual* on, 128–129
Protocols. *See* Treaties
Proximate cause, 207, 209–210
PSCs (private security companies), 17
Psychological and physiological effects, 157–176; cyberterrorism, lab assessments of psychological effects of, 167–171; implications for ethics and law, 171–173; noncombatant immunity, 158–161; overview of, 157–158; psychological harms, 84, 232n15; severe mental suffering, 164–167; terrorism and cyberterrorism and, 161–164
PTSD (post-traumatic stress disorder), 161–162
Public health and safety, 129–130, 131, 159
Publicity, for terrorist attacks, 230, 237
Public policy, norms in, 22
Public records, outcomes of cyberattacks on, 159
Punishment, 19, 58, 94–95, 94n26, 121, 129, 133; aerial bombardment type, 221–222; collective punishment, 126. *See also* Retribution and retaliation
Purposes of wars, 77
PUR (Principle of Unnecessary Risk), 103n54

QUANTUMINSERT (router hack), 211
Quester, George, xi

Radio, Hague Rules for the Control of Radio in Time of War (1923), 204n17
Raids, as medieval warfare strategy, 80–81
Rational choice theory, 243n43
Rationality of human agents, 179, 183–185, 194
Rawls, John, 101n49
Reagan, Ronald, xi
Realists, types of, 64n24
Real vs. absolute war, 86
Reciprocity, principle of, 242–246, 247, 254
Red Cross, false use of emblems of, 146, 204, 208

Red Lion and Sun, false use of, 204
Reengineering problem, 106–107
Reflexitivy, intelligence collection and, 254–255
Regime change, postwar, 121
Rehabilitation, 120–124, 130
Reiter, Dan, 221, 222
Reliability: of keyword searches, 265–266; of metadata analysis, 273; trust vs., 217n51
Reprisals, 94n25
Research on cybersecurity, 169
Reservations to treaties, 38
Resilience, nature of, 164
Restoration. *See* Postwar period
Restraining the winner, as justification for laws on war termination, 118
Retribution and retaliation, x, 120–124, 126, 130
Retrospective reciprocity principle, 242–244, 247–248
Reversal of cybertheft, 121, 131, 133
Reverse engineered cyberweapons, 105, 107
Revisionists on killing in war, 100n46
Rid, Thomas, 3, 19, 25, 75–80, 83, 84n39, 153, 205
Rights: infringements of, 259; political entities and, 252; rights violations, 100n48, 110. *See also* Human rights; Privacy, right to
Rights of Passage (ICJ case), 49
Robots, autonomous, war between, 89n3
Rodin, David, 100n46, 223
Roff, Heather M., 201
Rome Statute of the International Criminal Court (1998), 38, 204n17
Ronfeldt, David, viii, xi
Ross, W. D., 57n6, 102n52
Rousseff, H. E. Dilma, 50n91
Router hacks, 211–212
Rowe, Neil, 60n12, 68, 69n36, 207n24
RSA Security LLC, 190
Rules: reality of, 141. See also *Tallinn Manual*
Rules of law, 188–193. *See also* Customary international law; International law and legal norms; Law
Ruses, 203–204. *See also* Perfidy, ruse, and deception in cyberwarfare
Russia (Russian Federation): on armed cyberconflicts, regulation of, 40n32; Georgia, cyberattacks against, viii, xi, 4, 27, 29; as possible instigator of cyberattacks against Estonia, 4, 26–27, 77–78, 115, 124–125, 184; response to application of international law to cyberspace, 52; sabotage, preference for, 127

Sabotage, as act of war, 25. *See also* Cybersabotage
Sanctions, 121, 131–132, 133, 172
SCADA (Supervisory Control And Data Acquisition) systems, 213
Schmitt, Michael N., x, 28, 34, 60. See also *The Tallinn Manual on the International Law Applicable to Cyber Warfare*

Scholarship, immaturity on cyberwarfare, 5
Search phase (of keyword searches), 262–263
Second Lebanon War (2006), 163–164
Security standard (moral framework): coercive state actions and, 252–256; conclusions on, 275; keyword searches and, 265, 267; metadata analysis and, 272–273; warrant process, 270
Self-defense, right to, 42n43, 241
September 11, 2001, terrorist attacks, 230
Serb wars, 118
Shamoon cyberattack, x, 50
Shanghai Cooperation Organization, 39–40
Sharp, Tracy, 165–166
Shattered Assumptions Approach, 170–171
Ships, deceptive lighting of, 204, 206
Sieges, 80–81
SIGINT (signals intercepts): HUMINT vs., 266n24; SIGINT-prompted investigations, 263–268; types of, 256. *See also* Keyword searches; Metadata and metadata analysis
Signature strikes, by drones, 180–181
Singer, Peter, 28
Situationism, 194
Sixth Annual European Forum on International Cybersecurity (FIC), 18n15
Skerker, Michael, 251
Slippery slope arguments, 274
Smail, R. C., 80–81
Smart arms race, 146
Snowden, Edward, 20n21, 152, 211, 229, 233, 234, 251
Social capital, political capital vs., 216n49
Social engineering, 193
Social vs. political trust, 216n49, 217n52
Societies, trust in, 217
Soft law approach, to norms for cyberwarfare, 13
Software, weaponization of, 62
SolarWorld, 190
Soldiers. *See* Warriors and cyberwarriors
South America, trust of U.S. by democracies in, 248n52
South Korea, cyberattacks on, 167
South Ossetia, 2008 cyberattacks on, 76n6
Spamming, 203
Spanish Civil War, x
Spearphishing, 192–193, 210n32
Specially affected states, 49
Speed of cyber vs. kinetic attacks, 185, 194
Spies, 151–153, 190. *See also* Cyberespionage
SplashData, 107n63
State and cyberstate of nature, 241
State practice (*usus*), 46–47
States: authoritarian, retrospective principle of reciprocity and, 248; citizens, relationship to, 59; clustering of, espionage and, 247–248; coercive state action, 251–256; as creators of international law, 52; espionage by, 251; just liberal, patrol in, 259–260; as locus for international law, 19; positions on legality of cyberactivities, 50–51; private enterprises, distinction from, 191
Status quo ante, just war theory on, 58
Stealth technology, 212–213
Steinhoff, Uwe, 90n7, 91n10
Stores, as analogies for websites, 66–67
Strategic cyberwar, viii–ix
Strawser, Bradley Jay, 16, 106n60
Stress, 168, 169
Strikes (labor), 237
Stuxnet (worm): as act of war, 83; APTs and, 211; attribution of, 128, 182–183; collateral damage from, 231; description of, 27, 128, 229; difficulty of control of, 221; discrimination capabilities of, 96–97, 104; international acceptance of, ix, 50, 64; Jenkins on, 26n33; kinetic and cybereffects of, 192; Lucas on, 28n35; physical damage from, 3–4; proportionality of, 29; retaliation for, x; as sabotage, 78–79; size of, 97n35; trust as target of, 84n39
Subversion, 151
Success of cyberwarfare, 99n44, 108n65
Suffering, minimization of, 144
Sun Kailiang, 190
Sun Tzu, 201
Supervisory Control And Data Acquisition (SCADA) systems, 213
Supreme Court, U.S.: *The Paquete Habana* case, 46; on privacy of telephone metadata, 270n28
Surinam, 267
Surveillance. *See* Cybersurveillance
Sutton, Willie, 26
Symantec, 61
Syria, Israeli actions against, 4, 27, 79, 220n66, 229, 230

Taddeo, Mariarosaria, 15–16, 16n11, 65
Tallinn. *See* Estonia and Tallinn cyberattacks
The Tallinn Manual on the International Law Applicable to Cyber Warfare (Schmitt): on arms, 62n18; on attacks, 42, 43; on attribution, 126; on civilian contractors, 149n37; on collateral damage, 145; on collective punishment, 126; on computer network exploitation, 189, 236n29; on cyberattacks, 160–161, 165, 178, 214; on cyberespionage, 151, 189; on cyberwarrior codes of honor, 140; on damage from cyberattack, 3n7; description of, 5, 124–125; on espionage and cyberespionage, 148, 236n29; influence of, 44; Internet, lack of understanding of, 165; on *jus ad bellum* and IHL norms, 41; missing

principles of, 130–131; on noncombatant immunity, 105n58; on objects, civilian vs. military, 105n58; on occupation, 128–130; on perfidy, 154–155, 207–208, 214; post-conflict situations, rules possibly relevant to, 116; on postwar investigation, 126; principles of, and aftermath of cyberattacks, 124–130; process for development of, limitations of, 17–20, 30; on protected persons, 105n58, 128; purpose of, 148; reception of, 17; as source for international policy on cyberwarfare, 144–147; on technical experts, 147; on use of force, 41–42, 83. *See also* Codes of conduct
Tallinn 2.0 project, 52
Targeted sanctions, 132
Technical experts, *Tallinn Manual* on, 147
Telecom companies, 270
Teleological elements (in codes of honor), 143
Telephone metadata, 268, 270n28, 271
Termination of cyberattacks. *See* Aftermath of cyberattacks
Termination of wars, 117–118, 219–224
Terminology: for cyberwarfare, 35–37; state of flux of, 3n5
Territorial integrity, 232n13
Terror, description of, 161
Terrorism and cyberterrorism: bodily injury and loss of life from, 161–164; description of, 230; implications for ethics and law, 171–173; lab assessments of psychological effects of, 167–171; psychological harm from, 160; severe mental suffering from, 164–167
Thick theory #2 (rehabilitation), 121–122
Thick theory #1 (retribution), 121
Thin theory (overlapping consensus), 120–121
Third parties, human agents as, 179
Threats of action, Clausewitz on, 84n37
Threshold problem, 109n68
Tiel, Jeffery R., 152
Time period for state practices, 46, 47, 49
Tit-for-tat principle, 243
Totally reversible harm, 98
Traditional just war theorists, 99–100n46
Traditional warfare, 60. *See also* War and warfare
Transportation networks, 159
Trapdoors (cyberoperations), 210
Treaties: compliance verification of, 44–45; cyberconflict and interpretation of, 39–45; interpretation of, 39, 41–44; as statutes in international law, 63; treaty law, 37–39; Treaty of Versailles, impact of, 118; on weapons, U.S. nonparticipation in, 44n56
Trust: deception and, 215–224; between enemies, 201–202, 203; reliability vs., 217n51; ruses and, 204; social vs. political, 216n49, 217n52; as Stuxnet target, 84n39
Turing, Alan, 1

UAVs (unmanned aerial vehicles, drones), 62, 106, 109n68, 180–181
UK. *See* United Kingdom
Ukraine, nationality of rebels in, 189n41
Uncertainty, methodology of, 23–31
UN Charter: cyberconflicts, applicability to, 40, 41–42, 47, 63; on economic and political coercion, 63n20; as *jus ad bellum*, 36; states as possible violators of, 50
Unconstrained war termination, 118
Undisclosed acts, 49
UNIDIR (United Nations Institute for Disarmament Research), 17n13
Uniform law treaties, 45
Uniforms, false use of, 204–205, 208
United Kingdom: Additional Protocol I, interpretative declaration on, 39; assault, definition of, 84; classification of cyber attacks, 76n2; on cybertreaty norms, 45; Government Communications Headquarters, 211, 251; information sharing by, 247; Spanish Civil War, intelligence in, x; U.S., covert political action against, 237
United Nations: emblem of, false use of, 146, 204; General Assembly, 18, 48; General Assembly Resolution 3314 (1974), 90n7; Group of Governmental Experts, 40; Institute for Disarmament Research, 17n13; peacekeeping operations, 238; Security Council, 28, 42n43, 63. *See also* UN Charter
United States: on application of international law to cyberspace, 52; on armed attacks, 42; China, actions against, 4; covert political actions against, 237; covert political actions within, 239n34; Cyber Command, 60; cyberoperations, restraint in use of, 223; foreign trust of, 248n52; government–private enterprise distinction in, 191; information sharing by, 247; International Criminal Court Statute and, 38; Iran hostage rescue, 239n35; on military manuals as source of *opinio juris*, 48; nuclear weapons, use of, 24; reflexivity of intelligence collection in, lack of, 255; response to condemnation of Chinese cyberoperations by, 51; sabotage and cybersabotage by, 51, 127–128; spying on Germany, 219; Stuxnet worm, possible responsibility for, 79, 128, 183, 192; treaties, non-participation in, 37, 44n56, 46, 48; treaty-making power in, 38
United States Army, Field Manual, 143
United Steel Workers, 190
Unit 61398 (PLA), 19, 190, 234–235

Subject Index

Unjust aggression, 108
Unjust gains, 121, 131
Unjust harm, 95
Unmanned aerial vehicles (UAVs, drones), 62, 106, 109n68, 180–181
Unprincipled belligerents, 116
Unprivileged irregular combatants, 263n21
U.S. Code Title 10, 189
U.S. Steel, 190
US Air Force Research Institute, 16
Use of force. *See* Force
Usus (state practice), 46–47
Utilitarianism, 63
Uzbekistan's intelligence capabilities, 255–256

Valeriano, Brandon, 211n34
Verbruggen, J. F., 80–81
Verification, caution and care in, 145
Verizon, NSA collection of metadata from, 229, 233–234, 246
Versailles, Treaty of, 118
Victory in war, 119–120n18
Vienna Convention on the Law of Treaties, 37, 39
Vietnam War, vii
Vihul, Liis, x, 34
Violence: maximally discriminatory, 92–93; nonphysical, 84, 86; physical purpose of, in war, 77; of war, 91–92
Viruses. *See* Malware; Stuxnet
Vocabulary. *See* Terminology

Waismel-Manor, Israel, 157
Waldron, Jeremy, 172–173
Waltz, Kenneth, 64n24
Walzer, Michael, 63, 90n7, 94n25, 99n46
Wang Dong, 190
War and Self-Defense (Rodin), 100n46
War and warfare: absolute vs. real, 86; air warfare, Hague Rules on, 44n57, 204n17; Aquinas on, 148n35; armed interventions short of war, 238; civilian assets in, 105; civil wars, 219, 223n76; covert political actions and cyberactions, 228–250; cyberwarfare as war, 75, 85–87; cyberwarfare's differences from traditional warfare, 60–61, 69; description of, 2–3; ethics in, vii–viii, 63; Form of War, 91n15; humanization of, 95; ideal *jus in bello*, cyberwar and, 96–103; ideal war, possibility of, 90–95, 108; likelihood of, xi–xii; medieval warfare, 80–82, 85; modern conventional warfare and potential for evil, 105–106; naval warfare, 204; peaceful settlements of, 223n76; World War II description of, vii; worst possible, 91n11. *See also* Acts of war; Armed conflicts; Cyberwar and cyberwarfare; Just war theory; *specific wars by name*
War crimes and war crimes trials, 121, 126, 214
Warriors and cyberwarriors: codes of honor, 139–140, 141–143; cyberoperatives vs. cyberwarriors, 143n12; ethical principles for warriors, 139–140; moral justness of war and, 155n65; roles of cyberwarriors, 150; similarities and differences, 148–149. *See also* Codes of conduct
Water supplies, 159
Weapons. *See* Cyberweapons; Nuclear weapons; *specific types*
Web. *See* Internet
Wen Xinyu, 190
Werner, Suzanne, 222
Westinghouse, cyberattacks against, 190
Westphalian intermediate principles, 62, 63, 67, 68, 69
Whethan, David, 75
Whose Justice? Which Rationality? (MacIntyre), 23n29
Winners (of wars), restraining, as justification for laws on war termination, 118
Wired Science (PBS), episode on Estonian cyberattack, 14n3
World Conference on International Telecommunications, 38
World War I, 118, 122
World War II: German warfare techniques in, xi; as model of rehabilitative ending of war, 122; Treaty of Versailles and, 118n11; unethical practices in, vii
Worms (malware), 211n34
Worst possible war, 91n11
Wrongdoing, 232n15

Yahere (fict. country), 191
Yannakogeorgos, Pano, 22, 58n9
Yuen, Amy, 222

Zero-sum war, 222